TO SERVE CANADA

Aerial view of the Royal Military College, 1985

TO SERVE CANADA

A History of the Royal Military College Since the Second World War

Richard A. Preston

With a Foreword by
General A.J.G.D. de Chastelain

University of Ottawa Press
Ottawa • Paris

© Minister of Supply and Services Canada, 1991
Printed and bound in Canada
ISBN 0-7766-0327-2

Canadian Cataloguing in Publication Data

Preston, Richard A., 1910-

To serve Canada : a history of the Royal Military College
since the second World War

ISBN 0-7766-0327-2

1. Royal Military College of Canada — History —
20th century. I. Title.

U444.K5L1   1991   355′.0071′171372   C91-090285-2

UNIVERSITÉ D'OTTAWA
UNIVERSITY OF OTTAWA

Published by the University of Ottawa Press in cooperation
with the Department of National Defence and Canada
Communication Group — Publishing, Supply and Services
Canada.

Catalogue number: D61-1/1991E

Design: Anita Bergmann
Jacket photograph courtesy of P.D. Mansbridge.
All photographs courtesy of the RMC Archives unless
otherwise indicated.

Contents

Foreword

A former Dean of Arts at RMC, Professor George Stanley, in his book *Canada's Soldiers*, chronicles what he calls "the military history of an unmilitary people." Yet the history of Canada since Confederation has been one rich in the military achievements of Canadians in five wars, two of them World Wars. In the conduct of each of those wars, and in the periods before and between them, the role played by the Royal Military College has been significant, even pivotal.

In *Canada's RMC*, Richard Preston recounted the details of the first eighty years of RMC's contribution to Canada, from the college's foundation in 1876 until the period immediately following the Korean War. In *To Serve Canada*, he takes up the story of RMC as one of Canada's three military colleges and examines its development through the uncertain years of the Cold War, through the vagaries of public indifference towards defence, through the evolution of degree-granting status and the moves towards institutional bilingualism, and through the frequent Ottawa-directed re-evaluations of their roles.

Unique tensions inevitably exist in a bipolar military–civilian institution, both components of which seek the same goal – the production of effective military officers – but through the competing imperatives of the simultaneous development of military leadership and academic excellence. There is a potential for friction between professors and officers who must compete for cadets' scarce time, but solidarity unites military and civilian staff members in response to periodic instructions from outside to re-cast the direction in which the military training or academic programs are going.

To Serve Canada covers a period in the country's history when the pinnacle of peacetime military activity (a Permanent Force strength of 120,000 at the beginning of the 1960s) is eroded by a succession of circumstances: by anti-war sentiment spilling over from a Vietnam-conscious United States in the late sixties and the early seventies; by a population increasingly bereft of personal military experience – and hence, perhaps, lacking sympathy for military aims; by the public's diminishing belief in the reality of the threat to world peace, as nuclear stalemate proves effective in deterring world conflict; and by governmental concern over defence spending in the face of growing national indebtedness.

This is a period in which the changing nature of war is parallelled by radical changes to Canada's social fabric, both elements having a profound effect on the way the armed forces must view themselves in the light of their roles and traditions to this point. The rush to an increasingly sophisticated military technology has placed a premium on the production of "technologist" officers to fill the engineering branches, at the same time as traditional military wisdom continues to reserve the highest ranks for those in the fighting classifications. The demand for bilingual officers, in a country where bilingualism is now an official requirement, adds a new dimension to the training and education of officers who must communicate in order to be able to lead, but whose secondary education is directed by provincial governments little interested in teaching second languages at an age when that is most easily done. The changing societal and legislative norms that demand that women have the opportunity to serve in all walks of life, including on the bridge, in the cockpit, and in the trenches, have compelled the armed forces to make an introspective evaluation of their approach to the profession of arms and of the way their members perceive that profession and practise it.

This book will be valued by scholars of military history as well as by other observers of Canada's social development. Some would have it that the military forces of a country reflect the values that its society wishes to preserve, and few can deny that the motto "Truth, Duty and Valour" provides an admirable credo in which all Canadians can take great pride. By chronicling the development of RMC and its sister colleges from the post-Korea period to the present, Dr Preston has provided a valuable and entertaining addition to the historical literature of this country.

This foreword opened with a reference to the five wars that Canada has fought in in this century, and as this book goes to press the fifth war is only just coming to a close. The Gulf War was a strange war for Canada. The country's contribution was small but it was consequential and militarily effective. No Canadians were killed in it and none were wounded, but the war's significance at this time in Canada's history is striking if only for two reasons. The first is that Canada did indeed go to war again, despite the belief of some that Canada's military role, now and in the future, should be only as a "peacekeeper." The second is that much of the leadership in the war, from the Force Commander down to the most junior officer, was provided by the products of the Canadian military colleges, whose graduates quite clearly are still effectively educated and trained to serve Canada.

General A.J.G.D. de Chastelain
Chief of the Defence Staff

Ottawa
1 April 1991

Preface

The promotion of military professionalism is a more complex problem than professional development in other fields. It includes two elements – training and education – that are somewhat incompatible. Military training must develop an attitude of mind in addition to practical expertise: soldiers, sailors, and fliers must subordinate themselves to the service of the state as formulated by higher authority, even to the point of being prepared to sacrifice their lives. They must be bound psychologically to the good of the whole rather than to their own self-interest. In other words, they must learn to obey without question. Education, however, is necessary if an individual is to develop the ability to reason, a quality that is especially needed in persons of high rank. Military training and education, which must both begin early in a career, thus combine the promotion of military cooperation, expertise, and dedication to service with a conflicting objective – learning to think independently.

In areas like military and marine engineering, where technical competence is a basic requirement, this paradoxical combination of opposites in the process of producing military leaders has long been fairly well understood and achieved. It has not, however, always been realized that a preparation that not only trains and indoctrinates but also fosters the capacity to think is just as necessary for those whose duties in military, naval, and air service are primarily in operations. This need is made more imperative by the fact that the combat officer is often more likely than the technical officer to be appointed to high command, preparation for which must begin early. Hence, those cadets or young officers who, from interest or personal capacity, elect to follow operational rather than technical careers, including many who are less capable of the mathematical and scientific skills needed in a technological education, must be developed intellectually as well as practically and ideologically. The formative programs for this purpose may be built on various academic disciplines. Mathematics or the sciences are most frequently preferred, but other disciplines that appear to have less direct military relevance or seem liable to foster a critical rather than conforming mind may also be appropriate. In many military colleges and academies,* however, sound academic criteria have often been endangered by the priority given to military training and to the indoctrination of young officers.

Before 1939, and briefly after its reopening in 1948, the primary function of the Royal Military College of Canada (RMC) was to produce militia officers with little or no legal obligation to career service. Since it had to prepare a majority of its graduates for university degree courses and civilian careers, as well as for severely competitive entry into British technical corps, the prewar college had had more success in maintaining the quality of

*Americans use "academy" to mean a pre-commissioning training institution, where the British use either "academy" or "college."

the academic elements in its curriculum than was usual in other military colleges.

My earlier volume, *Canada's RMC: A History of the Royal Military College*, published in 1969 by the Royal Military College Club and the University of Toronto Press, was the first full account of RMC's history to appear in print. It told the college's story from its foundation in 1876 to its closure in 1942 during the Second World War, and then from its reopening in 1948 to the year 1959, when the Ontario legislature passed "An Act Respecting the Royal Military College of Canada" that conferred on RMC the power to grant academic degrees. The central theme of *Canada's RMC* was the historical basis of those academic qualities that justified the conferring of the degree-granting power.

Because pertinent records were not then available, the period from 1948 to 1959 was sketched only briefly in that book. Furthermore, by 1959 the new RMC had two features that sharply distinguished it from the old. First, it had been formally re-established in 1948 as one of two (later three) tri-service institutions to serve all three of Canada's forces (which it had actually done informally before 1942). Second, as was confirmed by the creation of the Regular Officer Training Plan (ROTP) shortly after the reopening, it had become primarily, though not exclusively, devoted to the production of regular rather than reserve officers.* These two innovations called for a new study that would relate them to the earlier development of the college.

On 30 July 1981, when the first printing of *Canada's RMC* was running low in stock, Brigadier-General John Stewart, commandant at that time, arranged for a second printing as a Regimental Institute project. At the same time he asked me to write another book that would tell the story of the postwar RMC. This book, *To Serve Canada*, is not the second of a two-volume history of the college but a completely new look at RMC's history, with particular attention to the period since the Second World War.

The theme in this new history is different from that

of *Canada's RMC*. It studies the problem of retaining and promoting the college's expertise in developing military professionalism despite the new stresses from academic requirements imposed by accelerating technologies and by the acquisition of degree-granting powers. It also examines another set of complications caused by several new factors: the existence of two sister colleges, each with its distinctive composition and aspirations; the unification of the Canadian Forces; programs of institutional bilingualism and biculturalism designed to preserve national unity; a social revolution that included youth and women's movements; and measures to extend RMC's offerings to serving personnel. Some of these innovations are especially important in this story because they have had a direct impact on the functioning of the Cadet Wing (formerly the Battalion of Gentlemen Cadets), the core of RMC's military training system, and therefore on its traditional method of developing professionalism.

Stress on military professionalism in RMC was not new in 1948. From its foundation, RMC graduates had competed for a limited number of commissions in the British Royal Engineers and Royal Artillery. For over half a century a number had also become career officers in the Canadian Militia (Permanent Force), in the Royal Canadian Mounted Police, and, on a smaller scale, in the Royal Canadian Navy and the Royal Canadian Air Force. Thus, in 1939, seventeen of the seventy-two permanent force Royal Canadian Artillery officers were RMC ex-cadets. (This was, however, probably a higher proportion than in any other branch or arm.) The success of RMC graduates in British technical corps and in the Canadian regular forces amply demonstrated the professional

---

*Cadets under the Regular Officer Training Plan receive a four-year education at public expense in either a services college or a university and are obliged to serve for a comparable period in one of the armed forces. Cadets under the Reserve Entry Training Plan pay fees, though most of them have scholarships, and can enter a civilian career on graduation as long as they serve for a comparable time in a reserve force.

quality of RMC's training and education. The ultimate tests came in the two world wars when RMC graduates, along with Canadian Permanent Force and Militia officers (and in the first war also along with officers and non-commissioned officers borrowed from the British army), provided a basis of military expertise for Canadian forces that were largely recruited from civilian life for the duration of the war. The record of Canada's contribution in both of those wars speaks for itself. It was founded, in part, on RMC's achievement in fostering military professionalism among its graduates.

It is now nearly forty years since the establishment of the new RMC. Many of the problems mentioned above have come to a head only within the past decade or so. Nevertheless, it is time to attempt to evaluate the college's postwar evolution and accomplishments. Lacking the supreme test of a major war, any assessment can only be tentative and inconclusive, for the circumstances of peacetime soldiering differ from those of war when other qualities move men to the top. Even so, it is possible to look at the record of RMC's postwar graduates as evidence of the success of the new structure. Ex-cadets of both prewar and postwar generations fought in Korea, winning a total of sixteen decorations, three OBEs, three MBEs, one DSC, and eight MCs; and four of these MCs went to postwar graduates who had gone straight from the classroom to war. Citations for some of the MCs particularly stressed qualities of leadership. This suggests that, in these cases at least, the college had given an adequate psychological preparation for military operations.

In peacetime soldiering, yardsticks for measuring the success of an officer-production program lack certainty. One of the more suggestive is the success of graduates in rising to high rank. The record of the new RMC can be compared with earlier experience. In 1914, neither of the two Canadian major-generals in the permanent force were ex-cadets, and out of eighteen colonels only five were from RMC, two who had graduated and three who had not. In 1939, three of eight major-generals, six

of fifteen temporary brigadiers, and six of forty colonels were ex-cadets. These relatively low RMC numbers, especially among colonels, may be explained in part by Canada's militia tradition: RMC's professional contribution did not become really effective until the war began.

By way of contrast, it should be noted that on 30 June 1987 the *Canadian Forces' List* showed that when there were 2726 RMC graduates listed in the Canadian Forces,

Cadets and staff form a maple leaf on the square, 1989.

Graduation parade, 1970

ception of CF Europe, were held by CMC graduates, as were 43 per cent of the higher appointments in National Defence Headquarters (NDHQ) and throughout the forces. General Fred Sharp, a 1938 RMC graduate, had been one of the first chiefs of the Defence Staff after the unification of the Canadian Forces. Since the time when postwar graduates could be expected to have reached the top, three out of the four officers appointed as chief of the Defence Staff (CDS) – General Ramsey Withers, General Paul D. Manson, and General A.J.G.D. de Chastelain – have been ex-cadets.

The quality of RMC's education can also be measured by the success of its graduates in non-military occupations, even though, unlike in the old RMC, only the first three classes, and a few recruits in the late 1960s and 1970s, entered on the Reserve Entry Training Plan (RETP) which allowed them to go directly to a civilian career. Many others, however, took up a second career after their military service, and their success there should be noticed. Graduates of the new RMC can be found in as wide a variety of important civilian positions as were held by their predecessors who graduated before 1939. R.V. Hession became president of Central Mortgage and Housing Corporation, deputy minister of supply and services, chairman of the Royal Canadian Mint, and deputy receiver general. A former CDS, Ramsey Withers, is now deputy minister of transport, and several other ex-cadets have become deputy ministers in provincial governments. Others have succeeded in various parts of the public service, including in the Department of External Affairs. Many more have flourished in large corporations and small business enterprises, even though they have come from a wider cross-section of Canadian society than their predecessors in the old RMC and so have fewer personal and family contacts on which to base a business career. Several have entered the academic profession in universities. It is also interesting that a number of postwar graduates have elected to take up medical education and have become medical officers in

the only general, three lieutenant-generals out of eight, ten major-generals out of twenty-seven, thirty brigadier-generals out of seventy-six, and 107 colonels out of 307 were graduates of the Canadian Services Colleges (CSCs), now Canadian Military Colleges (CMCs). And these figures may not tell the full story: inclusion of the letters "rmc" in the *Canadian Forces' List* occurs only when the individual supplies it. In that same year all the principal commands in the Canadian Armed Forces, with the ex-the services. Several have become clergymen, both Catholic and Protestant.

Another indication of success is the retention rate of officers, especially after the termination of their period of obligatory service but also in later career. The Canadian Military Colleges, and RMC in particular, did well in this respect in the 1970s and 1980s by comparison with other sources of officers. On occasion, however, there was complaint about the RMC product – about a lack of real military interest on the part of some graduates. That defect, created in part by the ROTP system that puts emphasis on a relatively cheap academic education rather than on dedication to a military career, serves to demonstrate, by contrast, the degree of success achieved by others. In another dimension, the high quality of RMC's technical education led some graduates to seek more profitable civilian jobs. Because of this criticism and also for financial reasons, there has been much discussion of the cost-effectiveness of the Canadian Military Colleges as against a supposedly cheaper program of producing officers by subsidization through civilian universities. RMC's greater rate of retention was one argument in its favour, and was undoubtedly a result of its success in stimulating military professionalism.

To Serve Canada provides an understanding of the problems in professional military development, solutions for which are vital to the maintenance of an effective Canadian defence. Because the problem of military education, as contrasted with military training and apprenticeship, which runs through the history of the college, is also important in officer-production systems everywhere, the introduction to this book, entitled "Tradition and Change," analyses other Western experience in this field.

Those readers whose interest is solely in the history of the Royal Military College of Canada and in its contribution to Canadian defence may therefore prefer to pass directly to chapter 1, which briefly retells the history of the old RMC and the establishment of the new college in 1948.

The main references used in preparing this book can be found in the Notes. A full list of sources will be deposited in the Massey Library at RMC. Ranks cited in the text refer to the appropriate time; those given in the biographical footnotes are the highest attained.

Acknowledgments

A number of individuals contributed to the preparation of this book by providing encouragement, support, and help. Major-General John A. Stewart, Major-General Frank Norman, Brigadier-General Walter Niemy, and Commodore Edward Murray, who were the RMC commandants while it was in preparation, all regarded a book on the history of the college since the Second World War to be essential for a better understanding of its problems and accomplishments. An ad hoc committee chaired by Dr D.M. Schurman, head of the Department of History at RMC, approved the project. Dr W.A.B. Douglas, director of history at National Defence Headquarters, proposed that this history be published by his directorate.

During academic sessions, Duke University provided me with office space as an Emeritus Professor; and during summer vacations the RMC history department gave me use of an office as honorary professor. One summer, when that space was not available, Lieutenant-Colonel D.A. Lefroy, head of the Department of Military Leadership and Management, filled the breach. Without those facilities my task would have been much more complicated.

Some of the costs of preparing a typescript were borne by modest grants from the Duke University Research Council. At a later stage, then Lieutenant-Colonel A.H.C. Smith, RMC's director of administration, arranged for a DND contract that partially covered other expenses.

At the outset Barbara Wilson of the National Archives of Canada and Dean William Rodney of the Royal Roads Military College made available a preliminary list of documentary sources that had been prepared for a history of Royal Roads. Many of those sources had relevance also for RMC. I am also indebted to Dr Keith Crouch, chief librarian, Major Alphee Bake, Mr Benoit Cameron, and Suzanne Burt of the Massey Library at RMC and to many reference librarians and others in the Perkins Library at Duke. Dr Steve Harris at the Directorate of History, DND, Captain P. Fortier, the RMC registrar, and Mr Logie Macdonald, the associate registrar, and their staffs gave me guidance and access to records in their custody. Captain J.R. McKenzie, curator of the RMC Museum, provided many of the illustrations reproduced in this book. Professor J.E. Pike and Lieutenant-Colonel P. Nation answered many questions, drawing on their vast knowledge.

Lieutenant-Colonel Don W. Strong, executive director of the Royal Military College Club, and his secretaries Joy Fehr and Linda McGinley, were especially helpful in providing information about ex-cadets, as was Brigadier-General R.T. Bennett as well. Lieutenant-Colonel R.C. Coleman (no. 7272) and Major G. Croutier of DND's Media Production researched biographical information for two civilian members of the Canadian Services College Advisory Board.

The Directorate of Professional Education and Development of DND provided access to the minutes of Advisory Board meetings and of sessions of the General and the Academic (Educational) councils.

Several individuals lent me documents or supplied information that supplemented those in official repositories. The most substantial collection was Brigadier-

General W.K. Lye's file of Miscellaneous RMC letters. Dr David Baird, Principal John Dacey, Dr W.F. Furter, Dr Yvan Gagnon, Rear-Admiral W.M. Landymore, Rear-Admiral D.W. Piers, Principal John Plant, and Major-General George Spencer also provided papers. Major-General John Stewart gave me access to a diary he kept during his term as commandant. Captain M.D. Fabro (no. 13868) granted me permission to reprint the cartoon, "Putting the 'M' back in RMC," he had published in the cadet magazine *The Arch*.

Information to set the history of Canada's RMC in perspective was provided by discussions with Dr David Chandler, professor of history at the Royal Military Academy, Sandhurst; Principal F.R. Hartley of the Royal Military College of Science, Shrivenham; Sir Michael Howard of Oriel College, Oxford; Dr Peter Dennis, formerly of RMC, Kingston, and now at the Australian Defence Force Academy; Richard Alexander, deputy director of studies, and Evan L. Davies, head of the Department of Strategic Studies and Internal Affairs at the Royal Naval College, Dartmouth; Commander P.J. Jewell, deputy dean, Commander M.J. Kitchin, director of undergraduate studies, and Lieutenant-Commander P.J. Payton, college historian, of the Royal Naval Engineering College, Manadon; Dr Geoffrey Till of the Royal Naval College, Greenwich; and many members of the American services who had knowledge of the United States Military Academy, West Point, the United States Naval Academy, Annapolis, and the United States Air Force Academy, Colorado Springs. Commodore Darroch N. Macgillivray of the Canadian Defence Liaison Staff in London facilitated arrangements for visits to the British naval colleges. C.A.M. Jones, director of studies at Dartmouth, and Captain G.C. George, dean at Manadon, facilitated my visits to those colleges. Conversations with RMC faculty members during my summer sojourns in Kingston contributed to a better understanding of developments there, especially after my tenure as professor ended.

My colleague at Duke, Dr Alex Roland, read the first rough draft of each chapter as it appeared and made many valuable suggestions. Parts or all of the early drafts were also read by Lieutenant-General W.A.B. Anderson, Faith Avis, Dr and Mrs Stuart Barton, Air-Commodore L.J. Birchall, Air-Vice-Marshal D.A.R. Bradshaw, Dr Pierre Bussières, Dr John R. Dacey, General A.J.G.D. de Chastelain, Captain (N) P. Fortier, R.B. Hamel, Commodore W.P. Hayes, Dr Barry Hunt, Brigadier-General W.K. Lye, Commodore Edward Murray, Lieutenant-Colonel P. Nation, Brigadier-General W. Niemy, Major-General Frank J. Norman, Rear-Admiral D.W. Piers, Major-General Roger Rowley, Major-General George Spencer, Brigadier-General W.W. Turner, and Major-General John A. Stewart. Dr W.A.B. Douglas, Dr Norman Hillmer, Dr Roger Sarty, and Brigadier-General R.T. Bennett all read later versions and provided extra information that saved me from many errors. The final version and the opinions expressed in this book are my own.

Gordon Smith, a research assistant provided by a federal government student employment scheme, checked references from the minutes of RMC's Faculty Council and Faculty Board during one summer. Mrs D. Eskritt, the RMC commandant's secretary, gave me valuable assistance in locating individuals. R.B. Hamel, second-language coordinator at RMC, took charge of the preparation and circulation of various drafts to readers.

I am indebted for secretarial help to Grace Guyer of Duke's Department of History, Karen Brown of RMC, and particularly to Vivian Jackson of Duke, who patiently typed the whole manuscript from very rough drafts. At National Defence Headquarters Annie Rainville and her staff transcribed the text on to a word-processing program. David Dunkley prepared the index, and Jennifer Wilson provided invaluable assistance in the preparation of the manuscript for publication. Finally, I am once again especially privileged to have had Rosemary Shipton as my editor.

Abbreviations and Acronyms

Adm   Admiral
ADM(Per)   Assistant Deputy Minister for Personnel
AFHQ   Air Force Headquarters
A/M   Air Marshal
B&B   Bilingual and Bicultural; Bilingualism and Biculturalism
B&B Admin   Bilingual and Bicultural Administration
BOTC   Basic Officer Training Course
BSM   Battalion Sergeant Major
CAS   Chief of the Air Staff
CD   The Canadian Forces' Decoration
CDEE   Canadian Defence Educational Establishments; also Commander, CDEE
CDS   Chief of the Defence Staff
CEF   Canadian Expeditionary Force
CEGEP   Collèges d'enseignement général et professionnel
CF   Canadian Forces
CFHQ   Canadian Forces Headquarters
CFMC   Canadian Forces Military College
CGS   Chief of the General Staff
CinC   Commander-in-Chief
CMBG   Canadian Mobile Brigade Group
CMC   Canadian Military College
CMR   Collège militaire royal
CO   Commanding Officer
COTC   Canadian Officer Training Corps
CNP   Chief of Naval Personnel
CNS   Chief of the Naval Staff
CP   Chief of Personnel
CPD   Chief of Personnel Development
CRAD   Chief of Research and Development
CSC   Canadian Services College

CWC   Cadet Wing Commander
DATES   Director (or Directorate), Administration, Training, and Education
DGITP   Director-General, Individual Training Programs
DGMT   Director-General of Military Training
DGRET   Director-General, Recruiting, Education, and Training
DHist.   Director (or Directorate) of History [DND]
DND   Department of National Defence
DNE   Director, Naval Education
DPED   Director (or Directorate), Professional Education and Development
DRB   Defence Research Board
DROTP   Director, Regular Officer Training Plan
DSC   Distinguished Service Cross
DSO   Distinguished Service Order
GOC   General Officer Commanding
GSO1   General Staff Officer, Grade 1
HQ   Headquarters
JCSC   Joint Canadian Services Colleges Committee
JSUC   Joint Services Universities Committee
JSUCC   Joint Services Universities Coordinating Committee
JUMAC   Joint Universities and Military Advisory Committee
MARCOM   Maritime Command
MBE   Order of the British Empire, Member
MC   Military Cross
MLM   Department of Military Leadership and Management
(N)   Navy (designating a naval rank when used with lieutenant or captain)

NATO	North Atlantic Treaty Organization	UNTD	University Naval Training Division
NCO	Non-commissioned officer	USAFA	United States Air Force Academy, Colorado Springs
NDHQ	National Defence Headquarters		
NORAD	North American Air Defence Command	USMA	United States Military Academy, West Point
NRC	National Research Council	USNA	United States Naval Academy, Annapolis
OBE	Order of the British Empire, officer	UTPM	University Training Plan, Men
OCP or OCTP	Officer Candidate Program, or Officer Candidate Training Program	UTPM(W)	University Training Plan, Serving Women
		UTPO	University Training Plan, Officers
ODB	Officer Development Board	UTPO(W)	University Training Plan, Women Officers
PMC	Personnel Members Committee		
POW	Prisoner of War		
PPCLI	Princess Patricia's Canadian Light Infantry		
PWOR	Princess of Wales's Own Regiment		
QC	Queen's Counsel		
QMG	Quartermaster General		
QR&O	Queen's Regulations and Orders		
RCA	Royal Canadian Artillery		
RCAC	Royal Canadian Armoured Corps		
RCAF	Royal Canadian Air Force		
RCCS	Royal Canadian Corps of Signals		
RCE	Royal Canadian Engineers		
RCEME	Royal Canadian Electrical and Mechanical Engineers		
RCHA	Royal Canadian Horse Artillery		
RCIC	Royal Canadian Infantry Corps		
RCN	Royal Canadian Navy		
RCNC	Royal Canadian Naval College		
RCR	Royal Canadian Regiment		
RE	Reserve Entry		
RETP	Reserve Entry Training Plan		
RETP(W)	Reserve Entry Training Plan, Serving Women		
RMC	Royal Military College of Canada		
RN	Royal Navy		
RNCC	Royal Naval College of Canada		
ROTP	Regular Officer Training Plan		
ROTP(W)	Regular Officer Training Plan, Serving Women		
RR	Royal Roads Services (or Military) College		
RRMC	Royal Roads Military College		
RSM	Regimental Sergeant-Major		
RUTP	Reserves University Training Plan (Air Force)		
SACEUR	Supreme Allied Commander, Europe		
SALT	Strategic Arms Limitation Treaty		
SHAPE	Supreme Headquarters, Allied Powers, Europe		
SLT	Second Language Training		

Tradition and Change in Military Education

Every country faces similar problems in training officers, and each responds to these problems in ways that are appropriate to its national history and situation. How, then, have other countries coped with the challenges that faced the Royal Military College of Canada? How have they accommodated the academic and military elements in their mandate, and how have they reacted to the social, technological, and other changes that have imposed new requirements on the military profession?

Colleges or academies that prepared young men for immediate commissioning in armed forces emerged in Europe in the seventeenth and eighteenth centuries. During the course of the nineteenth century they became a preferred mode of educating officers in many countries, but there were often other routes to a commission as well: in a regiment or fighting ship, in special training units and ships, and in establishments that provided short courses. All these other routes were varieties of apprenticeship. It was not until after the Second World War that Michael Lewis, Samuel Huntington, and Morris Janowitz* established convincingly what the military had known for over half a century through their war and staff colleges and their military academies – that a naval or military career is a profession with basic characteristics like other professions. It is safe to say that since 1945 the significance of a military education has been much better understood.[1]

Huntington's definition of the military as a profession laid down principles for application in military academies.[2] The characteristics he cited were, first, expertise based on historical understanding rather than on rote learning; second, corporate sentiment (especially important for effective cooperation in the military); and, third, social responsibility. Military academies had come to be the preferred mode of producing professional officers, he said, because they could provide not only the academic education essential to military expertise but also the environment most suitable for ensuring corporate sentiment and social responsibility.[3]

Like all other professionals, the military claim exclusive control of the skills of their craft. They act collectively to conceal internal conflict, but they can preserve public confidence only by responsible behaviour – by social responsibility.[4] In recent years, many military colleges have shed much of the cadet's routine training in skills in favour of academic education and professional development. They often leave military training (in Canada, classification training) to the summer months, away from the academy, or to early commissioned service. Hence, academies that emphasize military training

---

*Professor Michael Lewis, Royal Naval College, Greenwich; Samuel P. Huntington, professor of government and director of the Center for International Studies, Harvard University; and Morris Janowitz, professor of sociology, University of Chicago, and founding chairman of the Seminar on Armed Forces and Society.

or detailed military subject matter have been called "trade schools."[5] By contrast, those that offer a predominantly educational curriculum, even though it is all at the undergraduate level, can properly claim to be professional schools.

Military training and indoctrination is essential in officer production. "Obedience," Admiral Mahan once said, "is the cement of the structure . . . it is the life-blood of the organism."[6] To foster obedience, military academies pay much more attention to indoctrination and discipline than cadet training systems in the universities. Military academies stress motivation, dedication, loyalty, character, and leadership. Professional schools of religion preach service and social responsibility, but the military profession, more than the church, assumes the possibility of laying down life itself in the cause of duty.[7] Some military academies that overemphasize military virtues at the expense of academic elements have been labelled "seminary-academies."[8] Blind obedience, as Mahan goes on to show, can be self-defeating. It must be qualified by individual judgment, and individual judgment is promoted by education.

To inculcate conformity to accepted standards, military colleges mould character and shape minds by a wide variety of means. They segregate cadets, and especially new recruits, from civilian society. Their courses are longer than training courses. Military colleges are essential where there is no traditional military class on which to draw. They impose rites of initiation and passage. They often have distinctive, sometimes traditional, uniforms. They foster pride in their institutions and in the armed forces and country they serve. They hero-worship. They draft codes of conduct, teach them by precepts, catchphrases, and slogans, and enforce them by routine discipline and punishment. In so doing they exercise an abnormal control over individual behaviour. They also identify the individual cadet closely with his college, its sports, and his service. They expose him to experienced veterans. They use drill not merely to sharpen attention

or provide a spectacle, but for the more subtle psychological effect it has on the individual – to make him a willing and cooperative participant in an effective whole.* They give experience in command, in the exercise of discipline and authority, and in the managing of the interior economy of a military unit, which is quite different from anything in the outside world. Finally, they encourage self-confident belief in a military elite and in the superiority of the military spirit.

Although indoctrination and training are essential elements in academy life, the primary *raison d'être* of the institution is academic preparation for a professional military career. Before commissioned military officers were generally acknowledged as a distinct profession, it was the officers of the technical corps who were known as "professional officers," thus coupling them with civilian professional engineers. Significantly, the first military academies to become permanent were set up to provide the necessary academic foundation for these professional officers of the technical corps. In several countries, military colleges were the first engineering schools of any kind. They did not long retain their monopoly in producing professional engineers, and in time they also lost their leadership in technological education. But because of technical necessity stemming from the development of weapons and their use, they retained mathematical, scientific, and technological courses. For technical officers, a military education must necessarily be founded on academic sciences.

Sometimes these same military academies also produced officers for the non-technical combat arms – the cavalry and the infantry – and their curricula included

*The American historian W.H. McNeill noted this effect of drill in his classic book, *The Pursuit of Power: Technology, Armed Forces, and Society since AD 1000* (Chicago: University of Chicago Press 1982), 254. Some individuals are alienated by drill, however, and excessive amounts may demotivate young men of high intelligence and be counterproductive.

elementary mathematics. Even where the combat arms had their own pre-commissioning institutions, the curricula usually had a mathematical basis because it was believed to be relevant not only to tactics, field fortification, gunnery, and navigation, but also the best means of developing precise, logical thought patterns. All academies, especially those which took boys at an earlier age, also included some general education. Indeed, from the beginning, many naval training colleges, while placing primary emphasis on seamanship and navigation, mingled general and professional education. By the twentieth century a basic requirement of general education pertained everywhere, either for admission or in further courses necessary for graduation.

The primacy given to military training and indoctrination in many of the academies often undermined their standards in general education, and sometimes in technical education as well. Thus a series of British royal commissions and other investigations in the late nineteenth and early twentieth centuries severely criticized the failure of the Royal Military College, Sandhurst, to maintain academic standards. They held that this lowered military proficiency. The commissioners usually contrasted Sandhurst, the cavalry and infantry school, unfavourably with the engineering college, the Royal Military Academy, Woolwich; but at times that academy did not escape unscathed. Then, when the two institutions were amalgamated as the Royal Military Academy, Sandhurst, after the Second World War, the liberal arts courses at RMA Woolwich were found to have been superior to those at RMC Sandhurst, where primacy had been given to practical military training. Technical officers in Britain had thus been more broadly educated than, and were often intellectually superior to, those whose career was more closely related to combat.

It was not only in Britain that academic standards among combat officers were low. Republican France commissioned less on the basis of social position and wealth than in Britain, but, although academic standards were enforced by competitive examinations, the French also promoted on other considerations such as horsemanship. Prussia had prescribed academic qualifications for commissioning since 1808 and had a structure of junior cadet schools with examination hurdles and rigorous instruction. There were high academic qualifications for membership on the General Staff, but preference was also given to aristocratic birth in commissioning and promotion. When bourgeois candidates penetrated the Prussian officer corps, they aped the manners of their aristocratic fellow officers.

The American academies were sometimes accused of nourishing a pseudo-aristocratic elite of the sword like those in the Old World, but the American practice of preparing line and deck officers in the same academies as officers destined for technical duties, along with the prevailing American scorn for Old World class distinctions, helped check this trend and sheltered academic standards. Even so, when a new approach to military professionalism in the 1920s led to a revival of the humanities and social science courses at the United States Naval Academy (where they had declined since its early years), the midshipmen called them "bull" and slighted them in favour of seamanship and the sciences.[9]

As a consequence of social privilege, then, or of overemphasis on motivation and training, the military has too often assumed that intellectual effort and a general education are secondary in the production of officers, especially of those officers who do the actual fighting. Educational curricula and standards have consequently often received short shrift. In some quarters, general education is still assumed to be of lesser relevance. Addressing a British commission investigating military education, a speaker expressed the belief that, since some officer cadets prefer to get on with practical soldiering as soon as possible, general education should be kept to a minimum.[10]

One other circumstance impeded academic development. Before technology-inspired changes affected armies and navies in the twentieth century, military science only inadequately appreciated their significance. Five years after Ivan Bloch had published his prescient forecast of the effect of greater fire-power on the conduct of future wars, a popular academic historian, surveying the development of military science in the century just past, revealed the extent of contemporary unawareness of the coming danger when he made no mention of the vast changes that technology had already brought in war.[11] Surprisingly, most professional officers and general staffs also failed to realize how much warfare had been altered by weapons development until they learned from bitter experience in 1914–18. Even after the First World War, only a few lone prophets or practitioners, including Basil Lidell Hart, General J.F.C. Fuller, Charles de Gaulle, and General Heinz Guderian, and then in a partial and distinctly flawed manner, appreciated that the introduction of tanks and aircraft meant a return of campaign mobility.

After the Second World War, the American General James Gavin, one of the most articulate exploiters of greater mobility in operations, discussed the postwar introduction of intercontinental nuclear missiles in a chapter entitled "The Most Significant Event of Our Time." Arguing that the traditional military virtues are still paramount in officer training, he said that the "quantum jump" in warfare was mobility. He thus appears to have overlooked what was, when he wrote, the revolutionary effect of nuclear missiles and the vastly increased destruction now possible at a distance.[12] Even with the most perceptive of military thinkers, then, full comprehension may lag behind technical change. Consequently, as a legislator told the Howard-English Commission, a solution for the problems that technological change brings to military education is always likely to be controversial.[13]

Gavin was correct when he said that technology has not changed the qualities needed for leadership; but what he failed to add is that technology has changed the military leader's environment in at least four ways. It has

Hockey: RMC vs USMA, Kingston, 1990

made available much more complex and infinitely more destructive weapons; it requires personnel with more complicated skills; it has provided a more elaborate communications system that centralizes control of the battle; and, because of the potentially devastating effects of deterrent strategies with nuclear weapons, it has made necessary a closer relationship between the military command and the political leadership. All these elements are quite different from what prevailed, even in the most advanced countries, at the beginning of this century. Consequently, there is a professional requirement for greater competence, more specialization, and better teamwork.

Technological needs have also increased the educational levels and range of competence required of enlisted men. Educated persons can be more difficult to lead, but the results are better. Technology has strengthened the need for the officer to vary his leadership to fit new conditions. Leadership can still be either authoritarian or motivational, but effective leaders now use both kinds together. The basic task is unchanged – to organize, focus, and motivate people – but the qualities of an effective leader have become more elusive, more psychological, and perhaps more mystical. A recent academic study of leadership argued that the problem of combining academic standards and traditional military indoctrination can be "explosive."[14] Successfully managed, it can also be extremely effective.

The Soviet military faces these same problems. The Soviet army assumes that, along with the size of forces, equipment, and the ability of personnel to handle it, the determining factors in conflict are a combat spirit and military tradition; consequently, it puts considerable stress in its officer production on military training and motivation.[15] However, there are some signs of a different approach. Major-General G.I. Pokrovskii, a distinguished Russian physicist and engineer, maintained in 1959 in his book, *Science and Technology in Contemporary War*, that "to conduct military operations successfully a well-prepared, technologically literate, [as well as] physically tough, fighter is needed," yet in the light of contemporary development in military affairs, military pedagogy requires "constant and searching creative development."

Pokrovskii added that "each individual military man" needs a correct understanding of the laws of physics and mathematics, rather than a mere rote memorization, and also of the logical method of mathematical analysis, of proof, and of the theory of probability, of the use of calculating machines, of computers, and of linear programming. He said that physics unites quantitative mathematical methods with experimental research and is the foundation of technology for military affairs and new means of combat. He noted that mechanics, long used for the solution of artillery problems, has now been applied to the conquest of outer space. He also instanced

as useful for military personnel, gas dynamics, hydrodynamics, atomic energy, heat resistance in materials, chemistry, and the biological sciences.

Pokrovskii stressed that modern technology has greatly increased the need for scientific and technical education in Soviet military academies. He asserted that "the requirements levied on man grow in proportion to the growth of technology," but concluded, "no matter how much technology has developed, and no matter how much it has facilitated the solution of combat missions, it does under no condition lead to a reduction of the requirements, both physical and moral, of man in war."[16] Such a broad spectrum of scientific studies, although required by modern technology across a whole officer corps, obviously could not apply to each individual.

Like many in the Western world, Pokrovskii may have been overreacting to Sputnik. He was misleading in his scant reference to another well-known aspect of Soviet officer production – the political propaganda that is used to stiffen military indoctrination. But what was more serious was that, like many of his Western counterparts, Pokrovskii also failed to notice some effects of technological advance and politico-social change. Professional soldiers are primarily concerned with the use of arms, and therefore with their development. But technology can alter, and indeed has to some extent already altered, the way in which military force must be used to resolve international problems. An early popular misconception was that the possibility of universal catastrophe by nuclear power would mean the eventual end of the use of force in war and the introduction of a new international order to be maintained by international constabulary forces. That dream has long faded, but, as several scholars have argued, military force has become an increasingly doubtful tool for use in some aspects of international relations. The distinguished French philosopher Raymond Aron conceived a sophisticated version of this idea – that technology has brought the substitution of crises for wars between the superpowers.[17]

Cadet Squadron Officer Ramsey M. Withers, RMC, 1952

General R.M. Withers, CMM, CD, CDS, 1980–3

Thus, while it is true that wars have actually become more numerous since the discovery of the way in which nuclear fission and fusion can be employed, total global conflict was held off. It has been prevented by military power, but that power has been largely exerted not in operations or by concerted international deterrence, but by the more dangerous device of mutual deterrence. National military forces, using the highest possible application of technology, remained an indispensable mainstay of the international state system, but they are now generally applied differently, notwithstanding the wars in Vietnam, the Falkland Islands, and Afghanistan.

The functioning of deterrence to prevent war, and also of international peacekeeping forces to contain crises and ease tensions among smaller powers that are less restricted than the superpowers in their proclivity to resort to war, requires the maintenance of military forces with traditional skills and qualities as well as with technical competence. These instruments cannot be fully understood and satisfactorily applied by an officer corps trained and educated only to use the most effective means of destruction available. At the same time, an exclusively scientific foundation for all officers on Pokrovskii's model would restrict the possibility of sound military accomplishment by the professional cadre operating as a whole. Therefore, although modern military education must seek to preserve the old military virtues – courage, loyalty, and obedience – and must add the skills and knowledge required by the new technology, it must also lay basic foundations for broad knowledge and wisdom. Those foundations can best be derived from a general education including social and humanistic studies. Accordingly, military education in the West has tended to become more like education in civilian universities, not merely in its content of science and technology, where it was always similar, but also in the addition of non-scientific studies to give broader understanding.

Another major factor that has reinforced this development in the content of academy education is the changing relation of the military profession to the society it serves. The soldier in the ranks, whether a conscript or a volunteer, is no longer an uneducated socioeconomic misfit. He is drawn from society at large. Although always subject to the possibility of an emergency during long periods of peace, he often resides in a civilian community rather than on a remote military base. The tasks he performs appear to be similar to those of his civilian neighbours. He may even work on an eight-hour basis, but subject to overtime without extra pay and to posting wherever and whenever the service orders. "Civilianization" of this kind means more than using civilians to perform duties once undertaken by military personnel; and it has affected commissioned ranks because officers are no longer drawn exclusively from an aristocratic elite. Retaining their distinctiveness, they must now earn their commissions by meeting academic as well as military criteria. They have more contact with their civilian counterparts, and they must be able to deal with them on terms of equality to maintain the prestige of their profession.

Professional military education is following trends observable in other professions. Although occupational specialization has become more common at the undergraduate level in all fields (as it already had earlier in military academies), the better liberal arts colleges, as well as the professional graduate schools, increasingly recognize the need to prepare their students "to act intelligently in the broader contexts of life as in their own work."[18] But Edward Katzenbach, a former US assistant secretary of defense, in a foreword to the book from which this quotation is taken, William Simons's *Liberal Education in the Service Academies*, said that the modern military officer must have an even more basic general education than that deemed necessary in the other professions.[19] This opinion was substantiated by Oxford professor Sir Michael Howard, when he co-chaired the commission to investigate military education in Britain in the early 1970s. He stressed "the need for character formation and character training to turn the boy into a man and the man into a leader, the need for technical training to make him handle the growing complexity of technological tools which will be at his disposal during this career, and third, and not least, the intellectual and moral education of the young officer as a whole, *the stretching of his mind*." The Howard-English Commission recommended the establishment of a military university, but it was not implemented because of cost.[20]

General Andrew Goodpaster, the former American commander-in-chief of US forces and supreme allied commander, Europe (SACEUR), who was recalled to service to take charge of West Point after it was wracked by cheating scandals among cadets, explained the reason for broader and more intellectual studies for modern officers: "If our military establishment is to fulfill its assigned role and do so in ways acceptable to the parent society, it must meet demanding standards of performance. Extensive programs of precise, coordinated training and education in a wide range of studies are needed. Individual competence must be developed in many subject areas." He said that military personnel share the same educational needs as the civilian community in many areas, and he proposed that these studies should be "civilian-based."[21]

Goodpaster was referring here to military education over the whole range of military service. He was not challenging the West Point practice of using service personnel to instruct cadets in many fields of study, even in those that are obviously "shared" with the civilian community. "Shared" academy studies include not only the sciences, but also the social sciences and humanities, because they foster flexibility of mind and a capacity to grasp ideas. This is a facility the military profession needs as much as, or more than, any other profession. To quote William Simons: "Among military officers of today (and certainly of tomorrow) creative thought, social conscience, and broad vision are traits perhaps even more necessary than among leaders in other professions." Simons held that what was required was a "liberal education" – one that paved the way for a better understanding of the complexities of human problems.[22]

These trends in military education have been discussed more in the United States than in many other countries. They are often seen by contemporaries merely as a form of civilianization, with a convergence of the functioning and performance of the military with their civilian counterparts, and with greater interdependence and assimilation in society as a whole.[23] Similar trends have occurred in the relations between the military and civilian societies in all democratic states.*

From the military point of view, civilianization often connotes a tendency to apply civilian standards of work to its very different conditions and service. In Canada, as elsewhere, civilianization may seem to the military to weaken military professionalism and its capacity to carry out effectively its role as protector of society. For some service personnel, a liberal education appears to be incompatible with their concept of military training and duty, especially in the combat arms and the lower commissioned ranks. For these people, general education detracts from what they believe to be the important parts of preparation for the military profession. They would

Cadet Wing Commander Paul D. Manson, CMM, CD, RMC, 1956 (Photo courtesy of General Manson)

General P.D. Manson, CDS, 1986–9 (Photo courtesy of General Manson)

*David R. Segal argued that this convergence of military and civilian society does not necessarily mean interdependence, and that the ending of the draft in the United States would possibly cause the military to develop its own distinctiveness by the isolation of its personnel. "Convergence, Isomorphism, and Interdependence at the Civil-Military Interface," *Journal of Political and Military Sociology* 2, 2 (fall 1974): 157–72.

Conflict of opinion on this issue was a continuance of a difference of opinion between Huntington and Janowitz about the proper role of the military in a democratic society. Huntington held that the military should be politically neutral and isolated from society. Janowitz believed that the military was becoming, and should become, integrated with society. Arthur D. Larson, "Military Professionalism and Civil Control: A Comparative Analysis of Two Interpretations," *Journal of Political and Military Sociology* 2, 1 (spring 1974): 57–72.

prefer more military training and professional indoctrination.

Lieutenant-General Sidney B. Berry, superintendent
of West Point, speaking about USMA in 1976 at RMC's
Centennial Symposium of Military Education, placed
primary emphasis not on the academy's contribution in
education, but on its traditional function to "strengthen,
develop, and nourish in the cadets those values, attitudes, and qualities that prepare its graduates to deal
effectively and victoriously with the constants of the battlefield." He illustrated his point by a simple diagram
showing USMA as the connecting link between American
society and the battle. His concern was to indicate that in
the early years of a military career the young officer, especially in the combat arms, has more need for military
training than for a general education. Berry noted that
from 1964 to 1976, although the academy had altered in
many ways in response to military needs and social pressures, "the focus of military training of cadets tended to
change from preparation for generalship to preparation
for lieutenantship."[24]

These were the years of Vietnam when the immediate
American demand for junior officers, as is customary
in wartime, had risen sharply. In more normal times,
the problem of providing a training suitable for junior
officers, when some of them must be simultaneously educated for future higher command, is more of
a dilemma.

Supporters of a broader education in military academies contend that basic intellectual processes are not pertinent to particular professions but are general in
application. They argue that in preparation for lifetime
careers, a narrow professional education can mean rigidity and the numbing of originality. Nevertheless, when
John P. Lovell, in his book *Neither Athens nor Sparta*, presented a powerful case for more intellectual education
for soldiers along with professional development, he was
compelled to conclude that there is no consensus on what
the changes in military education should be.[25] The basic

unresolved problem is how professional indoctrination
can be preserved in an increasingly academic milieu –
and that is the theme of this book.

Other recent developments that affect military academies and the production of officers are the application
of behavioural and managerial sciences to organizations
and the use of electronic devices for computation and
analysis. The resulting concept of the cadet as manager-
in-training may sometimes be more appealing to the military than is the alternative concept of the cadet as
scholar. However, two former officers blamed the American failure in Vietnam in part on excessive resort to
managerial theory and techniques in the United States
army.[26] This was undoubtedly an oversimplification,
but it is true that courses on management techniques
and psychology have to some extent supplemented
or replaced earlier instruction in military colleges on
leadership.[27]

In some cases that trend may have gone too far. In
1980 the American Army Staff College's *Military Review*
devoted an issue to replies from officers in the field about
this leadership and managerial-sciences issue in the
United States. Most contributors pointed out that command and management are radically different. In the
following issue of the *Review* Dr Sam Sarkesian, a military
sociologist, agreed. He added that leadership has become
more complex as a result of technological development
and the switch from a military to a political-socio-military
environment: "there is continuing disagreement regarding the meaning of management and its relationship to
the military as an institution, and to the concept of leadership."[28] Although many Canadian officers are critical
of this debate, the Canadian Forces face this same dilemma in some degree.

Finally, a survey of universal trends in military education must necessarily take into account the attitudes of
the age group from which officers come – the eighteen-
year-olds and adjacent ages. The last half-century has
seen social and political revolutions of unprecedented

proportions, including the overthrow of colonialism. Many of these changes involve a greater sharing of benefits and a voice in decision-making. At the same time, traditional forms of authority have been superseded or weakened, and behaviour patterns relaxed. There has also been a great increase of political demonstrations, revolts, terrorism, guerrilla activity, and crime, and also of various forms of egalitarianism and "participatory democracy." No segment of the population has been as much affected by, and as much involved in, this worldwide phenomenon as the group in the high schools and universities. Throughout the West, students have demonstrated against war, militarism, and nuclear policies. Their pacifism reached its peak in the United States at the height of the Vietnam War, and the protest movement spread to Canada, even though few Canadians were directly involved in that war.[29]

In the United States the student protest movement declined with the ending of conscription. There and everywhere else the subsequent onset of economic adversity and a resultant concern for jobs has made youth protest somewhat less critical. It appears also to have increased interest in military academies. But the youth movement led to some visible changes in Canadian and other military colleges, including a liberalization of cadet lifestyles to allow cars, marriage, and bar privileges. The admission of women into the cadet academies may also be described as a yielding to external popular pressures from a protest movement. What the effect of the so-called reforms in cadet lifestyle will be on youth's willingness to embark on military careers is not yet clear. A significant decline in the pool of potential officer candidates in Canada may have been caused not only by this alienation of youth, but also by the absence of imminent danger for Canada and a need for military preparation to counter it.

Earlier, with the quickening of the Cold War, the need for officers had increased in all countries, including Canada. In order to meet greater demand for, and pres-

Hudson Squadron caught with its pants down, 8 May 1987

sure on, recruiting, which has been caused by new needs as well as by new academic standards, academies often resorted to offers of a free university-type education in return for a limited obligation of military service. Such offers were attractive because of the growing cost of university fees and because they provided alternative opportunity for social and economic betterment. But there have been serious consequences. First, it is still difficult to secure enough applicants who have the personal qualities deemed requisite for an officer but who can also meet the academic standards for entry. Second, the increased burden of academic work along with professional preparation and training falls heavily on those cadets who lack intellectual attainment or who may have entered an academy merely to get a cheap education. When there are shortages of academically qualified candidates for commissions, there is a temptation to reduce academic standards and barriers.

Sometimes a consequence of these various circumstances is said to be a decline of military professionalism, especially in Canada where military need is less clear. Some blame the growth of academic requirements and non-military content for allegedly diminishing cadet professional motivation and commitment and the supply of officers. Thus, American air force officers who have considerably more academic qualifications than their Royal Air Force counterparts are said to show a slightly less degree of commitment to a military career. But this "more purely heroic attitude" in the RAF may in fact be

Exercise Courageous Cub, 1985. Basic military training continues after arrival at the college.

due to what an American sociologist has called "family patterns based on elitist, ascriptive, and particularistic criteria."[30] It may also be due to the fact that lower academic qualifications mean less alternative opportunities and therefore a greater commitment to the career in hand. The Royal Air Force, a relatively young and glamorous service with a modern image, has perhaps been able to draw longer from the elite class that was formerly the mainstay of British officers. Another critic said in 1959, however, that the Royal Military Academy, Sandhurst, which combined the old RMC Sandhurst and RMA Woolwich and which could no longer depend on the elitist British public (private) schools, was designed to produce officers of the old type by methods that had not been adjusted to modern conditions.[31] This problem of adapting education and training to the society from which recruits come must be kept in mind when talking of military education in Canada and of its adaptation to Canadian needs.

Because tradition is an important element in every military academy, there is a tendency for military authorities to cover up weaknesses by asserting that the old standards and accomplishments have in fact been successfully maintained. John Lovell criticized such claims when they were made in relation to American academies in the immediate post-Vietnam era. He said "sensitive observers know better." But he also noted that even in pre-war years, indoctrination with military ideals was never as completely accepted by cadets as the authorities would have liked.[32] There is often much distortion in academy propaganda, and the effect of military academy preparation on cadets and on their degree of professional commitment is hard to establish with certainty. At West Point empiric studies have shown that, while there is discernible change in cadet attitudes towards superficial career patterns, there is also evidence that "cadets do not experience profound changes in the direction of conformity to a common mold" in their broad professional perspectives as a result of academy indoctrination.[33]

Cadet commitment to a military career and acceptance of traditional ideals of military professionalism presumably correlate to some degree with rates of career retention. It was once shown that in the United States, except at the Naval Academy, retention rates from the service academies had been lower than from the university ROTCs.[34] This suggests that the four years of socialization in the academies has not always been as effective as was desired or expected. It will be shown later in this book that Canadian experience is different. Furthermore, there is evidence that, in Canada, commitment to a career grows more out of years of actual service than from college indoctrination.[35] Here again, however, the rise or decline of possibilities for alternative employment must also be a factor that has to be taken into the reckoning. Moreover, if academy training prepares a cadet to cope better with future military service, it presumably also prepares him for a steady increase in the degree of his commitment as he climbs the promotion ladder. The contribution that an academy makes to career commitment is thus important, but its nature is not simple.

As a consequence of the impact of technology and socio-political forces on military colleges, and of the uncertainty about how best to adapt them to serve the needs of armed forces in a changing society, it is not surprising that the American academies have experienced problems with the maintenance of their traditional codes of conduct. Similarly, in the United Kingdom, the length and nature of educational training has been a subject of vigorous debate.[36] In Australia, the question of tri-service training and a university-type education has been explored at length.[37] In the United States, where public debate is more prolific and less restrained than elsewhere, there has been scholarly and media discussion about the merits of retaining academy training or substituting other forms of officer production – for instance, through university ROTCs.[38]

In Canada there has been a series of departmental investigations into the nature and degree of profession-

alism attained in the armed forces and the military college system. Questions about the merits of academy or university forms of officer production crop up from time to time. Concern is frequently expressed about the larger cost of the military colleges and whether the advantages they offer in indoctrination and professional training sufficiently offset the savings that could be made by subsidizing university officer-training corps undergraduates who would commit themselves to a military career. This question is becoming more pressing in view of the growing extent to which it is necessary to send commissioned officers later in their careers for further specialized education, frequently in civilian graduate schools.

This history of the Royal Military College of Canada since the Second World War as one of the Canadian Services (now Military) Colleges is designed to facilitate understanding of these important universal problems of military education. It must take into account the fact that,

Prime Minister Brian Mulroney visits RMC, 5 February 1991.

because of Canada's earlier relationship with Britain, officer education in the dominion was originally directed primarily towards a part-time militia, and only to a small extent towards a Canadian regular force and the British army. Now, as a junior partner in an alliance with the United States, Canada must still accept the need for co-operating in defence measures it could never undertake alone. Canada has always had less incentive for defence preparation, but greater freedom to experiment.

Furthermore, Canada's RMC is still affected by problems and developments that are peculiarly Canadian – for instance, the creation of the tri-service system and the integration and unification of the armed forces, the existence of three regional colleges each serving substantially the same purpose, and the growth of bilingual programs in an attempt to ensure the preservation of a country with two founding peoples. Measures of national indoctrination and nation-building have always been important in all military colleges everywhere. Canada's ethnic diversity, and Canada's need to preserve its identity and independence distinct from its neighbour to the south, make them particularly so in RMC. For all these reasons, the history of RMC may have something to teach other countries.

The graduating class of 1989 marches off the square.

CHAPTER 1

The Old RMC and the New Canadian Services Colleges

Military academies strive to promote military skills and knowledge, corporate sentiment, and motivation – the essential elements of military professionalism. In all countries, methods of pre-commissioning education and training vary according to the perceived role of the military, the social and political conditions, and the historical development of military education. The early history of the Royal Military College of Canada illustrates the point.

RMC's story down to its re-establishment after the Second World War was covered in considerable detail in *Canada's RMC*.[1] The final chapter sketched further development up to the acquisition of degree-granting powers in 1959, but only in broad outline. This book will expand on the last part of *Canada's RMC* before recounting the college's postwar history as one of three units in the Canadian Military (formerly Services) Colleges' system established in 1948. When RMC reopened in 1948, it maintained some traditions tested by time but introduced other innovative measures. These features included tri-service education and preliminary academic progress towards degree-granting power. Soon afterwards came another innovation – the Regular Officer Training Plan (ROTP) – whereby the college concentrated almost exclusively on the production of career officers.

To provide a background for this new story, it is necessary to review briefly the history of the old RMC. This outline will bring out more clearly the relation of subsequent developments to Canada's defence problems. It considers why RMC was originally set up in 1876 as a unique military college that educated for both military and civilian careers, how that dual capacity strengthened academic standards, and how from time to time changing defence needs or other problems affected the college's development. The story of the campaign to have RMC reopened after the war will also be recalled, with added information to remind readers that the old college was, in effect, re-established. The chapters that follow will explore the influence that the revival of the old college in 1948 had on RMC's further development to meet Canada's defence needs in the second half of the twentieth century.

Before the British North American colonies were granted responsible government and before Confederation joined them together in 1867, provincial militia acts had provided only for the registration of able-bodied males, sometimes with a little training, who would then serve in support of the British army in a major emergency or under their British governor and his military staff in lesser dangers. In 1867, when the memory of fear of an American invasion during the Civil War was still fresh in Canada, the new dominion parliament's first piece of legislation was a Militia Act to set up a Canada-wide military force under Canadian control. Sir George-Étienne Cartier (Macdonald's French-Canadian lieutenant, and minister of militia 1867–73), who introduced the bill, said that a defence force was essential to nationhood;[2] but since Canada still expected British military guidance and support in the event of serious danger, the new force was only a voluntary, part-time militia. In 1870 the British garrison began to withdraw from most of the country, and Canada moved involuntarily a step closer towards complete responsibility for its own defence. Then in 1874 the Royal Military College Act provided for the education and training of young Canadians to fit them to be officers, either in the Canadian Militia or in a Canadian Permanent Force that the college's founder, Prime Minister Mackenzie,* was later to claim he had contemplated. Even if that claim were an exaggeration, his measure was a significant move towards providing Canada with an increased military potential.[3]

The Canadian military college differed significantly from similar colleges of the period. British and European military academies usually produced only career officers for regular armed forces.** In the early days at the United States Military Academy (USMA) at West Point, the graduates followed either military or civilian careers, but after the Civil War its output became entirely military. The United States Naval Academy produced career officers from the first. When RMC was set up, Canada had no permanent force, apart from two artillery schools established in 1871. While the Treaty of Washington was being negotiated in 1870–1, Canadians began to think that the United States was no longer a threat. By the time the first RMC class graduated in 1880, prospects for a larger Canadian permanent force seemed dim. A few of the staff appointments in the new militia infantry and cavalry schools that opened in 1883 went to ex-cadets. By the mid-1890s, nineteen RMC graduates had joined Canada's Permanent Force, but most others had to seek employment in the private sector. Many hoped to be civil engineers, and the RMC curriculum was designed to make that possible.

One feature of RMC, borrowed from West Point, was vital for future success in this respect. The course was four years long, the same as in Canadian universities, compared with two-and-a-half years at the Royal Military Academy (RMA), Woolwich.*** At the outset, it was established that the four-year course would include compulsory mathematics, fortification, artillery, surveying, military history, law, strategy and tactics, French or German, chemistry, geology, and drills and exercises. There were to be advanced options in mathematics, fortification, chemistry, physics, languages, architecture, and civil engineering. Because many of these subjects could serve either a military or a civil purpose, cadets were able to prepare for civilian careers with the options available.

To open the Canadian RMC in 1876, Lieutenant-Colonel E.O. Hewett of the Royal Engineers,† who was appointed as its first commandant, recruited two British Royal Artillery officers as professors. Like him, they had graduated from Woolwich. He also secured the appointment of a Canadian civilian to teach German, French, and English. During its first two decades, the RMC "superior" staff usually had up to fifteen members, ten of them professors. One or two of the professors were British officers seconded from the British army; five or six were civilians; and the remainder were Canadian Militia officers. After 1882, two of the militia officers were ex-cadets, who assisted in teaching several subjects and doubled as company commanders.

In the early 1880s, one of the professors was a leading Canadian chemist, Dr H.A. Bayne. Later, another was a research-oriented physicist, Dr J.A. Waddell. A third was a well-known Canadian painter, Forshaw Day, who taught

*Alexander Mackenzie, second prime minister of Canada 1873–8.

**France's École polytechnique, which simultaneously prepared some students for non-military government service, was one exception.

***After 1882, the RMA course was only two years. F.G. Guggisberg, "The Shop": The Story of the Royal Military Academy (London: Cassell 1900), 153–4.

†Lt-Gen. Edward Osborne Hewett, commandant 1875–86, later superintendent of the Royal Military Academy, Woolwich.

sketching.* By 1890 four civilians were teaching arts subjects, one instructed in civil engineering, and another in the sciences. That year a seventh civilian was appointed to teach the mathematics hitherto taught by an officer. From 1892 to 1897 there was a professor of literature as well as a professor of English. Clearly, education at RMC had become more civilianized, with greater emphasis on the arts; but the college still had a program and a faculty that could also prepare cadets for a military career. Three professors with military connections taught subjects that had military connotations. Other individuals who contributed to laying a foundation for military professionalism included a retired British officer as commandant (Major-General D.R. Cameron),** a Canadian staff adjutant (Lieutenant-Colonel S.C. McGill)*** who had served in the British army but was now in the Canadian Militia, and a British regimental sergeant-major (J. Morgans),† an internationally known athlete, who taught drills and exercises.

The four-year course had been adopted not merely to equate with Canadian universities but also because, like the United States, Canada had no traditional military aristocracy or "leisured class" as was found in Europe. Time was needed to make up for the lack of a Canadian military tradition. For political reasons, the college had to be open to a broad stratum of society across a large and still undeveloped country. Virtually no colonists had yet settled on the prairies, and British Columbia was only sparsely settled. Few Maritimers or French-speaking Canadians enrolled. The majority of the early cadets came from well-to-do English-speaking families in Ontario and Quebec, and even they, including some who had been at private or residential schools, needed the four-year socialization process to shape them in a military mould. They were kept isolated from civilian society, especially in their first year. More liberal than West Point in this respect, however, RMC adopted university terms, gave its recruits Christmas leave, and closed for two months' vacation in the summer. RMC gentlemen-cadets were treated as rankers (the West Point method) rather than, like British cadets, as potential officers.[4] The difference was important. Life under strict military discipline enforced conformity on a diverse group of unmilitary colonial boys.

What was most effective in this respect was the deliberate subordination of recruits by seniors for the whole of the first year. "Recruiting," as this practice was known at RMC, claimed to inculcate the military virtues emphasized in Hewett's college motto, "Truth, Duty, Valour." Common misery under subjection bound the recruits into a tightly knit group that later merged its corporate sentiment in loyalty to the college and in support of Canadian military preparation.

This military-civil educational system was designed to build a reserve of officers for possible emergencies in a country where defence was not an immediate problem. In later years other British colonies in similar situations showed interest in the Canadian example. The states of Victoria and Queensland in Australia, the Dominion of New Zealand, the Union of South Africa, and the Commonwealth of Australia all considered its use. None of them adopted it, but in most cases that was because of the cost of a military college rather than because they found the system unsuitable. When, in 1910, Australia decided to establish Duntroon, its planners opted for a college that would produce only regular officers.[5] The Royal Military College of Canada remained unique.

*For the careers of Bayne, Waddell, and Day see Preston, *Canada's RMC*, 69, 83, 143.

**Maj.-Gen. Donald Roderick Cameron, RA, had a brilliant record as a young officer on operations in India. After retirement from the army he was engaged in several international diplomatic missions. He was commandant 1888–96.

***Lt-Col. Sydenham Clitheroe McGill joined the Royal Canadian Rifle Regiment as an ensign in 1859. He was staff adjutant 1883–1900. See Preston, *Canada's RMC*, 111n.

†For Morgans see ibid., 147–8.

To promote the military tradition, RMC quickly built up its own pantheon of heroes, beginning with the "Old Eighteen," the first class that had voluntarily submitted itself to some of the rigours undergone by later recruit classes. In the course of time the names of ex-cadets who had earned distinction in distant imperial wars were added, though hero-worshipping was never done as systematically as at West Point and Annapolis, perhaps because the RMC heroes were not national figures. RMC had, however, rapidly developed a group military spirit. This was shown when the cadets volunteered to go as a unit with the militia force sent to suppress Riel in 1885. The offer was politely refused, but it was solid evidence of military zeal in the college.[6]

A year earlier a group of ex-cadets had planned the Royal Military College Club to further the ideals they had assimilated during their four years at RMC. Ex-cadets retained, and used, their college numbers along with their surnames as a mark of identification among themselves. They became a distinct group with a sense of close identity. In their four cadet years under military routine, they had gained an insight into military professionalism as they steadily exercised more authority and learned something of the substance of a military unit.

The RMC Club was not a professional military organization. Most of its members, despite their military training and interest, were civilians. Unlike their fellows in Britain and the United States, Canadian ex-cadets could not merge their RMC-inspired military identity in a greater whole because Canada had almost no army. Their professionalism had, in effect, been cut short at graduation. The RMC Club was, consequently, a fraternal organization, with many members prominent in Canadian life. It was devoted to the ideal of military preparation for Canada and the British Empire. Carl Berger has shown that in the minds of Canadians of their class at that time, these two interests coincided;[7] and it was to serve those interests that the members directed their efforts towards maintaining the college.

Even though membership in the RMC Club was open to all ex-cadets whether or not they had graduated, the total number eligible for membership in March 1885 was no more than 116.[8] Down to the end of the nineteenth century the college had room for sixty to seventy cadets in single rooms. The size of this contribution to defence compared favourably with that of West Point in the United States, where the population was ten times as large. Enrolment at West Point was only 252 in 1892.[9]

It disturbed some Canadian anti-imperialists that the college had sent more graduates to serve in the defence of the empire than in Canada. Of approximately two hundred who had graduated by 1895, when eleven were in the Canadian Permanent Force, six in the North-West Mounted Police, and about fifty-seven had militia commissions, at least sixty-three were in the British army.[10] In the two decades after RMC was founded, except for the Riel rebellion in 1885, Canada's need for defence had seemed to be declining. RMC graduates who wanted a military career had to turn to the British army. Some allegedly sold their swords elsewhere.*

A substantial number of RMC ex-cadets were in the British army because, before the first class graduated in 1880, the British War Office had agreed to give the Canadian college four commissions annually. At least one of these commissions could be in the prestigious Corps of Royal Engineers, entry to which was normally by academic competition from RMA Woolwich. In years of imperial crisis, more than the four commissions were granted. RMC graduates normally went to the School of Engineering, Chatham, or, if destined for the Royal Artillery, to its school at Shoeburyness – the same path as the cadets from Woolwich. They compared well. At first it was proposed that, before Canadians could receive their commission, they would require additional training

---

*The House of Commons was told that a number of ex-cadets from RMC commanded troops on both sides in the Sino-Japanese War. *Debates*, 1896, II, 6761. This allegation cannot be confirmed.

to make up for experience not available in Canada[11] – for instance, they should witness the manufacture of ordnance. This proposal was never adopted. The longer Canadian course, with its military and civilian content, apparently compensated for lack of practical experience with troops and lack of visits to major industrial plants in Britain. The academic training at RMC appears to have taken its graduates at least as far as Woolwich-trained officers.

RMC cadets in the British army had remarkable careers. In 1899 Sir Charles Tupper* drew the attention of the House of Commons to several who had already done well. He and the Liberal minister of militia, Frederick Borden,** named Percy Girouard, Philip Twining, Van Straubenzee, and Huntley Mackay, all technical officers.***[12] From the first, then, RMC turned out excellent "professional soldiers" for the technical corps.[13]

RMC ex-cadets in the British army also distinguished themselves as "professional soldiers" in the combat corps. Several modern scholars have argued that most British non-technical officers in the nineteenth century did not measure up as military professionals by standards already established in Europe. Although Britain had abolished commissioning by purchase a few years before the Royal Military College of Canada was founded, entry to cavalry and infantry regiments was still by invitation, and an officer usually needed a private income to supplement his pay. After the abolition of purchase in 1871, the War Office experimented briefly with direct commissioning from the "public" (private residential) schools, with a brief period of regimental service followed by a post-commissioning course at Sandhurst. The experiment failed, because officers did not study. Sandhurst soon restored the cadet system, but the qualifying examination for admission as a cadet at Sandhurst continued to be low. British officers were in effect gallant amateurs, cultivating a *credo* that all they needed was the natural leadership conferred by breeding and courage. Safe behind the Royal Navy, Britain could afford this type

of army officer, and in Britain's little colonial wars they performed satisfactorily.[14]

In comparison, RMC graduates commissioned in non-technical regiments showed up well, but because those commissions were much in demand the War Office was reluctant to open the doors too wide to colonial applicants. By the end of the century, however, several Canadian ex-cadets had gained distinction in imperial military operations in addition to what was achieved in technical capacities – for instance, by the engineers Huntley Mackay, W.G. Stairs, W.H. Robinson, and K.J.R. Campbell in Africa.† H.E. Wise and H. Freer,†† who had taken infantry commissions in the British army but were

*Sir Charles Tupper was high commissioner in London 1884–96 and briefly prime minister of Canada.

**Sir Frederick Borden, a medical doctor, was minister of militia 1896–1911. A militia enthusiast, he instituted reforms and expanded the service.

***Col. Sir Edouard P.C. Girouard (no. 147) was director of military railways in South Africa 1899–1902, and governor of northern Nigeria 1908 and of East Africa 1909–12. During the First World War he was director-general of munitions supply in Britain. See Preston, *Canada's RMC*, 110n.

Maj.-Gen. Sir Philip G. Twining (no. 88) was an instructor and professor in military engineering in Britain 1895–9. He became director-general of fortifications and works in the War Office 1918–20.

When Tupper spoke, three ex-cadet brothers named Van Straubenzee had served in the British army: Sir Casimir (no. 162), who became a major-general; Col. A.H. (no. 23), who became commanding royal engineer on Salisbury Plain; and B.W.S. Van Straubenzee (no. 100).

Capt. Huntley Mackay (no. 39) became the acting administrator of the British East Africa Company.

†William Grant Stairs (no. 52), William Henry Robinson (no. 62), and Kenneth J.R. Campbell (no. 81) all served in Africa. Stairs was chosen by the American Henry M. Stanley for an expedition to rescue Emin Pasha. Robinson was the first ex-cadet to be killed in action. Campbell was deputy commissioner and vice-consul in the Oil Rivers District on the coast of Nigeria.

††Wise and Freer were no. 2 and no. 3, respectively, members of the "Old Eighteen." They both joined the infantry in the British army. Freer had served in Egypt, and Wise later served in India.

on half-pay in Canada in 1885, performed invaluable service closer to home as staff officers with the Canadian Militia force that went to suppress the Riel rebellion.[15] Several ex-cadets became British colonial administrators. The RMC experience had laid solid foundations for a military professionalism that compared favourably with that of the average British cavalry and infantry officer.

After 1888, the college began to decline. The War Office complained that ex-cadets commissioned in the British army were not up to the standards of their predecessors. Enrolment fell away and excessive "recruiting" attracted the attention of the Liberal opposition in the House of Commons. It was said there that the commandant, General D.R. Cameron, was out of touch with modern military developments, and that few of the graduates were serving Canada in the permanent force.[16] Opponents argued that RMC, where the fees had risen to $400 a year from the original $100, was being improperly maintained by the government to give rich men's sons a civilian education that, according to the BNA Act, should be given in provincial universities where it would cost less.

The college had turned more towards producing civil engineers, had expanded its staff of arts professors, and had appointed a civilian, Iva Martin,* to teach mathematics. In 1895, however, the Board of Visitors reported that some civilian professors were ineffective teachers. Lieutenant-Colonel F.C. Denison,** the father of one of the cadets who was a member of a noted militia family and a member of parliament, blamed Cameron and introduced a motion in the Commons to reduce the RMC appropriation by the amount of his salary.

Denison's amendment was defeated, but when in 1896 the House debated the condition of the militia, the RMC question came up again. Major-General Ivor Herbert,*** appointed GOC Canadian Militia in 1891, had already begun much-needed militia reform. The government dismissed Cameron and asked the War Office to send a British regular officer to command RMC. Possibly because no engineer or artilleryman was available, it sent Colonel Gerald Kitson,† an infantry officer. Kitson purged the staff, dismissed the civilian professors who taught civil engineering, French, English, and drawing, tightened discipline, restored morale, and reduced the course to three years.

Four years later, the war in South Africa marked a turning point in Canada's military history. Previous British colonial wars had involved only a few individual Canadians, but this was different. The Afrikaaners were a European people with modern weapons. Instead of Canada depending on Britain for defence, Britain now called on Canada for help. Many French Canadians, empathizing with the problems of another ethnic group in the empire, were skeptical that the empire's survival was sufficiently threatened to warrant Canadian participation. English-speaking Canadian imperialists forced Sir Wilfrid Laurier's hand, however, and he agreed to send a special force to South Africa.

Kitson obtained a British offer of extra commissions in the army for RMC cadets, but most of the senior class he had seen right through the college preferred to take civilian jobs. Disgusted, he resigned to become military attaché in Washington. Later, however, RMC cadets and ex-cadets amply made up for this initial failure of the military spirit.[17]

The war had aroused great patriotic excitement in English Canada. A call for volunteers from the permanent force, in which Herbert had organized an infantry

---

*Iva Edwin Martin, professor of mathematics, mechanics, and astronomy from 1890, director of studies (DOS) 1917–22.

**Lt-Col. Frederick Charles Denison, Conservative MP for West Toronto 1887–96. See Preston, *Canada's RMC*, 100n.

***Col. Ivor John Caradoc Herbert, Grenadier Guards, later Baron Treowan, GOC Canadian militia 1890–5 with the acting rank of major-general. His attempts to eliminate political patronage led to his dismissal. See ibid., 167.

†Maj.-Gen. Sir Gerald Kitson, King's Royal Rifles, commandant RMC 1896–1900. See ibid., 153–70.

regiment in 1893, and from the militia, which a successor, Major-General Edward Hutton,* was in process of developing into what he called a "Canadian national army," had brought quick response. Altogether, 113 RMC ex-cadets served in South Africa, but many of them were already in the British army. Nevertheless, some senior ex-cadets played important roles, especially in the military railway construction that made a vital contribution to victory. Lieutenant-Colonel Percy Girouard, who had built Kitchener's railway in the Sudan in 1896, Lieutenant-Colonel H.S. Greenwood, Captain Joly de Lotbinière, and Captain H.C. Nanton built and operated military railways with great efficiency.** RMC's education of civil engineers had thus paid off in war. It was to do so again in a greater conflict in France.[18]

After 1891, three GOCs, Herbert, Hutton, and Lord Dundonald,*** worked to reform and expand the Canadian Militia, yet each left without completing his tour of duty – mainly because of conflict with the Canadian government about political interference and patronage. At the same time, the South African War and the increasing tension in Europe made the militia problem urgent. Minister of Militia Borden, although he had dismissed two of the GOCs in order to assert Canadian control, worked conscientiously to achieve ends they had sought.

Britain's withdrawal from the western Atlantic pointed to the need for greater Canadian self-defence. The first step in 1905 was an arrangement to take over the garrisoning of Halifax from the British. To do this, the size of the permanent force had to be doubled to 4000. Although the idea of a possible danger from the United States was maintained for some time longer as the pretext for militia development, there was an unspoken realization that Canada might again be called on to send an expeditionary force overseas. A subtle change was made in the Militia Act with that possibility in view.[19] There were thus more openings for military careers in Canada. At the same time, although RMC continued the

military-civil system of education with Kitson's reforms, the commandants who followed him increased the emphasis on military training to prepare cadets for commissions. After 1910, when international tension was growing, all cadets were required to attend one militia camp in order to qualify for graduation, and they had to take up either a permanent force or militia commission. If not attached to a unit, they had to parade with the militia as supernumeraries.[20]

Between 1911 and 1914 the output of RMC graduates could have filled most of the vacancies in the permanent force, but only twenty-three of its 127 new commissions went to cadets: the remainder were filled from the militia. During the same period, nineteen ex-cadets went into the British army. Many young Canadians seeking a military career with good prospects thus continued to find a British commission more attractive than one in the Canadian Permanent Force. The remaining graduates took civilian jobs. Many members of the Canadian professional classes still sent their sons to RMC to prepare them for civilian employment. Once it was shortened to three years, however, the RMC course was no longer adequate for either military or civil purposes. In 1910 the RMC Advisory Board recommended that the course should include more advanced military training and that there should be an alternative option with less emphasis on engineering and more on general education. These

---

*Lt-Gen. Sir Edward Thomas Henry Hutton, GOC Canadian militia 1898–1900. A vigorous reformer, he was dismissed for intriguing with Canadian imperialists to involve Canada in the war in South Africa.

**Lt-Col. H.S. Greenwood (no. 57) commanded the Central South African Railway Volunteer Corps in South Africa. Maj.-Gen. Alain Chartier Joly de Lotbinière (no. 69) served in Africa and was commanding engineer with the Anzac Corps at Gallipoli and in France. Brig.-Gen. H.C. Nanton (no. 78) served in the North-West campaign, in India, and in charge of railway operations in the South African War.

***Lt-Gen. Douglas M.B.H. Cochrane, Earl of Dundonald, GOC Canadian militia 1902–4.

two changes, one professional and the other academic, could not be effected simultaneously unless the four-year course was reinstituted.

By 1914, RMC's enrolment had grown to 128 – nearly twice what it had been ten years earlier. Apparently quality had been restored sufficiently to satisfy the majority of applicants who sought civilian careers and, on the military side, the education given at Kingston was later said to have been "undoubtedly of a higher standard and more thorough" than that at either Woolwich or Sandhurst.*[21] In some spheres related to military technology, Canada now had a measure of sophistication. In 1903 Captain W.B.M. Carruthers,** an ex-cadet who had served in South Africa, organized a signals formation in the Royal Canadian Engineers and, after 1905, signalling was taught at RMC.[22]

In the First World War, in a country notoriously unprepared for military ventures, RMC proved its worth in Canada's surprisingly large war effort. It has been claimed that 980 ex-cadets, all that are known to have been able and available, served in the war. Most were in the Canadian Expeditionary Force (CEF) or in the Canadian Permanent Force. Another 390 served with various British forces. Although only 2 per cent of all CEF officers throughout the war were ex-cadets, 22 per cent of the command and staff of the first contingent was from the college when it moved overseas in 1914. RMC graduates received a large share of wartime promotions to high rank and decorations for valour or distinguished service. Thus, although not designed to produce officers for a huge wartime army, RMC justified Mackenzie's initiative when the country faced its biggest military challenge to date. It provided a degree of military professionalism that was rare in Canada.[23]

The Great War is often said to have brought Canada recognition as a nation. It certainly brought the dominion greater control of its own forces, and it also paved the way for RMC's Canadianization in 1919. Lieutenant-General Sir Archibald Macdonell,*** who had commanded the First Division of the CEF in France, was the first officer of the Canadian permanent force to be RMC's commandant. The military staff now also came from the same source. Henceforward, British officers who came to RMC to maintain the traditional British imperial tie were usually employed as directing staff in courses for the militia. From the time the college opened in 1876, Canadian civilians had been needed to teach academic subjects when competent military officers were not available. Most of the new postwar appointees were also civilians, but they were usually ex-officers with war service.

This preservation of RMC, now Canadianized, was especially important because it was against the drift of the times. Whereas before 1914 Canadians had been merely uninterested in, or indifferent to military affairs, there was now a strong antipathy towards any future overseas involvement that might become another slaughter like that on the Western Front. RMC thus became one of a few small islands of military interest in a pacifically inclined Canada.

Macdonell restored the four-year course from its wartime two years by keeping a senior class for three years. He permitted entry with junior matriculation from provinces where schooling did not go beyond grade 12. He

---

*The son of one of the first British officers to teach at RMC argued later that RMC graduates "were much better equipped for life than those who went through the colleges at home." G. Walker, "The Royal Military College," RMC Review 15 (June 1927): 24–7.

**Maj. W. Bruce M. Carruthers (no. 82) was at RMC 1879–83. Commissioned in the British Hussars, he later enlisted to serve in South Africa as a sergeant in the Royal Canadian Regiment. He became inspector of signalling in the Canadian Permanent Force, and he is revered in the Royal Canadian Corps of Signals as the "father" of the corps, which claims to be the oldest in the British Commonwealth. John S. Moir, ed., History of the Royal Canadian Corps of Signals, 1903–1981 (Ottawa: Privately printed 1962), 1–5.

***Lt-Gen. Sir Archibald Macdonell (no. 151) graduated in 1886 and served for twenty years in the Royal North-West Mounted Police. He was GOC, 1 Division, in the CEF, and commandant RMC 1915–25. See Preston, Canada's RMC, 226–50.

believed that RMC should not carry its cadets through to university degree standards. Instead, he negotiated for the accreditation of the RMC diploma by universities and professional societies. He thus maintained the system whereby cadets could receive education for a civilian career while simultaneously receiving military training. He planned to expand the college to 300 cadets, more than twice the size before the war. These plans made possible the organization of the growing cadet body on the lines of an infantry battalion. Macdonell also introduced field operations and taught trench-warfare tactics based on his experience in France. Since he believed the college should prepare for all arms, at times it paraded for meals as an artillery battery or as a cavalry regiment; there was also instruction in those arms.

Macdonell aimed to produce generalists trained in all branches of military service and with a general basic education in all branches of engineering. After 1921, RMC permitted specialization in civil or mechanical engineering only in the fourth year. Specialization in chemistry was at first permitted in the third, but in 1925 that too was eliminated. At the same time Macdonell put so much stress on military training that Iva Martin, retiring as director of studies, complained that the commandant had seriously compromised academic standards. This may have merely been personal pique. Although Colonel H.J. Dawson,* Martin's successor, remained director of studies, Macdonell made sure that the next appointee, Colonel E.J.C. Schmidlin,** was downgraded to senior professor, the level beyond which Martin had been promoted. By 1932 there was a "unified syllabus." Like West Point, RMC then offered only what was assumed to be the ideal basic professional preparation for all military officers.[24]

The military aspect of the college was also strengthened when RMC cadets again began to use part of their two months' summer vacation for military training. The RCAF introduced the practice in 1922. Other cadets jeered at classmates selected for flying training, asking whether they had not had enough "playing soldiers down here without spending the summer at it."[25] But from the next year, until the onslaught of the Depression in 1929, cadets began in increasing numbers to train during the summer with the militia, the RCN, or the RCAF.

In the first decade of the restored four-year course, while two-thirds of RMC graduates took up civilian professions on graduation, the rest went into either the Canadian or the British services. The War Office, although noting they were still of excellent quality, commented that RMC cadets commissioned in the British army now came from lower down the graduation list than before the war.[26] One reason for this was that many cadets were seeking commissions in Canadian technical corps in preference to British.

Outside factors soon brought a restoration of more advanced academic levels in the college. The Depression brought an increase in the number of applicants for university admission, including RMC. Some universities adopted entry by senior (grade 13), instead of junior, matriculation. Because RMC had to meet conditions in all provinces, it retained the technical requirement of junior matriculation but added a requirement of further study in certain subjects, to be tested by an RMC entry examination. The Depression also postponed the construction of the dormitory needed for expansion to the 300 level and, as the number of applicants grew, there was increasing competition for the limited number of places.

But the Depression also cut down employment opportunities in engineering. Parents who wanted the RMC course for their sons to prepare them for civilian careers

---

*Col. Herbert John Dawson, director of studies from 1922 until his death in 1926.

**Maj.-Gen. Edward James Schmidlin (no. 600), BSM 1904–5 and governor general's gold medalist, became senior professor in 1926. He was QMG in 1940 and head of mechanical engineering at Queen's University 1942–6.

now turned more to law and business. Many thought that the four-year RMC course was too long and too mathematical for either of those careers, and some members of the RMC Club pressed to have it cut down to three or even two years. By the 1930s, however, Woolwich and Sandhurst had improved to the point where the diluted RMC four-year course was no longer as competitive academically for entry to the British army as it had once been. Brigadier Harold Matthews,* appointed commandant in 1935, Colonel Schmidlin, now director of engineering in the Canadian army, and General Andrew McNaughton,** a chief of the General Staff who became president of the National Research Council, therefore succeeded by the late 1930s in retaining the four-year course against attackers and reintroduced an engineering specialization to meet standards in Canadian universities and the British army.[27]

General McNaughton held that a thorough and specialized scientific training was necessary for modern war and that routine drill and mechanical exercises that could be learned on the barrack square should be dropped in favour of general education. He was also in favour of adding constitutional history and international studies, and, shortly before the Second World War, more social studies appeared in the curriculum. A proposal in the House of Commons that RMC should give its own degree was dropped, however, when the leader of the opposition questioned whether it would be constitutional to give federal support to what would become a "purely civil institution."[28]

In the 1920s and 1930s Agnes Macphail,*** a pacifist member of parliament, regularly moved the reduction of the college's annual budget on the grounds that it favoured a privileged class at public expense. Yet, despite this outspoken critic, and even though the college's development was slowed by financial stringency, there was general support for RMC. Because of that support, ex-cadets were available in significant numbers in 1939 as one of the three interlocking foundations on which

Canada could build its war effort. The other two were the permanent forces (three under-strength permanent battalions of the militia, a brigade of artillery, and two cavalry regiments as well as a tiny navy and an air force) and the militia, navy, and air force reserves, equally starved for funds and equipment.

In the Second World War Canada once again mobilized large forces. RMC provided 50 per cent more officers to Canadian and imperial forces than it had in the previous conflict, and it had a bigger share of staff appointments and high ranks, including the army commander, General Harry Crerar.† Some British military leaders criticized the quality of Canada's top generals[29] – an interesting contrast with 1914–18 when Lloyd George declared that Australia's Monash and Canada's Currie were superior to his British army chiefs – but the Canadian officers who led the thousands who volunteered learned their trade effectively in war conditions. RMC graduates, especially those with permanent or reserve militia experience, played an important part, and Guy Simonds,†† later the CGS, was picked by Montgomery as one of the outstanding tacticians of the war. During the

---

*Maj.-Gen. Harold Halford Matthews, British Columbia Horse, CO 8th Canadian Infantry Battalion overseas, commandant 1935–8, adjutant-general 1938–40.

**Gen. Andrew George Latta McNaughton commanded the Canadian Corps artillery in the First World War, CGS 1929–35, president of the National Research Council 1935–9, GOC 1 Division, 1939, GOC-in-C, First Canadian Army Overseas 1942–3, minister of national defence 1944–5, president of the Atomic Energy Control Board of Canada 1946–8.

***Agnes Campbell Macphail, MP 1921–40, MLA (Ontario) 1943–51, a feminist and anti-militarist.

†Gen. Henry Duncan G. Crerar (no. 749) graduated in 1909. After serving in France in the First World War, he was professor of tactics at RMC 1928–30 and commandant 1938–9. He was GOC-in-C of the Canadian army 1944–5.

††Lt-Gen. Guy Granville Simonds (no. 1596), RMC 1921–5, was associate professor of artillery and instructor in tactics 1938, GOC, 2nd Canadian Corps, 1944–5, and CGS 1951–5.

long period of military neglect, RMC's military-civil education had helped to prepare key personnel to lay a foundation for Canada's contribution to victory.

Before the war ended, Canadian defence chiefs began to plan a postwar defence establishment. They assumed that, in order to back its new international status and to make a contribution to the preservation of peace, Canada would need larger professional armed forces than before. In June 1945 they estimated a total requirement of 105,788 men in all three services, with 50,500 "auxiliaries" in training and 227,396 reserves, mainly in the militia. They assumed there would probably have to be conscription.[30] The government quickly ruled out conscription and cut down the proposed regular force establishment to between 45,000 and 55,000 men.[31] That number was still four or five times larger than the permanent force in 1939. It would require a more certain and continuous supply of officers than before the war. The General Staff's postwar plan H, dated 20 September 1945, stated that before a decision could be made about RMC, further consideration was needed on several questions. For example, how much of the education should be at state expense? What should be the conditions of entry, the minimum age, the length of the course, and, on graduation, the possibility of having all RMC graduates in the permanent force?[32]

In 1919 all that had been necessary to put RMC on a peacetime footing to produce officers had been to extend the wartime course to the prewar length or more, and to reintroduce general and technical courses dropped during the war. But in 1942 the cadet college at RMC had been closed and its buildings used to accommodate various wartime courses, including a staff college that would still be needed for the new peacetime army. In the meantime, the Royal Canadian Navy had in 1942 opened HMCS *Royal Roads*, a college to train its officers. It planned that when the war ended it would revive the old Royal Naval College of Canada to ensure a supply of officers in peacetime. The army had shown no similar

concern for the future. The RMC Club executive therefore feared that government's dislike for military expenditures might prevent RMC from reopening. It began to work to keep the idea of the college alive.[33]

In Kingston in the winter of 1945–6, a group of ex-cadets met regularly to plan a campaign for the reopening of RMC. They included Brigadier "Ben" Cunningham,* commandant when the college was an army staff college during the war and president of the RMC Club in 1946, Colonel W.A.B. Anderson,** and Colonel W.R. Sawyer.*** Colonel S.H. Dobell† and E.W. Crowe†† in Montreal were also active. The RMC Club then set up a special committee chaired by Dobell, which approached two successive ministers of defence. The first,

*Brig. D.G. Cunningham (no. 1841) graduated in 1929. A militia officer in the Princess of Wales's Own Regiment, he served in the Dieppe raid, commanded a brigade in Normandy, and then, because of his war experience, commanded the staff college in RMC. After the war he returned to his law practice in Kingston.

**Lt-Gen. W.A.B. Anderson (no. 2265) graduated from RMC in 1936 and was commissioned in the Royal Canadian Horse Artillery. After serving in Europe, where he was personal assistant to General Crerar, he was director of military intelligence in Ottawa during the period when college reopening was being discussed. He commanded RMC 1960–2 and then became adjutant-general and deputy chief of reserves. He became commander, Mobile Command, after unification. After retiring, he was co-chairman of the Ontario-Quebec Permanent Commission for Co-operation and then secretary of the Management Board of the Ontario cabinet. For further details see Preston, *Canada's RMC*, 338n, and R. Guy Smith, ed., *As You Were: Ex-cadets Remember* (Kingston: RMC Club 1984), II, 273.

***For Col. Sawyer see page 26 below.

†Col. Sidney Hope "Choppy" Dobell (no. 1230) graduated from RMC in 1918. He was CO of the 6th Field Regiment, RCA, in 1942. He became comptroller of McGill University in 1947 and in that year was also president of the RMC Club.

††Capt. Ernest W. Crowe (no. 1542), BSM and silver medalist, graduated 1924. An actuary, he was a captain of the McGill Canadian Officer Training Corps 1941–6 and president of the RMC Club 1950.

The Honourable Brooke Claxton, KC, minister of national defence, founder of the tri-service Canadian Service College system

Douglas Abbott,* appointed an official committee under Brigadier Sherwood Lett,** but took no further action.[34] The second, Brooke Claxton,*** states in his unpublished memoirs that he gave more attention to officer training during his eight-year term of office than to any other subject. He felt strongly that "whatever service college we had should be for all the services as RMC had been to some extent before the war, that it should preserve the old traditions and avoid some of the bad precedents." He sent Air Vice-Marshal E.W. Stedman† to visit and report on West Point, Sandhurst, Cranwell, and Dartmouth. The chiefs of staff, Stedman, and Claxton then worked out a plan for reopening the college.[35]

Each of the three services had its own ideas about what was required for the preparation of permanent peacetime commissions. The navy looked at the USNA's four-year integrated academy course for all branches, but it preferred that its own prospective executive officers have two years of education and training at Royal Roads, followed by sea service and a six-month course at the Royal Naval College, Greenwich, England. Engineer officers would go to the RN Naval Technical College at Keyham.[36] The RCAF believed that aircrew for modern aircraft should start their flying training as young as possible, but soon realized that senior officers need a sound academic grounding. When it found that an independent air force college was unlikely because of cost and that the navy was afraid it might lose Royal Roads because of small numbers, it came to the conclusion that its requirements were sufficiently like those of the navy for it to ask to share Royal Roads. It felt that use of a service college, rather than of the universities, would make earlier military indoctrination possible. Its condition for sharing Royal Roads was that entry should be raised to senior matriculation as in most Canadian universities. On that basis, arrangements were made for a joint two-year RCN-RCAF college at Royal Roads to commence operation in 1947.[37]

Meanwhile, the RMC Club had secured vague promises that RMC would reopen, but not what form it would take.

In the army there was general agreement that all officers should have some general education, but not how much. Some thought that Canada should aim at the baccalaureate standard for all officers, as in the United States. Royal Canadian Engineers insisted that technical officers should have an honours degree in engineering from a civilian university. The army's Chesley†† Committee considered various methods of training officers. Plan A, which came to receive the support of General Charles Foulkes,††† the chief of the General Staff, was a revised university Canadian Officer Training Corps (COTC) that would offer military training to university undergraduates and send them on to a two-year course at RMC after graduation. Figures of cost that purported to show that plan A was cheaper than reopening RMC as a cadet college were, however, successfully challenged by the RMC Club executive as gross underestimates. The club also argued that, if university OTCs were the sole source of regular officers, the scheme could be unsafe: pacifist political action might find it easier to close a university

*Douglas Charles Abbott served in the First World War. He became minister of national defence (naval services and army) in 1946, minister of finance 1946–50, and a justice of the Supreme Court 1954–73.

**Brig. Sherwood Lett served in both world wars and became deputy chief of the General Staff and justice of the Supreme Court of British Columbia 1954–63.

***For Brooke Claxton see page 25.

†Air Vice-Marshal Ernest Walter Stedman served in the RNAS and RAF in the First World War and came to Canada in 1920 to be technical director of the Aeronautical Board. Later he was member of the Air Council in charge of research and development. He retired in 1946.

††Brig. Leonard McEwan Chesley (no. 1210), RMC 1916–17, RGA in the First World War, was in business in Montreal 1919–39. He served on the Directorate of Staff Duties in the Second World War and was director of staff duties and deputy CGS 1946–8.

†††Gen. Charles Foulkes, educated at the University of Western Ontario and commissioned in 1926, commanded the 2nd Infantry Division until 1944, and then the 1st Canadian Corps in Italy and Northwest Europe. CGS in August 1945, he was later chairman of the Chiefs of Staff. He retired in 1960.

scheme down, and undergraduate trainees would be exposed to anti-military propaganda in the universities. Plan B, however – to reopen RMC with an obligation to serve in the active force – would require a substantial enlargement of the college to produce all the officers needed. The solution, therefore, was to adopt a third option – a combination of both plans.[38]

By the time this decision was reached in 1947, prospects for military development in Canada had reached a postwar low. Prime Minister Mackenzie King wanted to return to the "old Liberal principles of economy, reduction of taxes, and anti-militarism."[39] The strength of the armed forces had now sunk to 32,612, less than three-quarters of what the cabinet had proposed in 1945.[40]

What had facilitated the decision to open the RCN-RCAF college at this unpropitious time was that Brooke Claxton,* the new minister of national defence, saw a program of service integration to reduce costs and interservice friction as part of his mandate.[41] He now agreed to reopen RMC in 1948 as a tri-service college associated with Royal Roads that would also accept army as well as navy and air force cadets.[42]

Meanwhile, the RCAF was satisfied it could secure enough applicants for short-term commissions to meet its immediate flying needs, and had come to support the idea that a four-year educational foundation was necessary for long-term commissioning.[43] Claxton arranged for army and air force cadets from Royal Roads to finish their four-year course with two years at RMC. Naval cadets would go to sea after two years as before.[44] The idea that university cadets would come to RMC after graduation for military training was dropped. Instead, Canadian Services Colleges and university cadets would train together during three summer vacations with the service of their choice. Finally, graduates of both RMC and university plans could choose between regular and reserve commissions.

Brigadier D.R. Agnew,** an ex-cadet and prewar instructor, was appointed commandant to re-establish the college. The new tri-service RMC, with an establishment of 400 cadets, twice as large as in 1939, inherited the structure as well as the traditions of the old. As had been the case immediately before the war, cadets had to pay fees of $550 in the first year and $300 thereafter, as well as a recreation club fee of $30.[45]

In the past, in order to maintain its accreditations, Canada's RMC had had to guard its academic reputation more jealously than military colleges in other countries and had by and large succeeded. Emphasis on academic standards in the new RMC was therefore not new. Most of the features of the academic program had been introduced before the war. Specialization had begun before 1939 in several engineering disciplines, and some arts courses had been added about the same time. Finally, the tri-service system simply formalized the practice at RMC, where cadets had been able to enter one of the three armed forces.

But some things pointed to the future. Fees were low and there were numerous entrance scholarships in addition to the dominion cadetships for the sons of deceased veterans and officers and men with fifteen years service.[46] Summer training, now obligatory, which extended through most of the vacation in every year and was paid at the rate of a second lieutenant, made it possible for cadets to finance their way through to graduation after the first year.

Brigadier D.R. Agnew, CB, CD, LLD, commandant, 1947–54

*Brian Brooke Claxton served in the 10th Siege Battery in the First World War. A lawyer, he taught at McGill, was elected MP in 1940, and shortly afterwards was made parliamentary secretary to Mackenzie King. As minister of national defence 1946–54, he supervised the rebuilding of the Canadian Armed Forces during and after the Korean War.

**Brig. Donald R. Agnew (no. 1137) was at RMC 1915–16. He served in the RCA in the First World War and was instructor in artillery at RMC in the 1930s. After serving again in the Second World War, he was commandant 1947–54. He died in 1968, one year after retiring as commissioner of the Commonwealth War Graves Commission. See Richard A. Preston, "Brigadier D.R. Agnew," in Smith, *As You Were*, II, 324–5.

Colonel W.R. Sawyer, OBE, CD, psc, BSc, MSc, PhD, LLD, D.ScMil, FCIC, vice-commandant and director of studies 1948–67

The "New One Hundred": Opening ceremonies, 20 September 1948

A significant change was that the "New One Hundred," as the first postwar RMC class named itself, drew only 22 per cent of its members from private and residential schools compared with 83 per cent in 1919. In subsequent years the percentage would decline still further.[47] The private schools were now sending more of their students to universities; indeed, people in the professions and other relatively well-off parents were seeking occupations for their sons that were more remunerative than military careers. Furthermore, since

entry to RMC was based on the same senior matriculation requirement as university engineering, as well as RMC entrance examinations in mathematics and English, this eliminated some weaker students. The "New One Hundred" in 1948 came from a broader cross-section of Canadian society. Although a disproportionate number was still from Ontario, many cadets came from other parts of the country. Twelve per cent of the first class was French-speaking, more than before the war but far less than the proportion of French Canadians in the country as a whole. Applications for admission in 1949 showed the same distribution. Out of 447 applicants, 195 came from Ontario, 125 from all the western provinces, 28 from the Maritimes, and 88 from Quebec. Forty of the Quebeckers were English-speaking.[48]

Colonel W.R. Sawyer,* an ex-cadet and prewar RMC professor of chemistry, had been deeply involved in planning the college's reopening and was, almost obsessively, devoted to its development. He was appointed director of studies and charged with recruiting staff for the college and drafting a curriculum. Believing that the military profession must be able to conceptualize and communicate, he was convinced that the modern officer must be both technically and generally educated. He was therefore determined to make RMC more like a university.

Sawyer stipulated that specialization in civil, mechanical, and electrical engineering should start, not in the final year, but in the third, and he also introduced an arts option. He organized departments in the various engineering specialties, and also in mathematics, physics,

*Col. William Reginald Sawyer (no. 1557) graduated from RMC in 1924 and then attended Queen's University, McGill, and Harvard. He served on the RMC staff in the Department of Physics and Chemistry 1935–9 and became GSO1, Chemical Warfare, with the Canadian army in Western Europe. After the war he was director of weapons development in Ottawa 1945–8.

and chemistry. He added four arts departments – history, English, French, and political and economic science – and gave them equal status with the sciences to provide courses for all cadets in the first two years and for the arts option thereafter. To ensure that instruction was of university calibre, he ruled that the optimum qualifications for RMC faculty would be the PhD and a continued engagement in research. As few Canadian service officers at that time could meet that requirement, most of the new appointments came from the universities, although many were former wartime officers. They brought with them their university concepts of standards and quality.[49]

In one important department, military studies, the conditions of faculty appointment were different. Brigadier Agnew asked for serving officers loyal to their own service but experienced in interservice cooperation, who would be capable of influencing cadets and who were also first-class lecturers and fluent writers. Otherwise, he said, "their teaching will suffer by comparison with the civilian academic staff."[50] The course in military studies presented topics common to all services and also introduced cadets to the basic features of the service of their choice. To give it standing in the predominantly academic curriculum, it was placed under the director of studies. However, a cadet's military indoctrination would depend not so much on this academic course in military studies as on his exposure to the four-year military routine and on his summer training with a particular service.

More innovative than the greater emphasis on academics and the expanded summer training was the creation of the tri-service colleges system in which one college fed into the other. This necessitated a correlation of curricula. Preliminary arrangements for correlation were made by the two commandants and a Joint Services Colleges Committee established on 8 January 1948. Renamed the Canadian Services Colleges Co-ordinating Committee on 26 January, when the system as a whole was named the Canadian Services Colleges, the committee was to operate only until the colleges got under way. At that point they would come directly under the defence secretary, who would be responsible for them to the Defence Council.[51] Further curriculum correlation was arranged directly between the two colleges, with RMC, as the recipient of all cadets after their first two years, exercising a leading role. For example, Dr G.F.G. Stanley,* appointed to head the history department at RMC after 1949, arranged for more emphasis on New France; he also informed Royal Roads that more attention would be given to military history in RMC history courses than in university departments of history.[52]

Within RMC, academic matters, although ultimately the responsibility of the commandant, were to be administered by the director of studies. The RCN and the RCAF wanted a service officer of lieutenant-colonel's rank as vice-commandant, but the chief of the General Staff, Lieutenant-General Charles Foulkes, ruled that as only 15 per cent of the course was military, he would not spare an officer at that level. This arrangement was to be reviewed after one year's experience.[53] Because of the large part he played in organizing the college, Sawyer, although now an unattached reserve officer, was appointed vice-commandant and was called out on active duty whenever the commandant was absent. Another departure, one adopted at Royal Roads also,[54] was the appointment of a registrar. At RMC this was Colonel T.F. Gelley,** a former RMC faculty member who had been associated with Khaki University in England. He oversaw

*Dr George Francis Gilman Stanley (no. H8829) was with the Historical Section overseas in the Second World War with the rank of lieutenant-colonel. He was head of the history department 1949–67 and became chairman (later dean) of the Division of Arts. He resigned to become director of Canadian studies at Mount Allison University and was lieutenant governor of New Brunswick 1982–7.

**Col. Thomas Fraser Gelley (no. H6888) was registrar from 1947 to his retirement in 1957, and secretary-treasurer of the RMC Club from 1957 until his death in 1968. See Preston, *Canada's RMC*, 233n.

routine academic administration in place of the staff adjutant who had been responsible for academic records before the war.[55]

For the first year of operation, to advise him on policy and to discuss and approve cadet grades, Sawyer set up an Academic Board which consisted of the single professor teaching the first-year course in each department. At its first meeting on 15 November 1948, he made it clear that grades were to be assessed by the judgment of the instructor and that there would be no predetermined percentage of failures. He also informed the board that in the last ten years of the old college the wastage rate in the first year had been about 25 per cent, including both academic failures and withdrawals for other reasons.[56]

Although to some people the new RMC appeared to have become a replica of a civilian university, it was really quite different. Its curriculum included only a small number of the disciplines normally found in a university. Subjects taught had been selected for their relevance to the task of preparing military officers. Since classroom content was left to individual departments and instructors, teaching could be affected by a professor's research or other interests. Sawyer accordingly prescribed that RMC courses should be made appropriate for a military college by the inclusion of content and examples with military relevance. This could be done more easily in some departments than others – for instance in history or political and economic science rather than in language courses. Science courses could similarly be adapted to military pedagogical use, though here again some were more suitable for military purposes than others.

Vital decisions about the establishment of the tri-service Canadian Services Colleges and their curricula had been made at a time when it was assumed that Canada's regular forces would remain small and that a surplus of their graduates would go into the reserve forces. The restored RMC, then, had been built on the lines of the old. Prime Minister Mackenzie King, addressing the RMC Club at its annual meeting on 30 September 1950 in Kingston, summed up broad reactions. He said that it is "always difficult to feel altogether happy about major new developments in any institution to which one is attached. But from what I can learn, most of what is best in the old traditions is being carefully – one almost may say lovingly – preserved. At the same time new life is being added to those traditions in a manner no old cadet will ever need to feel ashamed of."[57]

The "new" RMC, despite its association with Royal Roads in the Canadian Services Colleges system, its tri-service educational function, and its increased concern with academic standards, was in fact the old RMC writ large. Some serving officers feared, however, it was not what was required to produce officers for modern military and naval professional forces. Yet in 1950 RCN ships, officered in part by RMC and Royal Roads's ex-cadets, sailed for the Korean war zone within three days of the government's announcement that Canada would participate in the conflict; and army members of the first post-war graduating class went directly to the war. Ex-cadets won four Military Crosses in Korea and, in two cases, the citation made particular mention of "powers of leadership."[58] This was a tribute to the quality of the new RMC's professional training. At least twenty from later classes also served in Korea after the armistice, and some were decorated.*

The war in Korea marked the last time when the RMC system of education and training without commitment to a regular force career, as restored in 1948, would be tested in war. At the same time, NATO, the Korean War, and NORAD brought changes in Canadian defence policy that would lead to radical developments in the Canadian Forces and the services colleges.

*See appendix M.

Winds of Change:
The Regular Officer Training Plan
and Collège Militaire Royal

After the Second World War Canada's Armed Forces, pared to a skeleton and settling down to peacetime soldiering, were left with inadequate means to maintain the military prowess they had displayed in Europe.[1] Three broad objectives officially defined their tasks: the defence of North America in cooperation with the United States; the provision of administrative, training, and operational staffs as well as reserves to make possible an expansion like those of 1914 and 1939; and, in keeping with the recent creation of the United Nations, cooperation with other free nations to preserve peace and restrain aggression.[2] Exactly what this international cooperation would involve was nowhere made clear. None of the three functions aroused a sense of urgency about military preparation. A lack of clear political direction threatened the military's capability and sense of purpose.

When Mackenzie King appointed Brooke Claxton minister of national defence in December 1946, it was for the express purpose of cutting down military expenditures. Nevertheless, despite King's primary intention, Claxton began to restore the image of the military and to refill some of the deficiencies in its strength, though without any significant expansion. He approved the establishment of the tri-service Canadian Services Colleges (CSC) on the lines of the prewar RMC – there was no obligation for graduates to take regular force commissions. The government and people of Canada assumed in 1948 that the threat of war was at least as

remote as they had thought in the 1920s and early 1930s.

Before the college reopened in 1948, the Czech coup and the Soviet blockade of Berlin reminded Canadians of a need to maintain a military potential, but there was no immediate growth of defence forces. Canada's assumption of a share in responsibility for the defence of Western Europe in the North Atlantic Treaty Organization alliance in 1949 also brought little change in military development. Claxton called NATO "a pact for peace," and said that "the final result will not be to increase the expenditure which every nation on our side must take." By the time the treaty was signed in April, the strength of the three Canadian Armed Forces had been rebuilt only to about 40,000 from the postwar low of 32,000 in 1946. In September it was still less than 46,000.[3] RMC's first postwar graduating class would not be available until 1952. How the need for normal officer replacement was to be met, let alone how future expansion was to be provided for, was not indicated.

Having decided to reopen RMC, the government gave little further thought to officer production before the Korean War. Claxton said there were 5000 candidates in training for Canadian commissions, and boasted that was "in greater proportion to the population than in any country."[4] Major-General W.H.S. Macklin,* the army ad-

---

*Maj.-Gen. Wilfrid Harold Stephen "Slim" Macklin served in both world wars and was adjutant-general 1949–54.

jutant-general, advised the Cabinet Defence Committee, however, that "for professional reasons the majority of the officers required by the services should have a university degree or equivalent," and that the services were short 1870 officers with that qualification. They had already found it difficult to attract suitable candidates of that calibre.[5] What Claxton did not indicate was that two-fifths of the candidates were only training for short-service commissions that offered little prospect of a full military career.[6] Some of them had only a junior matriculation academic qualification, one year short of high school graduation in Ontario and some other provinces. In 1947, when the RCAF needed 165 officers for various appointments as engineers or other technical duties, it offered commissions to ninety-five veterans graduating in pure and applied sciences, but only six accepted.[7] The other services had similar problems.

The minister also did not mention that the remaining 3000 officer candidates in training, many of whom did have the desired academic qualifications, were not committed to join the regular forces. Nearly 2500 were in voluntary unpaid university units of one kind or another, such as the Canadian Officer Training Corps, from which most would take only a reserve commission. Graduates of the services colleges were similarly obliged to take only reserve commissions. By 1950 the arrangements Claxton had made for officer production did not satisfy current needs. The Korean War soon created a serious shortage of young officers.[8] All three services resorted to the subsidization of university cadets for all or part tuition for their final year and also gave them living allowances, if they would agree to take regular commissions.[9] The RCAF went a step further. It proposed to subsidize some university cadets after their first year, a pointer to the future.[10]

The Canadian Services Colleges, which at that time had only about one-tenth as many cadets as the university units, had been set up in large part as a more reliable source of technical officers. But here also acceptance of a regular commission was optional, and RMC graduates needed a further year at a university to qualify for many technical appointments in the services. It was in fact expected that, out of the hundred RMC cadets who would graduate annually from 1952 on, only about sixty would join the army active force or the RCAF.[11] (Naval cadets would have gone to training and sea service with the Royal Navy after their second year.) Although the Canadian Forces were still small, the output from the services colleges would obviously not go far towards meeting the need for long-service officers.

The new RMC, now the heart of the officer-production system, was under closer government scrutiny than before the war because it was authorized by the National

RMC in the 1950s

Princess Elizabeth and Prince Philip, 12 October 1951

Defence Act of 1950* rather than a separate RMC act. Since it was tri-service, it was only administered, and not governed, by the army. It came directly under the Department of National Defence. Command was to rotate among the services. Control passed from the ad hoc Canadian Services Colleges Co-ordinating Committee to the defence secretary, who was to report to the Defence Council.[12] In practice, however, supervision was exercised by the Personnel Members Committee (PMC), which reported to the joint Chiefs of Staff Committee. A joint committee for selection and entry of cadets, created to select recruits for the two colleges for the second year of operation, worked so well that Instructor Lieutenant-Commander J.C. Mark,** speaking for the director of naval education (DNE), recommended to the PMC that a continuing committee should be established to collate college regulations, work out selection procedures, and prepare for selection and entry in 1950. On 13 October 1949 the Canadian Services Colleges Committee met for

the first time. This was a first step towards the exercise of a closer control of the two colleges from Ottawa. It was a significant change from the prewar RMC. Brigadier Agnew and Colonel Sawyer, fearing that closer control by the Department of National Defence might have the effect of reducing attention to RMC's self-perceived problems and needs, urged the reconstitution of the prewar RMC Advisory Board; but the Canadian Services Colleges Advisory Board was not set up until 1955.

Meanwhile, the Canadian Services Colleges Committee provided a channel for the expression of concern about shortages in officer production. Two days before its creation, the Defence Council had considered a radical proposal that probably came from army sources – that at the end of their first year all CSC cadets should be required to give an undertaking to join the active forces for a definite period after graduation.[13] This proposal was referred to the Canadian Services Colleges Committee, which discussed it at its second meeting. Brigadier T.E. d'O. Snow, the deputy adjutant-general, then reported back from the committee to the PMC. He said that the army active force would have an annual requirement of 150 officers from 1950 to 1970, but only one-third of the RMC graduating class of about one hundred would be likely to take a commission in it. His committee therefore recommended that the PMC should consider exacting the proposed obligation from CSC cadets at the end of their first year.[14]

The proposal was referred to the commandants of the two colleges. Brigadier Agnew contended that obligatory service would lead to a lowering of standards. The director of naval education, Instructor-Commander M.H. Ellis, disagreed. The RMC brief also added that, since

*See appendix A.

**Instructor Lieutenant-Commander James Clarence Mark enlisted in the RCN Reserve in 1942, served in teaching positions during the war, and was CSC liaison officer, Department of Naval Education, 1946–51.

The new RMC in Korea. Four recent graduates won the Military Cross in Korea for gallant and distinguished services in action:

Lieutenant C.D. Carter, MC, Royal Canadian Engineers (Photo courtesy of C.D. Carter)

Cadet Squadron Leader H.C. Pitts, MC, Princess Patricia's Canadian Light Infantry

there were many scholarships, the present system imposed no financial obstacle to recruiting. Again DNE disagreed. He recommended a lowering of fees for the first year and a drastic application of academic standards at the end of that year. He also said that, although an attempt to enlist boys as cadet recruits for the forces might not have legal force because they were minors, routine enrolment in the services at the end of the first year of college, when they were older, would work.[15] What was not stated was that a program of that kind might become a dead end for those cadets who decided after one year not to give an undertaking to go regular. The proposal went no further at that time.

In 1950 the Chiefs of Staff Committee was informed that out of 374 applicants for the colleges, seventy-five had been rejected on academic grounds. The PMC had recommended that twenty-eight of the failures should be interviewed again. They were mainly from the Maritimes and western provinces, and their inability to meet the required standards was due to the lower educational

Lieutenant D.G. Loomis, MC, Governor General Vincent Massey, Lieutenant A.M. King, MC, and the Honourable B. Claxton, following the presentation of the decorations on the RMC parade square (Photo courtesy of A.M. King)

standards in those provinces. The Chiefs of Staff Committee ruled that there should be no reduction of the standards of entry.[16] Claxton added that the rejections drew attention to provincial disparities which the provinces themselves should remedy.[17] In September 1950, because of the acute shortage of junior officers for the army and the RCAF, the Chiefs of Staff Committee discussed various expedients, including commissioning university students at the beginning of their final year, requiring CSC cadets to opt for regular commissions after their first or second year, and giving preference to those cadets who would sign for a permanent force commission.[18] None of these proposals was introduced, but in May 1951 a conference of commandants and directors of studies of the services colleges, RMC and Royal Roads, discussed the possibility of admitting junior matriculants in order to tap provinces that had no senior matriculation. RMC members countered by suggesting that the college should obtain degree-granting powers to make it more attractive to recruits. They said that the curriculum and the staff already assembled warranted such a move. Further investigation of that counterproposal was postponed until the following year when the first batch of cadets had passed through the system.[19]

In 1950 the army had turned its attention from the production of peacetime officers to the raising of a special force for Korea. The 25th Brigade's personnel, although embodied as part of the active force, was enlisted for only eighteen months' service. Its officers were volunteers, mainly veterans of the Second World War. Some came from the regular units of the active force. This step, which was at first regarded as a temporary expedient, came to have lasting significance when, in 1951, after the government decided to order the special force to Korea, it became necessary to raise another brigade, the 27th, for service with NATO in Europe. The 27th was composed at first of militia, and later of regular units of the army active force. When the first contingent of the special force was brought back from Korea in 1952, it too was incorporated into the permanent active force, as parachute battalions of existing regiments. Henceforward regular battalions of the active force went to Korea. For the last two years of the Korean commitment the Canadian brigade, like that with NATO, was made up of career soldiers in units of the Canadian army active force.

As a result of foreign and defence policies that had abandoned Canada's tradition of isolation in peacetime, Canada needed to be able to maintain larger regular forces at home and overseas. Inevitably this had to affect the system of producing officers through the Canadian Services Colleges. Not merely was there now a bigger demand, but it also put a new emphasis on professionalism. In his official history of Canadian operations in Korea, Colonel Herbert Fairlie Wood noted that there is no substitute for battlefield experience as a basis for military proficiency. What he did not add was that war experience soon fades and, for a peacetime force with garrison duties, battlefield experience is not available.[20] There must be an effective program of military education to develop professionalism.

Because many officers had been taken from the active force for the special force, the army's difficulties in filling its officer requirements had been greatly exacerbated. The other services were similarly affected, though not as seriously, probably because their need for expansion was not as urgent. An immediate expedient was that all three services stepped up their commissioning by direct entry for short-service commissions. Then, in May 1951, Lieutenant-General Guy Simonds, newly appointed chief of the General Staff, introduced a new plan to recruit officers on a bigger scale for the army's non-technical corps. Candidates would need only junior matriculation and would be given short-service commission after training in a forty-two week course at Camp Borden in Ontario.

Simonds was also anxious to increase the army's intake of long-service officers. In the fall of 1951 he proposed two alternative schemes for the Canadian Services Colleges; either they should continue to admit with senior matriculation and cut the course to two years, or they should admit with junior matriculation for three years. He said that the latter plan would bring in more applicants from Quebec and from the Maritimes, which had no grade 13. The chief of the Naval Staff, Vice-Admiral Harold T.W. Grant,* supported him. Junior matriculation entry was what the navy had originally wanted for the short-lived RCN/RCAF college at Royal Roads.

The RCAF was adamantly opposed to any such change. It could always get plenty of applications for flying training with short-service commissions, and it had set its sights on the Canadian Services Colleges as a source of supply of better educated regular long-service officers. Notes of a conversation between the chief of the Air Staff, Air Marshal W.A. Curtis,** and the vice-chief, Air Vice-Marshal F.R. Miller,*** tell the story. It was probably Curtis who was speaking:

> I said I would oppose this – I want no part of it. In my view the present courses [in the Services Colleges] are perfect. While we have no graduates yet, we are likely to have a few in the next few years. The thing will build up and we will, after a time, get a number of graduates. I feel we are cutting off our noses to spite our faces in changing the present course. I suggested if the Army and Navy want to change things, they should move out of RMC and set up their own colleges, and the Air Force would carry on with things as they are – otherwise we will lose our best [RMC] instructors. The output at RMC would fit into the Air Force very well. We would be glad to take it over. CGS [General Simonds] said this

*Vice-Adm. Harold T.W. Grant, educated at the Royal Naval College of Canada, chief of the Naval Staff 1947–51.

**Air Marshal Wilfred A. Curtis, chief of the Air Staff 1947–53, first president of York University.

***Air Chief-Marshal Frank R. Miller, vice-chief of the Air Staff 1951–4, deputy minister of national defence 1953–60, chairman Canadian Chiefs of Staff Committee 1960–4, chief of the Defence Staff 1964–6.

would happen over his dead body. I said that was fine – we'd still take it over. Dr. Solandt [chairman of the Defence Research Board and member of the Chiefs of Staff Committee] is behind us 100%. General Foulkes [chairman of the Chiefs of Staff Committee] says we are on the right track also, and if we hold on the Minister [Claxton] will have a reason for turning down this request.

As the Army and the Navy are the ones who have not solved their problems and who are dissatisfied, I feel that they should be the ones to get out and establish their own schools. The RCAF would then take over RMC and continue with it as it is now.

We have a minimum [expectation] of 60 [officers per annum] and we can play with anything more than that up to 150 – as much as RMC would put out . . .

Every educational man in Canada will be against us if we change our present way. Also business people who know anything about it will object. I feel it is unsound business to change the role of the colleges at this time – just as we are about to get something out of them.

Dr. Solandt will back you up on this. Also the Chairman [General Foulkes] is sympathetic to our arrangement and would like to keep the present arrangement although he may not be able to state his view.[21]

The reason for the RCAF's different stand was that it regarded high command as a prerogative of its general (ie, aircrew) list and wanted to ensure that some aircrew officers had a thorough academic grounding. Its technical officers were more remote from actual combat than were their equivalent in the other services, and it wanted them to develop in a military environment. What made it possible for the RCAF to satisfy these two objectives in the CSC four-year course was that it had less difficulty than the other services in recruiting.

The army had quite a different problem. It was more anxious to get young officers than to ensure that some of them were prepared for high command in later life. When Simonds's plan was discussed at the Chiefs of Staff

Committee in November, each service wanted its own college. The chairman, General Foulkes, said that neither government nor the minister would approve a third college at this time.[22]

The navy's problem was a shortage of Canadian ships for sea training. It relied on the Royal Navy, but that led to trouble. On 20 April 1950 General Foulkes told the chief of the Air Staff that Claxton had said that naval cadets "must go four years."[23] He was reacting to the 1949 report of the commission under the chairmanship of Rear-Admiral E.R. Mainguy,* which had attributed disturbances in RCN ships in part to the British training of RCN officers that separated them from their Canadian ratings.[24] The navy, while it clung for a time to its practice of sending its deck officers to train with the Royal Navy after two years in the services colleges, had now to develop a program to produce technical officers in Canada. Some time after Simonds made the proposal to cut the length of the RMC course, then, the Chiefs of Staff Committee instructed the PMC to study an RCN plan to secure a supply of technical officers by subsidizing cadets right through a four-year university course if they would accept a commitment to take a regular commission on graduation.

On 6 March 1952 the PMC reported back on the navy's university plan. It said that if that scheme were implemented on a scale sufficient to meet 50 per cent of the long-term requirements for technical officers in all three services, it would necessitate a substantial revision of the operation of the system, including the services colleges. Most candidates who intended to join any of the Canadian Forces would obviously prefer to be subsidized in the universities; they could then live at home, rather than pay their tuition in the CSCs where they must submit to discipline and where they would need an extra year to

*Rear-Adm. E. Rollo Mainguy, RCN, educated at the Royal Naval College of Canada, chief of the Naval Staff 1951–6.

earn a degree. Few would be willing to attend the services colleges unless they could do so on terms of equality with cadets in the universities. The PMC, therefore, proposed to approach the same problem from a different angle – to use the services colleges "to provide a sufficient standard of education for officers who did not require technical or professional standards." Clearly the word "professional" was used here to mean professional engineers rather than professional sailors and soldiers. But the PMC noted that the present four-year science and engineering course in the services colleges must be maintained in order to meet the RCAF's requirement to produce its technical officers there. However, Simonds's plan to cut down the present four-year CSC academic course for army line officers to two years had now been linked to the navy's proposal to produce its technical officers in the Canadian universities. This obviously conflicted with the RCAF's desire for a four-year course in the services colleges for all its long-service officers.

The PMC had ruled earlier that long-term planning for permanent commissions should be in terms of obtaining 50 per cent of requirement from the services colleges; the remainder should come from the universities and from upgrading from the ranks. It had now agreed that, broadly speaking, those commissioned from the universities would make up that part of the officer corps in the army and navy that required technical professional qualifications. Others would come through the CSCs. It said that adopting these proposals would require an increase in enrolment in the services colleges from 650 to approximately 1325. When that could be achieved, the university part of the plan could be reduced. But the chairman of the Chiefs of Staff had instructed the PMC that present planning should not involve new construction at the colleges. The PMC therefore recommended that the army-navy scheme should be adopted as an immediate measure to remedy shortages. It could be modified later to divert more recruits to the services colleges when further construction was authorized and com-

pleted.[25] If this scheme had been put into effect as suggested, RMC would have become a two-year course for regular army and navy candidates for line and executive commissions, and at the same time a four-year course for prospective aircrew, air force technical officers, and reserve-entry cadets of all services – a confusion of aims and programs.

Foulkes found a way to compromise between these conflicting service views on officer production. He associated it with a different but related question – the under-representation of French Canadians in the services generally and in the officer corps in particular. Lack of French-speaking officers was believed to be partly the reason why francophones were reluctant to enlist in the ranks. The question had become politically sensitive when the Progressive Conservative opposition in the Commons criticized the failure of the Canadian Services Colleges to recruit French Canadians. They said there should be a French-speaking college.[26] Simonds was given the task of investigating ways of increasing the proportion of francophone cadets and, at the same time, of remedying the officer shortage.

Within a week of the first Progressive Conservative statement in the House of Commons, the army had inspected its old cavalry barracks of St-Jean, Quebec, as a possible site for a one-year preparatory school to bring graduates of Quebec's classical colleges* and junior matriculants from other provinces up to the level of senior matriculation.[27] Shortly afterwards, Foulkes received a report on numbers. The population of Canada was 27.5 per cent French Canadian, but only 2 per cent of naval

---

*At that time the Quebec system of tertiary education was quite different from that of English-speaking Canada. Derived from pre-revolutionary France and primarily philosophical, it mainly prepared for the law and the church. A four-year course at the classical colleges covered the age group which in English Canada, as in the United States, was in the last two years of high school and the first two years of university or college.

officers, 12 per cent of army officers, and 4 per cent of air force officers were French. The report said that to "remedy the imbalance," French-Canadian intake in the services colleges should be higher than 27.5 per cent.[28]

On 23 April 1952 Claxton submitted an outline of a "Plan for Production for Officers" to the Cabinet Defence Committee. This plan said that if there were an emergency within the next two years it would be possible to call up veterans of the Second World War and Korea; but, after 1954, wastage (the British-Canadian term for loss by retirement or any other cause) would be about 1500 a year, of whom 560 would be long-service officers. It asserted that the Canadian Services Colleges and the university training plans were working well and should not be changed, but it proposed modifications that would bring Canada's officer-training arrangement into line with those in Britain and the United States, "where all officer candidates are members of the armed forces from their entry into the training establishment, and all who qualify enter the regular forces." It proposed to reduce the optimum standard of academic training for line officers from four years at university standards after senior matriculation to two years, but to maintain the present standards for those officers who required technical qualifications. It recommended a two-year course at the Canadian Services Colleges, and also two-year courses at certain universities where cadets could be accommodated under military supervision. There should also be degree courses at some universities to produce technical officers who could not be accommodated at the services colleges. Finally, there should be a one-year preparatory school, preferably in Quebec, to raise students from Quebec and the Maritimes from junior to senior matriculation. Cadets in the plan would be paid. After graduation they would be obliged to serve for five years. This would result in the Canadian Services Colleges housing 585 potential regulars and sixty-five "civilian," or reserve entry, graduates by the fourth year of the plan's operation.[29]

On 26 May the minister received a progress report on this officer-production plan that said that all the universities approached, except Laval,* had agreed to offer the four- and two-year courses. Laval had replied that French-Canadian families would not be interested in a two-year course without a degree, and that graduates of the classical colleges and the Quebec provincial system could not fit into the proposed two-year course. Laval was also not interested in introducing a military preparatory course.[30] The implication was that the preparatory course should not be conducted by a university but under service auspices as Claxton had suggested. Laval indicated that such a course would be attractive to Quebec students and that it would be on a sound academic basis.

When Claxton announced this new officer-production program to the House of Commons on 12 June, it was substantially in the same form as he had presented it to the Cabinet Defence Committee except that the period of obligatory service had been reduced from five years to three.[31] The following day he instructed the deputy minister, C.M. Drury,** to prepare the necessary publicity for opening a new Canadian Services College (the preparatory course) in the fall – the Collège militaire royal (CMR) in St-Jean, Quebec.[32] On the same day the PMC met in a special session to work out details. The minister had ruled that, since the army and navy administered the other two services colleges, RMC and Royal Roads, the new college would be administered by the RCAF, thus putting it firmly into the tri-service system.

But instead of establishing the proposed two-year course for non-technical officers, the PMC now inserted merely that, "at the discretion of the services," cadets

---

*Université Laval and Université de Montréal were Quebec's only two French universities with graduate schools.

**Brig. Charles Mills Drury (no. 2082), RMC 1929–33, served with the RCA in the Second World War, and was deputy minister of national defence 1948–55 and minister of defence production 1963–8.

from services colleges could be transferred to general service – presumably with commissions – after two years.[33] The Defence Council noted that all naval cadets not in the Regular Officer Training Plan (ROTP) in the services colleges or the universities could already choose to join the RCN after their second year, that the army would institute similar arrangements for certain branches, but that the RCAF would not give long-service commissions in any branch before completion of four years of higher education.[34] These exchanges on officer production show that the Canadian Services Colleges were not to become entirely short-term military training schools on Simonds's lines. The four-year course culminating with RMC graduation would be retained for all RCAF cadets, possibly for some army classifications, and also for the reserve entry cadets of all services.

One consequence of the introduction of ROTP had been recognition of a need for further development towards centralized control of the Canadian Services Colleges from Ottawa. However, a little earlier an announcement in the *Canada Gazette* had provided a system of government that mingled prewar precedents and contemporary usage in Canadian universities. The minister of national defence was named president of RMC and there was to be a Canadian Services Colleges Advisory Board for the three colleges. The board's members, appointed by the minister, would include representatives for each province and from the RMC and Royal Roads clubs. They would meet annually to advise the minister but would have no executive power.[35] Since the minister would usually be little more than a distant figurehead, the commandants and directors of studies in each college thus appeared to retain considerable freedom of action. They could presumably use the Advisory Board, when it met, to influence departmental policy.

The introduction of ROTP and of a third college in Quebec necessitated an amalgamation of the Canadian Services Colleges Committee with a dormant Joint Services Universities Committee (JSUC), which had been set up to supervise the Canadian Officers Training Corps and other university training units and which had lapsed for want of authority. A new Joint Services Universities Coordinating Committee (JSUCC), established and named in January 1952, was now empowered to consider interservice aspects of training in university units and in the services colleges, the integration of military studies courses, and other matters affecting administration, personnel, and materiel. It was to take necessary action and to recommend to the PMC policies affecting the services programs at the universities and the Canadian Services

Visit of Field Marshal Viscount Montgomery, 17 April 1953. The snow fell for only a short while.

Colleges. It was also to act as the liaison between the Department of National Defence and the Military Studies Committee of the National Conference of Canadian Universities.[36] This JSUCC began to deal not only with such matters as a services colleges' form of drill to accommodate different practices in the three services, but also with wastage rates and the proportion of academic education to military training.[37] It thus had great potential for shaping the future development of RMC and the other colleges. In 1954 that potential was strengthened by new terms of reference; and from 1955 to 1957 JSUCC was aided by a Joint Universities and Military Advisory Committee (JUMAC) that included civilian appointees from the universities.[38]

Meanwhile, when the proposals for ROTP and CMR were still under discussion, the JSUCC and the two existing colleges had proceeded in the summer of 1952 with recruitment for the first class that was to include ROTP cadets. The prospect of a four-year course was still being offered to candidates. In August, when selection was in the final stage, the JSUCC recommended to the PMC that all scholarships, except the dominion cadetships for the sons of certain veterans, should be either cancelled or discouraged. But the PMC decided not to take that step unless it appeared that the scholarships operated against the functioning of the ROTP system.[39] For legal and technical reasons all 305 new recruits were admitted in September as reserve entry (RETP) – the old system – but ninety-eight were subsequently voluntarily transferred to ROTP, backdated to 12 September 1952. Others followed before the end of the year. In January 1953, to increase the output of the colleges, the government approved in principle an addition of two years at CMR to follow the preparatory year already operating there. Foulkes had instructed the PMC to study the benefits of putting Royal Roads on the same basis as CMR for entry – that is, to have a preliminary year there also. But that innovation would have created a serious bottleneck in the third year, when RMC would have to absorb the output

of CMR as well as of Royal Roads. So the proposal was shelved.[40] Space at RMC would be at a premium until construction was completed. Meanwhile, when the problem of a great disparity between CSC graduates and the products of the officer candidate schools for short-service commissions was brought to the attention of the Chiefs of Staff Committee in March 1953, it laid down as the desirable minimum standard for all Canadian officers either first-year university or three years of education beyond junior matriculation.[41] The idea of a two-year course at RMC was thus apparently still alive.

The second class to include ROTP cadets, and the first to be recruited as such, was admitted to the three colleges in the fall of 1953. The army and navy, still anxious to increase their enrolment of junior officers, again brought up the idea of increasing the number who could enter with junior matriculation. Simonds commented on the large numbers of junior matriculants who had applied for CMR. He argued that maintaining the senior matriculation standard at RMC and Royal Roads excluded many good candidates, especially as there was considerable disparity of standards between the provinces. So he proposed that the three colleges should be put on the same basis with the same courses. He believed that three years after junior matriculation was sufficient for executive, line, and aircrew officers in all three services. He said that an officer who needed a degree in the interests of the service could get one in a university in one year after graduation from the colleges. What he did not apparently know was that this could only apply to those who had taken a four- or five-year CSC course. Simonds argued that his plan would get a young officer into the services at an earlier age, which would be "much to the good . . . the better rooted will be his general military education." A cut in the number of years in RMC from four to three would make additional places available in the colleges: "Every year of delay . . . projects further into the future the time when the numbers trained will really meet our needs." It was necessary, however, to keep faith with cadets already in the system, and existing courses should therefore continue to function. Possibly that was an oblique reference to the RCAF's insistence on a four-year course following senior matriculation.[42]

In the following month, Foulkes circulated around the chiefs of staff schemes submitted by the various separate services to suggest how the output from the Canadian Services Colleges and the ROTP could be increased. The navy now proposed that junior matriculants be given two years of academic training to bring them up to senior matriculation level – a process that was already being achieved in one year at CMR. The army offered Simonds's scheme. Foulkes then asked for a study of numbers and of the effect of introducing the junior matriculation entry.[43]

These new attacks on the four-year academic system, aimed at abolishing or restricting its third and fourth years before the third batch since the reopening in 1948 had passed through them, soon reached the ears of the RMC faculty. By now most of the teaching staff necessary for a four-year course at RMC had been appointed. It seemed as if professors, who had been recently appointed with the prospect of teaching in a four-year academic college based on senior matriculation, might now find they had been misled. However, on 12 March 1954 Sawyer took pains to tell the faculty that the rumours they heard were untrue. The minister had reassured him that the four-year course was here to stay. Claxton had said that Ralph Campney,* the associate minister of national defence and, "so far as the Minister is aware, also the Chiefs of Staff," agreed. Claxton had also informed Sawyer that the tri-service system should no longer be considered experimental.[44] Brigadier Agnew was equally reassuring.

In fact, the minister's support for tri-service military education, coupled with the RCAF's insistence on a four-year course for both aircrew and technical commissions and with the continuance of reserve entry, had effectively blocked Simonds's plans to change the Canadian Services Colleges into a two- or three-year training program for line officers, diverting the production of technical officers to the universities. In the background there had also been considerable support for the new RMC from the RMC Club. When a former member of the RMC faculty, toasting the "Old Brigade" – the class that had graduated twenty-five years earlier – at a club banquet, praised the products of the old college and disparaged the new system, its "doctors," and its "Faculty Club," E.W. Crowe, an ex-cadet club's president, feared that the attack might make Ottawa think that the club was not fully behind the new college. He pointed out that only 42 per cent of that class had joined the regular forces, and he warned that intemperate criticism of the new system, which was designed to improve on that record, might get the college closed down.[45]

Since only about one-quarter of the first four annual entries to RMC before ROTP was introduced had taken up regular commissions, the number was clearly insufficient to meet the need for officers. Enrolment in ROTP had been offered first on a voluntary basis to cadets already in the colleges, and then to high school graduates. This would still not bring in enough. In April 1954, therefore, the PMC decided to limit future entry almost entirely to those cadets who would undertake to join the regular forces on graduation. The exceptions, the dominion cadetships, were few in number.[46] This meant a radical change in the composition of the cadet body. The plan was vigorously criticized on the grounds that it excluded the sons of middle-class families who did not need to make up their minds to join the forces when leaving high school but who might do so while in the services colleges. Critics also argued that in the place of those displaced

---

*Ralph O. Campney served in the First World War in a hospital unit, in the infantry, and as a pilot. He was associate private secretary to Prime Minister King 1925–6, parliamentary secretary to the minister of national defence 1951, associate minister 1953, and minister 1954–7.

would come many students who merely wanted a cheap education and who had no long-term military career interest. As it happened, the wastage rate of ROTP students, when they took up all the available places in the colleges from 1954 on, did indeed prove to be higher than for the two preceding years, when there had been mixed regular and reserve entry. At the same time, the services colleges' wastage rates with ROTP proved to be substantially lower than those of ROTP cadets in contemporary university schemes.[47]

The almost complete elimination of reserve entry in 1954 appeared to open the way for a system similar to the solution Simonds had sought so persistently without success – the restriction of the Canadian Services Colleges to a truncated course for non-technical officers selected on junior matriculation, with the diversion of potential technical officers to the universities. Under his scheme, line officers in the army would have completed their "military education" on the job with the troops – a late kind of apprenticeship. However, and as CMR's experience was to confirm, academic wastage is much higher in the junior matriculation year than in later years. Simonds's plan to recruit at that level would have transferred that year of high attrition from provincial and local budgets to the federal budget. It would not have produced the number he hoped for, unless academic requirements were lowered to keep the colleges full. It is hard to avoid the conclusion that at this time some senior army and navy officers, including Simonds, believed that for navy executive and army line officers, intellectual ability and application were less important than military indoctrination and training.

But any possibility of converting the services colleges to military training schools with a minimal academic and intellectual content had in fact been frustrated by the air force. The RCAF had not gone along with the plan. Air force candidates for both the general (aircrew) and technical lists were applying in fair numbers; and for those cadets the four-year CSC courses, with the third and fourth year continued at RMC in either science or arts, was what the RCAF desired. It would have been hard for the other two services to recruit on less favourable terms.

A Two-Year Military Training Course or a Degree?

In 1951 the RCN was short of senior officers and did not want to take its turn commanding RMC. It therefore recommended that Brigadier Donald R. Agnew should continue as commandant, taking over the navy's turn. When Agnew resigned in 1954 to become Canada's representative on the Commonwealth War Graves Commission in Europe, the RCAF succeeded to the command. Under Air Commodore D.A.R. Bradshaw,* the continuance of the four-year course was reinforced.

Bradshaw had graduated from RMC in 1934 and had been commissioned in the Royal Canadian Dragoons. He transferred to the RCAF and had a distinguished career in the war as a pilot and squadron commander. With his RMC background and his experience of two services, he felt he was well qualified to command the new tri-service college, but he was conscious of the fact that he was the first non-army commandant. Although he came directly under the Personnel Members Committee (PMC) to serve all three services, he encountered some petty obstruction from the army. Being a member of another service, however, also had advantages. He recalls that on one occasion General Simonds, the CGS, phoned to tell him to send the Leinster Regimental plate (which was deposited at RMC for safe-keeping and was then mostly in storage) to Ottawa for the use of his new Canadian Guards regiment. Bradshaw replied, "No, Sir." An army commandant, possibly dependent on the CGS for future advancement, could hardly have been so blunt. Bradshaw was similarly in a position to resist Simonds's pressure for any reduction in the length of the course.

Bradshaw was impressed by what the director of studies, Colonel W.R. Sawyer, had achieved in academic standards, and also by his plans for future development at RMC. Agnew had supported this trend because he felt his own academic career had been cut short by the First World War. Now Bradshaw was interested in maintaining high educational standards for the services, especially in the field of technology. Men in the ranks were better educated than in earlier days, and Bradshaw believed that officers must have proportionately more education in order to lead them. Moreover, he found that many permanent officers, especially those who had come into the service during and after the war, including naval officers on his own staff and elsewhere, favoured the four-year Canadian Services Colleges (CSC) course. Some members of the RMC Club also told him they wanted it. When he visited the large private schools to discuss recruitment, headmasters told him their boys wanted a degree before they were commissioned. Some senior naval

Air Commodore D.A.R. Bradshaw, DFC, CD, commandant, 1954–7

*Air Vice-Marshal Douglas Alexander Ransom Bradshaw (no. 2140) was at RMC 1930–4. Officer commanding 420 Squadron, RCAF 1942–4, he became director, Air Operations, Ottawa 1945, and then commanding officer, RCAF Trenton. After retiring from the RCAF, he headed the administrative services at York University.

officers like Admirals Hugh Pullen and Nelson Lay,* however, wanted the continuation of the same kind of training they had had themselves. In response, Sawyer held that the ultimate objective should be an RMC degree for all CSC cadets in all three services. He persuaded the new commandant to "battle for it."[1]

Before the degree question could be settled, Simonds's plan for a two-year course for non-technical officers still had to be put to rest. Since all applicants for the first year in a services college, except the preparatory

Inspection by Major-General G.R. Pearkes, VC, minister of national defence, accompanied by CWC Peter Meincke, October 1956

year at Collège militaire royal (CMR), had to have a senior matriculation that qualified them for engineering, almost all recruits initially entered as potential four-year students on the technical side. They did not have to make a firm commitment to a particular service until they had embarked on their first year or, if they were army cadets, to choose a branch until their second year. If found unsuitable for continuing in the sciences, or if they found their interests lay in other directions, they had alternative possibilities in the humanities and social sciences. Simonds had proposed to commission those cadets who dropped out of science as soon as they passed their second year, but most of the weaker ones did not get that far. Others who passed their second year on the condition they major in arts courses wanted the four years of education they had been promised. Only a few second-year cadets who could not go on in arts opted to take a commission at that stage – and they were accepted only if they seemed to have good officer potential. What Simonds had failed to realize was that it was the prospect of four years of tertiary education that had attracted many recruits to the college. Without fulfilment of that promise, recruiting would decline. His plan was doomed to fail because it would not have produced the numbers he sought.

Simonds probably did not have full support from his subordinates in the army. His tour as CGS ended in 1955. In October 1957 Colonel N.G. Wilson Smith,** director

*Rear-Adm. Hugh F. Pullen and Rear-Adm. H. Nelson Lay were graduates of the Royal Naval College of Canada who began their service careers as midshipmen with the Royal Navy. Pullen served with the RCN in the Second World War, was flag officer, Pacific Coast 1955–7, and then of the Atlantic Coast. Lay commanded a Canadian destroyer and an RN escort carrier in the war, and rose to be vice-chief of the Naval Staff in 1954.

**Maj.-Gen. N.G. Wilson Smith, Royal Canadian Infantry Corps, was directing staff at the Staff College 1945–8 after serving overseas. He became assistant adjutant-general in 1950 and a major-general at CFHQ.

of infantry, said he disagreed with the shortened-course policy Simonds had advocated: "My experience is that the best educated officer normally makes the best infantry officer." He added that the difference between Officer Candidate Programme (OCP) officers and graduates of the Canadian Services Colleges was "quite startling." He said that both began their careers fairly evenly, but within a few years the university-trained man had forged ahead and was a better officer in all respects. "The officer who is poor in his academic studies is normally weak on the military side as well." He noted the high failure rate among cadets who had been academic failures in the services colleges and who had obtained a commission through the OCP. By offering university training to infantry cadets, more and better cadets would be induced to opt for that arm. "Our Services Colleges are educational institutions first and military colleges second."

Smith also pointed out that the simplest and easiest channel to high positions of responsibility in the army was through the fighting arms, and it was a sacrifice of talent to put the best educated and trained men into the technical services. Summing up, he said that the best officer cadets would go where they could get the best education, that unless the infantry could offer that best education it would get the poorest product, and that the best cadets would then go into the technical services – where, however, they would be at a disadvantage in getting to high ranks.[2] This statement marked a revolution in the army's approach to combat-officer production, a rejection of the position long advocated by General Simonds. It confirmed the policy that had prevailed at RMC since 1948. The four-year course after senior matriculation was there to stay.

But there also remained in question where army and navy technical officers would be educated. In October 1954 the Joint Services Universities Coordinating Committee (JSUCC) had discussed the way in which CMR's output after three years could be disposed. To make room for more first-year cadets it had considered a bleed-off

either to the universities or to the services at the end of the CMR second college-level year. But transfers of technical students to a university would have the disadvantage of taking them out of a military environment at an important point in their training. The JSUCC then learned that Canada's university engineering faculties would be crowded in 1956 and that they would take few, if any, CSC cadets at the end of their CSC second year. It followed that the services colleges must accommodate most or all third- and fourth-year CSC technical cadets for the army and navy as well as the air force.

The JSUCC also pointed out that adding a further two years at CMR to absorb the surplus would please French Canada. RMC responded, however, that the opening of CMR had stopped the flow of French Canadians into the first year at RMC and terminated what had for three-quarters of a century been a contribution to national unity – the presence of francophones in RMC. It also suggested that keeping French Canadians on longer at CMR would impede their bilingualism. The JSUCC studied other expedients, such as putting all recruits and second-year cadets into CMR and Royal Roads, and all third- and fourth-year general (arts and sciences) course cadets in CMR. Its ultimate proposal, made in December 1955, was to accommodate all third- and fourth-year technical and non-technical streams from all the colleges for all the services at RMC. This could be accomplished by reducing the intake of recruits at Kingston and by building another barrack block like Fort Haldimand.[3]

Fort Haldimand had been built in response to expectations that by 1952 the strength of RMC would be 400 cadets. In March 1954 RMC had suggested that a third dormitory planned for CMR should be built instead at RMC, so that the expected third- and fourth-year cadets could be accommodated. The strength of the college would then be 550. Agnew had arranged for Fisher and Tedman, a firm of Toronto architects, to assess short- and long-term building needs at RMC. Their report, submitted in October 1954, recommended, in addition to the dormitory, an officers' mess that would capitalize on the unique features of its Point Frederick location looking out over historic and scenic Navy Bay to Fort Henry. The architects also recommended a sergeants' mess and improvements to the existing married quarters, quartermaster's stores, and college grounds. They advised that long-term development should include replacing the temporary buildings, presently used as engineering laboratories and for other purposes, by stone buildings architecturally consistent with the general appearance of the college. Finally, they suggested creation of a "north campus" enclosing the Mackenzie and Currie buildings, a new library and educational building, and the old artillery gun-shed now used by the electrical engineering department.[4]

By the time the report was submitted, Agnew had left RMC. It therefore fell to Air Commodore Bradshaw to prepare and submit an RMC development plan to meet RMC's new needs in the expanding CSC system. One of his first steps was to inquire about a long-range station development plan to accommodate the expected increase. He was told there wasn't one.[5]

Bradshaw was an administrator and planner. Whereas Agnew had called for an outside expert study, the new commandant formed and personally chaired the College Development Committee. He found that prewar facilities had been designed for 200 cadets and a civil engineering curriculum, but by 1948 the forecast was for 400 cadets and a curriculum that included a choice of engineering courses as well as a general arts course. The extra facilities approved thus far included a dormitory, an RCE workshop, and a Nissen hut to house a jet engine. Existing buildings had been adapted for academic accommodation, including old boiler rooms and coal cellars, hay lofts, storage facilities, and corridors. The college's accommodation was inconvenient and overcrowded. It lacked adequate facilities customary in military organizations, including messes, recreation rooms, and chapels. To accommodate the intake from CMR and

Royal Roads, RMC would now require immediate accommodation for 462.

Bradshaw's development plan had two phases. Phase 1 would meet the immediate needs for the 462 cadets, while phase 2 would provide for 650 – allowing for the full output of CMR (100) and Royal Roads (65). Phase 1 included a library and educational building as well as an addition for civil engineering, an officers' mess, artificial-ice rink, central heating plant, quartermaster's store, boat house, two chapels, and two tennis courts. Phase 2 would be the science and engineering building, a dormitory, gymnasium, and additional permanent married quarters.[6]

The Canadian Services Colleges Advisory Board discussed Bradshaw's development plan at its annual meeting in March 1957. It agreed there were difficulties in accommodating the second-year output of the three colleges at RMC, but warned there were disadvantages in sending the surplus to universities: the universities might be willing to accept only the best students, and French Canadians going to French universities would lose their competence in the use of English. When General Foulkes remarked that the chiefs of staff believed it would be difficult to obtain government approval for new construction at RMC, Paul Hellyer,* parliamentary assistant to the minister of national defence, commented that any calculation of the comparative costs of a services college and a university education should take into account all the costs to the Canadian public, provincial as well as federal. From the financial as well as the military point of view, he said, the development of the CSC system was to be preferred, even if it meant overcrowding until new accommodation was available.[7] Nevertheless, the department refused to approve implementation of Bradshaw's

---

*Paul T. Hellyer trained as air crew in the RCAF in the Second World War but was then involuntarily transferred to the army. Elected to Parliament in 1949, he became parliamentary assistant to the minister of national defence 1956–7, associate minister 1957, and minister 1963–7.

long-range plan. Henceforward it was submitted piece-meal from year to year, however, and Bradshaw's skill in planning thus had a long-term effect.

While these discussions about RMC's function and the accommodation it needed were proceeding under Agnew and Bradshaw, RMC's internal administration was being amended to ensure that it was appropriate for RMC's new status. At the beginning of the second year of operation, 1949–50, discussion of academic policy had been transferred from the Academic Board, consisting of the whole faculty, to a Faculty Council composed of heads of departments, the registrar, and the staff adjutant. The director of studies presided, and the council reported to the commandant. This left to the Academic Board, now renamed the Faculty Board, only the task of supervising mid-term, Christmas, and final grades. In 1952 some members of the board, while recognizing that in a military college ultimate authority must rest with the commandant, had protested that this rearrangement had deprived most faculty members of any input into policy, and even knowledge of decisions, except by courtesy of their department chairmen.[8]

This request for a voice in discussion of academic policy was partly a result of a belief that the prewar RMC had been run more as a school than as a university. It also indicated concern that undue military interference in academic matters might sap the college's academic credibility in the eyes of Canadian university faculties. An incident in the first term in 1948 had shown that this was a possibility, when a service officer teaching physics complained that cadets had been withdrawn summarily from his class for a clothing parade. The Academic Board ruled that this was a breach of college policy,[9] and henceforward the faculty kept a close watch on intrusions on class time.

At the root of the problem was competition among the military wing, the academic departments, and sports officers for the cadets' time. Two late afternoons a week were set aside for remedial and tutorial instruction, while another was allocated for drill for ceremonial parades when necessary. The staff-adjutant complained that the one-hour drill a week allocated for him for second-, third-, and fourth-year cadets was insufficient to prepare for major ceremonial parades and he sought to use the tutorial times.[10] Yet when members of the faculty reported that sports were interfering too much with study time, even the intellectually sympathetic Agnew was moved to say that a sports program of two hours was not excessive compared with twelve hours of evening academic work.[11]

Competition for the cadets' time was closely related to the problem of "wastage" during the college years. One source of difficulty was that cadets were involved in both intervarsity and intramural competition. Like parades, this competition was important for the development of a corporate military spirit, but the net result was a heavy burden on cadets who worked in a university-type academic program in a military institution. Since RMC was a small college competing with much larger universities, there was pressure on cadets to take part in intercollegiate athletics. The faculty insisted that cadets who were academically weak must not be coerced to participate in games if they risked failure.

Now that the government, through the Regular Officer Training Plan (ROTP), paid the full cost of educating and supporting most cadets and the college retained cadets in all fields for four years, there was greatly increased concern by both the military staff and the academic faculty about the possible effects of this conflict of interests on "wastage" – the loss of cadets from the system before commissioning – and "retention" – the ability to keep officers after the end of three years of obligatory service. Wastage at the college level had three different causes: voluntary withdrawal by cadets who found RMC military routine intolerable, academic failure, and elimination for lack of what were called at RMC "officer-like qualities." Inadequacy in academics was often accompanied by lack of potential in other respects, and, since it was easier to measure, elimination was most commonly asserted on that ground.

Before ROTP, academic failure in the first year at both RMC and Royal Roads had run at about 30 per cent of the class. This was in keeping with experience in university engineering courses, but from the first this failure rate had aroused much discussion at RMC because the college faculty and staff felt responsible for meeting a need for military graduates. Since RMC was small, grades could be examined and assigned collectively by a Faculty Board meeting of all concerned staff. They were then subject to ratification by the Faculty Council of heads of departments, where there was further discussion, and then to confirmation by the commandant. In practice the council dealt only with marginal cases and with an overall general review. The commandant's confirmation was usually a routine approval of the council's decisions, sometimes with comments on individuals and occasionally with a reference of a particular case back to the council. It was thus generally accepted that academic decisions were within the purview of the director of studies and the faculty. Since the faculty was concerned to maintain academic credibility in the university world, it was suspicious that there might be military pressure in favour of individuals who were weak academically but who had physical or other qualities that suggested they would be good officers.

After ROTP was introduced, the question of wastage had become even more serious. The director of studies circulated a memorandum to the faculty and the military staff to establish policy on extracurricular activities. He noted that RMC must have wide athletic and other interests, but added that group activities in the evening study periods must not take place without the permission of the director of studies or the registrar, and that cadets must not miss classes to engage in extracurricular activities without permission.[12] Later, the registrar had to remind the faculty that in a military college they must not unilaterally cancel classes, as often happened in universities.[13]

These measures demonstrate significant differences between cadets and university students. University students are free to study or not as they please. Their results are solely their own responsibility. Although military academics aim to develop a similar individual initiative in relation to studies, the cadet is not as much in control of his own time. Other duties, obligations, or interferences may infringe on his studies. The problem was, therefore, something more than a competition for the cadet's time. It raised basic questions about the value of a general academic education for professional development, as well as about the authority of the military and the academic staff. The most immediate issue at RMC in these problems was cost-effectiveness. Contemporary use of the term "wastage" forcefully illustrated the fact that the military authorities were concerned about a satisfactory return for money expended on the officer-production program.

In his unpublished autobiography, Claxton stated that CSC wastage was too high and therefore threatened his creation of the tri-service four-year military college system as a source of officers.[14] In Agnew's time, on 23 October 1953, Claxton had attended a meeting of the RMC Faculty Council to say he was worried by the high wastage rates, especially for the army. In view of the present need, it was necessary to produce and keep as many officers as possible. He asked whether the workload of the first-year cadets was too heavy, and said that anybody who hazed them should be "fired out of hand." Sawyer replied that the RMC curriculum paralleled those in universities, but that the cadet also had drill, compulsory sports, and other extracurricular activities. In a follow-up letter to Agnew, Claxton repeated his concerns, asked whether the RMC course was precisely related to what was needed for the services, but also said he thought that RMC was on more correct lines than Sandhurst, where the academic program was less important.

Agnew had replied that high wastage occurred partly because, in addition to their academic work, cadets had extra training for management and leadership. He noted that in the early years of the new RMC, some ex-cadets had criticized the course as being too narrowly academic; but RMC's record on the sports field had refuted that charge. Intramural sports were voluntary, but there was no shortage of aspirants. The good record of postwar RMC graduates in ten universities proved that the academic program was also sound. He produced figures to show that RMC's wastage was not much higher than elsewhere, especially when it was remembered that many of the college's recruits from some provinces could not have gained admission to leading Canadian engineering schools. Drawing from all Canada, RMC could not expect to get the same overall results as a provincial university which drew mainly from one source.[15] While 80 per cent of all applicants wanted to take a course in engineering, one-third of them did not have the necessary aptitude.[16]

Under Bradshaw, this problem of wastage continued to receive attention. At a mid-term marks meeting in November 1954, the Faculty Board noted that first-year averages were about the same as the previous year, even though the matriculation grades of entering recruits had been higher. Cadets from Nova Scotia, however, lacked proper preparation, especially in mathematics, and some French-speaking recruits were in difficulty because they had poor command of English. On the whole, the faculty believed that the fall term's full-scale athletic program had made it difficult for cadets to settle down to study.[17]

A year later, with an even higher average on entry, the mid-term marks of the recruit class were the lowest on record. The board noted that the high entry-grade average was possibly attributable to excessively high grading in Saskatchewan's matriculation, which had qualified a disproportionately large number of recruits. Members of the Faculty Board claimed, however, that there had been an increase in compulsory sports, including oblig-

atory attendance of spectators, and also of punishment drills. Both had interfered with studies. Moreover, cadets were coming less for tutorial assistance. These problems were referred to the Faculty Council for investigation.[18]

The JSUCC had already submitted the problem of high wastage rates to the newly created Joint Universities and Military Advisory Committee (JUMAC), which included appointed civilian academics. It asked JUMAC to investigate the ratio between academic and military training, but JUMAC declined to respond until JSUCC provided it with a complete analysis of results.[19] In the college itself,

Recruit arrival, 1958: Until the early 1970s new recruits arrived at RMC direct from home.

Bradshaw warned the cadets they should not have to be prodded, pushed, or urged by the academic staff but must rely on themselves. He told the Faculty Council that, despite its complaints about the number of failures and their causes, the faculty did not appear to have recommended the withdrawal of a single weak cadet. Money spent on cadets with poor academic performance was money wasted, he said, and cadets who failed seven out of ten papers at Christmas, or were below 40 per cent overall grade average, should be eliminated.[20] In fact, such cadets were usually eliminated later at the end of the session or were allowed to repeat a year at their own expense if their failure was not too severe.

One explanation of the admission of a number of academically weak cadets to RMC was disparity among various provincial educational systems. With the introduction of ROTP, the RMC qualifying examination, which had been administered to candidates from all provinces for purposes of comparison, had been discontinued. In 1955 JSUCC considered restoring the examination but did not do so, perhaps because the number of applicants was now so much larger.[21] The high wastage rate of first-year cadets from Saskatchewan, however, alarmed the minister.[22] Recruits from Alberta and Manitoba were also poorly prepared in mathematics and physics.[23] RMC found a remedy for these disparities by using a mathematical formula for the admission of recruits that adjusted matriculation marks in accordance with the performance of cadets from each province in RMC's first-year examinations over the previous five years.[24]

Disparity of provincial standards was only part of the trouble. Cadet academic performance was also affected by the competing pressures exerted to promote overall military quality. In January 1955 the registrar, Colonel Gelley, summarized the board's discussion of the continued decline of academic results by saying that leadership training in RMC appeared to be limited to athletics and that this concentration could lead to false values. Observance of sports programs three afternoons a week

meant that all other compulsory parades and activities were shifted to the two remaining afternoons, to the detriment of tutorials and cultural activities previously scheduled for those days. Furthermore, lack of emphasis on academic standards caused cadets to assume that all that was needed was a passing standard. Gelley said that cadets had no accepted criterion of excellence in intellectual and cultural activities, and that a heavy workload led them to use the study period as time for relaxation.[25]

Bradshaw also presented an oral brief on failure rates during a visit to Ottawa. He mentioned the variations in provincial standards, and that recruits entering CSCs had lower averages than students enrolling in engineering at Queen's University. But he attributed much of the problem of wastage to ROTP, which, over a three-year period, had altered the composition of the college and had changed cadet attitudes. Before ROTP, 81.9 per cent of cadets had come from urban centres, compared with 65.8 per cent now. Furthermore, ROTP applicants who lived in cities where there was a university usually gave it as their first choice so they could live at home. Agnew's annual report for 1953–4 had therefore recommended preferential selection of applicants in favour of CSCs.

There had also been a steady increase in the number of students for whom financial implications were the deciding factor. Many recruits frankly admitted they had applied for ROTP only for the purpose of getting a free education. When priority allocation into the CSCs was introduced in 1955, one recruit withdrew after the Christmas break and others were disgruntled because they had been sent to RMC, with its restrictions, instead of to a university where they could live at home, have personal freedom, and obtain their degree in four years. To get a degree at a CSC would take five years, and then only if the services sent them on to university. Bradshaw concluded that ROTP had led to mediocrity in academic achievement, a decline in motivation, higher first- and second-year failures, and suspected deliberate failures to avoid the repayment penalty.

Yet Bradshaw reported that RMC's "yield" of 55.4 per cent (as against DND's expected retention of 55.6 per cent) was more than 10 per cent above the Queen's University figure and slightly better than that of the prewar RMC. He said that in terms of output, the CSC system had been meeting its obligations because of "an excellent staff, a disciplined body of cadets, and more individual attention to a cadet's performance than is usually the case in a university." "Our staffs have laboured hard and with sincerity to produce the best results. I am convinced that they have more than carried out their responsibilities. Their results are excellent." To get even better results, Bradshaw advocated continued use of the "weighting factor" for provincial grades in determining entry, the introduction of 15 per cent reserve entry to tap a middle-class stratum of society now discouraged, rewards for academic achievement in service seniority as given in the RCN, restrictive qualifications for entry into engineering, and restrictions on the entry of Ontario matriculants who bolstered their school results by taking the grade 13 examinations over a two-year period.[26]

The Department of National Defence continued to worry about the CSCs' cost-effectiveness. In March 1957 Ralph Campney, now the minister, told a meeting of the Advisory Board that the increasing cost of military equipment made a new look at officer production necessary. ROTP was costing about $5 million a year. It cost $45,000 to educate a services colleges cadet. From the department's point of view it was more expensive to train cadets in the colleges than in the universities: $16,600 per annum for a CSC cadet compared with $6000 for each university cadet. He again raised the possibility that "general services officers" might be limited to two years in the colleges after senior matriculation.[27]

In August 1957, on the recommendation of the Personnel Members Committee (PMC), the Chiefs of Staff Committee set up an interservice committee to study ROTP. The only member of that committee who was, or had been, connected to the colleges was its chairman,

Captain W. Landymore,* RCN, an ex-cadet, who suspected that the committee had been struck to abolish the services colleges.

In fact, from the evidence placed before it, the committee arrived at a quite different conclusion. Landymore received a great deal of information from Lieutenant-Colonel Gordon Carington Smith,** permanent chairman of the Joint Services University Canservcols Committee. Carington Smith was also an ex-cadet, and belonged to a family that has now sent six young men to RMC and one to Royal Roads. Landymore found that the CSCs were not attracting enough applicants from successful matriculants motivated at the outset for service careers, and that the objective of many recruits was a degree rather than service in the armed forces. He assumed that 1500 officers were needed for replacements each year and that 356 of these should come from ROTP. His committee calculated that at present only 11 per cent of a smaller intake of officers came from that source. It concluded that, to meet the goal, recruiting must be increased.

Contradicting what the minister had said in March about costs, the Landymore Committee calculated that it would cost $24,500 to produce an officer through the university plan, compared with $24,700 through RMC, if the college were used to its full capacity. It argued there

*Rear-Adm. William M. Landymore (no. 2399) left RMC in 1936. He trained as midshipman with the Royal Navy, was deputy director of weapons and tactics at Naval HQ Ottawa, and attended the Naval Staff College and the Joint Services Staff College in England. He was director of manning, and then commander, Canadian Destroyers, Far East, during the Korean War, before attending the Imperial Defence College. Flag officer on the Atlantic Coast, he resigned from the navy in protest against Hellyer's program of unification.

**Lt-Col. Gordon Carington Smith (no. 1758) graduated from RMC in 1927. He was one of the founders of the Royal Canadian Armoured Corps and became staff officer, Manning, at CFHQ before retiring in 1958.

were advantages in the better quality of military indoctrination and training in a service institution. Landymore recommended against increasing junior matriculation entry, because greater wastage at that level made it more expensive, and also against the Reserve Entry Training Plan, because the quality of those cadets was only 1 per cent higher in academic grading than it was for ROTP cadets. Furthermore, he noted that the reserve forces,

The RMC–USMA hockey game had traditionally been one without penalties called. This gentlemanly practice eventually led to abuses from both teams and the first penalty in the series was called in 1954 against the USMA goalie, Dirk Lueders.

for which RETP was designed, no longer had the importance they once had had. Turning to the degree question, Landymore said that the practice of sending eighty-five RMC graduates on to university "just to get a degree," as was the case that year, would cost about $300,000 a year in pay, subsistence, and tuition. This was wasteful. He therefore recommended that RMC should apply for a charter to grant degrees. He also found that the services were sending some officers to universities to get postgraduate degrees that could be offered by RMC.

Landymore noted that the colleges were under the control of several different committees for different purposes. He recommended that, to simplify organization, there should be a director of the ROTP (DROTP), with a seat on the PMC for ROTP matters. His proposal was that the commandant of RMC should double as DROTP, and that his director of studies should be relieved of his task as vice-commandant and made director of education for ROTP. He ruled against a proposal that RMC be restricted to third- and fourth-year cadets, not because it would militate against training and academic development, but because CMR and Royal Roads could not produce enough qualified cadets for entry to RMC's third year. He proposed that graduates of RMC should be allocated to corps and branches according to service needs and cadet aptitudes, rather than on a basis of personal choice. He suggested that a few graduates should be permitted to serve in the Defence Research Board.[28]

Commenting later on the Landymore Report in the Advisory Board, the commandant of Royal Roads, Colonel P.S. Cooper,* said that the motivation of young men for military careers was beyond the control of the services. The problem lay in the home and in society

*Col. P.S. Cooper, Royal Canadian Infantry Corps, commandant Royal Roads 1957–60. GSO1 in 1947, he was attached to Canadian Defence Liaison Staff, London, in 1948.

generally. The emphasis should therefore be on inspiring service motivation after the acceptance of ROTP obligations.[29]

The Landymore Committee thus gave the colleges, and especially RMC, a strong endorsement. The PMC recommended to the chiefs of staff that a proposal to give degrees at RMC in arts and applied science should be approved. But the chief of the Naval Staff, Vice-Admiral H.G. DeWolf,* and the chairman of the Chiefs of Staff Committee, General Foulkes, both argued that a BSc degree could not be done at RMC in four years. Foulkes also said that an arts degree was "worthless" for the services. He added that what was being considered was a BSc in military studies. The Chiefs of Staff Committee then approved in principle a plan to approach the provincial government for a degree-granting charter for RMC. That committee also said that a DROTP should be appointed as soon as possible, but that he should be an officer in headquarters.[30]

Meanwhile, the first year's Christmas results at RMC were down again, the lowest in eight years and the third lowest in ten.[31] Analysis made by the JSUCC had in fact now established that CSC results compared favourably with those of civilian colleges and universities, but it still considered the academic failure rate in the colleges too high. At RMC the failure rate in the first year was 39.7 per cent, at Royal Roads 40 per cent, and at CMR 24.4 per cent, where there had already been a loss of 30 per cent in the preparatory year. It was 37 per cent in the university ROTPs where, however, entering quality had been lower. Further study of the problem was requested.[32] A year later, all the colleges reported improvement in the quality of applicants, with a lower wastage rate, much better than the university ROTP. The three commandants said this was due to an increase of numbers from which they had been able to select recruits.[33]

Nevertheless the RMC Faculty Board continued to worry about numbers and standards. It again complained about interference with study time, on this occasion as a consequence of the "excessive powers" of summary punishment exercised by cadet officers. These were said to be even greater than those of a subaltern in the service. It had also heard that seniors were judging cases in which they were personally involved. The Cadet Wing commander replied that records of all punishments were kept in the Orderly Room, and that the amount of punishment any individual cadet could receive was limited. "Circles," a form of extra drill imported from Royal Roads, were now restricted to recruits.[34] Concern about wastage had been partially eased in 1958 when Commodore Desmond Piers,** Bradshaw's successor as commandant, was able to report that RMC's first-year failure rate was now down to 16 per cent compared with a norm of 24 per cent.[35]

Meanwhile it was still being asked whether the academic quality of RMC graduates should now be recognized formally by the grant of an RMC degree. Several years earlier, in January 1953, when Agnew was commandant, Army Headquarters had asked for recommendations for sending RMC graduates returning from Korea to universities to take an arts degree. The college had responded that it was senseless to send RMC general course graduates to take a university pass arts course because in four years at RMC they had already covered at least as much as they would in three years in a university.[36] Despite this reply, the army had sent some ex-cadets who had served in Korea to universities immediately after graduation. The results were unsatisfactory

Commodore D.W. Piers, DSC, CD, commandant, 1957–60

---

*Vice-Adm. H.G. DeWolf was educated at the Royal Naval College of Canada and served in the RCN before and during the Second World War. He was chief of the Naval Staff 1956–60.

**Rear-Adm. Desmond William "Debby" Piers (no. 2184), RMC 1930–2, trained with the RN. During the Second World War he had destroyer commands on the Atlantic and Russian supply routes. Senior Officer afloat 1956–7, he was commandant 1957–60. Later he was director of naval plans and operations, and then chairman of the Defence Liaison Staff, Washington. After retiring, he was appointed agent-general in London for Nova Scotia 1977–9.

for those who took pass degrees because there was no challenge and little to be gained academically.[37]

It was the Army Headquarters' decision to send those general (non-technical) course graduates to university that revived discussion of degree-granting powers for RMC.[38] The Defence Council had rejected the idea in 1952,[39] but a year later the army's interest persuaded Agnew to send a memorandum to the PMC proposing that RMC should itself grant degrees, though only to general course cadets, in order to equate with the practice "elsewhere" (he presumably meant at West Point). He said this would improve RMC's appeal to high school students and would give CSC ROTP graduates the same qualification as was being earned by students with lower academic qualifications who went to universities under the ROTP scheme.[40]

The 1953 meeting of the Faculty Council attended by Defence Minister Claxton had discussed the question. Claxton had expressed surprise that a degree charter would need an act of the Ontario legislature. However, he did not think it "below the dignity of the federal government" to make such a request. In his follow-up letter he said he would approve the request provided provincial susceptibilities were not offended.[41] W.F. Nickle,* Kingston's representative in the provincial legislature and a member of the Ontario cabinet, confirmed that Ontario would respond favourably to a request. In an amendment to the speech from the throne in the Ontario legislature on 9 March 1954, Nickle recommended that Ontario give RMC a charter to grant degrees and honorary degrees.[42]

In the same year, a report prepared for the federal Civil Service Commission noted some skepticism within the federal service concerning the value of the services colleges to the military forces. It said, on the contrary, that the humanities did have a direct and practical application for military education: the value of languages was obvious, military history was needed for tactics and strategy, and English was an essential tool for military communications. In a modern society, soldiers had a multiplicity of tasks and needed an education equal to that of their fellow citizens. The report went on to quote the 1944 Guy Report on Sandhurst and Woolwich and to refer to West Point and Annapolis, to show the emphasis on general education in other countries.[43]

On 23 November 1955 Bradshaw sent the secretary of the Department of National Defence a formal proposal that degree-granting powers be obtained. After outlining the history of general education in the RMC, he noted that industry and the professions were now demanding higher qualifications and that the university degree was an important standard. He outlined the advantages that would accrue to the services from an RMC degree – namely, a lowering of cost, increase of prestige for the college, and removal of a "stigma" that was a handicap in the recruitment of staff. Bradshaw said that an RMC degree would attract more applicants because cadets entering the service wanted assurance of a degree before being commissioned. It would thus remove a disadvantage of the services colleges compared with ROTP in the universities. He stated there was no intention to ask for degree-granting powers in engineering in the near future, but noted that the RMC arts and general science curriculum now had to be designed to meet the requirements of the universities and that, with an RMC degree, it could be shaped to meet the needs of young officers.[44]

The RMC curriculum had already been tailored to meet university requirements. A reading course in medieval English history had been added to meet requirements for admission to the MA program at Queen's University to accommodate two RMC cadets who wanted

---

*William Folger Nickle served with the Princess Patricia's Canadian Light Infantry overseas during the First World War. Elected to the Ontario legislature for Kingston in 1951, he became provincial secretary and minister for planning and development in 1955.

to go on to graduate work there, and there was a proposal to add engineering physics so RMC graduates would be accepted into the fourth year of the Queen's physics option. The English department at RMC had introduced Old and Middle English literature in its third year for cadets who might want to go on to a degree in English in another university.[45]

In February 1956 the commandant presented a paper about his request to the Canadian Services Colleges Advisory Board, which had been set up by order-in-council in 1955. The board at that time consisted of the chiefs of staff, one appointed member from each province,[46]* and representatives of the RMC and Royal Roads ex-cadet classes. The board "made no serious objections" to the proposal for an RMC degree.[47]

Support for an RMC degree also came from an unexpected and formerly hostile source. After the Mainguy Report on the incidents in RCN ships, a commission headed by Rear-Admiral E.P. Tisdall** had studied the education of naval officers.[48] The RCN then presented a paper on its findings to the PMC. The Tisdall Commission declared that higher education was needed to qualify officers to use sophisticated technical weapons. Basic to this was a sound knowledge of the humanities and the sciences. RCN officers required the equivalent of a BA, with mathematics and physics as major subjects, as well as professional training. Some officers would later go on to graduate school. Tisdall said the RCN often had to send officers bound for graduate and professional schools to obtain preliminary bachelors' degrees in universities. At the same time, the Navy Board ruled that RCN executive officers would soon take the four-year course in the Canadian Services Colleges. The RMC course could serve the RCN's purposes for preparation for graduate school, but a degree was essential. It would therefore be an advantage from a naval point of view if RMC obtained a charter. After studying Bradshaw's degree proposal, the RCN informed the PMC it was strongly in favour because of the need for "high quality officers."[49]

On 15 February 1957 Sawyer tabled a paper on the degree proposal in Faculty Council. He said the plan would "provide for world-wide recognition of intellectual achievement at the university level at no extra cost." He argued that the National Defence Act placed the content and direction of training at the minister's direction. The Faculty Council decided to explore the possibility of giving engineering as well as arts degrees at RMC.[50]

But before any decision was reached on that question, Dr Percy Lowe,*** professor of mathematics and a wartime director of naval education, proposed that the RMC request for degree-granting powers should include a general science degree which, without specializing in any particular aspect of science, would give executive naval officers third- and fourth-year mathematics and physics at RMC to meet the RCN's requirements. Meanwhile, the Executive Council of the RMC Club, noting that its member on the CSC Advisory Board had been the one to introduce the question, said it wanted to go on official record as favouring a degree for RMC.[51]

Up to that time, specific proposals for degree-granting powers for RMC had applied only to the general course, which covered at least as much ground as a university pass degree. Cadets who followed the engineering option at RMC would still apparently have to go on to university for the final year to get an engineering degree. But meeting requirements for several different universities in various branches of engineering had imposed

*Later Ontario and Quebec each received an extra member and the Yukon and the Northwest Territories alternated in being represented.

**Rear-Adm. Ernest Patrick Tisdall, a cadet at RNCC in 1921–2, served with the RN in 1924. He specialized in gunnery in both British and Canadian ships and commanded HMCS *Skeena* and *Assiniboine* during the war, and the cruiser *Ontario* in 1951. He retired as vice-chief of the Naval Staff.

***Dr Percy Lowe, instructor in mathematics at RMC from 1921, was commissioned during the war and became director of naval education in 1944. Head of RMC's Department of Mathematics from 1948, he was chairman of the Division of Science from 1956 until his death in 1959.

strains on the RMC curriculum. That had been partially relieved by the introduction of a short spring term after final examinations. Sawyer and the RMC faculty had hitherto assumed that an RMC engineering degree would be allowed five years. This assumption proved to be wrong when, on 13 February 1958, the PMC approved RMC's application for degree-granting powers, including engineering, without provision for a fifth year.[52] RMC engineering students would therefore have to complete all requirements in four years.

Admission to a Canadian graduate school, when the applicant does not have an honours degree, requires preliminary make-up courses. It was therefore desirable that suitable CSC students should take an honours degree in arts or science at RMC. In July the JSUCC asked Sawyer to report on the possibility of introducing honours degree courses in the science option.[53] In December he told the second annual meeting of the PMC and CSC commandants that RMC was undertaking an intensive study of the equipment and staff needed for the granting of degrees. In February 1959 an ad hoc committee on honours courses recommended to the Faculty Council that honours courses should be prescribed by each department. Honours students should normally take six such courses each year in addition to military studies, one more than in the general course. An average of 66 per cent should be required for admission to honours and a minimum of 66 per cent should normally be maintained in honours courses thereafter.[54] Honours degrees in engineering would require the general use of the spring term after final examinations.

Before these details were settled, the Ontario legislature passed the Royal Military College of Canada Degrees Act in 1959. The first degrees in arts and science were given in that year.[55] Ex-cadet Alan Kear* had conducted an active correspondence campaign to have the award of degrees made retroactive for all ex-cadets who had completed a four-year course at RMC, and a clause to that effect was included in the RMC Degrees Act.

In 1960 it was estimated that the introduction of degrees in engineering would save $468,500 per annum. This figure was included in a Chiefs of Staff Committee submission to the Defence Council about the possibility of RMC granting honours degrees in sciences and engineering. Their memorandum stated that the Board of Examiners of the Engineering Institute of Canada had certified that RMC's courses were "entirely adequate." The committee had found that four more weeks were available for engineering courses at RMC than at Queen's University and that a more relaxed education schedule in the college permitted reduction in the intensity of instruction and a liberal allowance of tutorial periods. Ten to eleven weeks still remained for summer training. The humanities offered at RMC were largely in place of minor engineering studies at Queen's and were substituted because they produced more broadly educated engineers. If cadets were over-fatigued, this was not due to their academic schedule but to pressures imposed on them by parades, the Canadian Services Colleges Tournament, guards of honour, and other extracurricular activities, and the commandants should be instructed to reduce such pressures. The introduction of honours science and engineering degrees would, the committee reported, entail only a "moderate" increase of $59,000 in capital costs and an increase of ten faculty members at RMC, two at Royal Roads, and one at CMR. This it found "reasonable." This submission was considered (and presumably approved) by the Defence Council on 2 May 1960.[56]

The introduction in the Canadian Services Colleges of the same kinds of degrees as in Canadian universities – the BA and BSc, honours and general – instead of some kind of military degree, focused attention on a question

---

*Alan R. Kear (no. 3062), RMC 1949–53, an officer in the RCA Reserve, was deputy city clerk in Kingston from 1955. He joined the political science department of the University of Manitoba in 1969 and became professor in 1973.

that had been raised before and was to be increasingly debated in future years. Were the Canadian Services Colleges merely replicas of civilian universities, and how far did they serve a military purpose? Some senior officers strongly objected to RMC's conferring academic degrees, for instance Lieutenant-General Guy Simonds, who later refused an offer of an honorary degree.[57] At the Advisory Board meeting of 1973, General W.A.B. Anderson said he thought the degree qualification might be over-emphasized and what was actually required was "a mature officer." When told that continuing education, not the degree, was the object for serving officers, he said he was "terribly reassured."[58]

There was, however, an underlying belief that a basic education is the same for all fields, all professions, and every educated person, with the important additional feature for military colleges that they educate in a military environment and develop special attitudes and skills. The nature of the academic instruction in the services colleges was still an open question: How far was it, or could it be, adapted to the ultimate military objective of creating professional officers? On this there was no consensus.

In RMC, professional instruction given by the Department of Military Studies had at first been classed as an academic course under the director of studies. At the beginning of the 1952–3 session, Colonel Gordon Fawcett,* head of the department, submitted to the Faculty Council a complete breakdown of his course of instruction. Topics ranged from service etiquette and customs to administration, tactics and strategy, and amphibious operations. Unlike the university training units, where the military instruction was restricted to the service to which a cadet belonged, RMC provided some tri-service instruction in military studies. Altogether, the services colleges offered 20 per cent more military lectures than the universities.[59] After ROTP was introduced, the military studies department courses, which had not been very satisfactory, needed to be coordinated throughout the

An early tri-service Department of Military Studies: Major Gordon Fawcett, RCOC, Lieutenant-Commander Pat Nixon, DSC, RCN, and Squadron-Leader Alex Jardine, AFC, ARAeS, RCAF

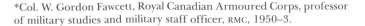

*Col. W. Gordon Fawcett, Royal Canadian Armoured Corps, professor of military studies and military staff officer, RMC, 1950–3.

CSC system. In May 1955 the CSC commandants' conference decided that RMC would be responsible for reorganizing the military studies curricula in all three colleges, subject to NDHQ approval.[60] Military studies at RMC was no longer in the Division of Arts and under the director of studies. Two years later a revised program introduced the important innovation that all cadets would now take all lectures in common – a significant step towards the principle of services integration.[61]

Some RMC courses had particular military relevance, and other academic courses also introduced material with military interest by way of illustration. The third and fourth year history courses, for example, included a seminar on the study of war in relation to society. A textbook based on this course at RMC was published in the United States and has been widely used there and elsewhere for over twenty-five years.[62] When naval cadets began to take four years at RMC, the history department added a course on naval history for them.[63]

Another military input came from the employment of service officers in departments other than the Department of Military Studies. At the outset in 1948, along with the appointment of many civilian professors who had wartime service experience, it had been necessary to second a few regular officers for temporary duty until permanent faculty could be secured. On 28 May 1954 the PMC approved the principle of permanently filling some RMC positions with service personnel.[64] Two years later Bradshaw wrote to the secretary of DND to say that eight new lectureships for serving officers would bring the proportion of military officers to civilians into a better balance. He said this would increase cadet motivation and would strengthen the colleges' links with the services. In 1957, however, because of a shortage of technical officers, the Rank Structure Committee rejected a request for ten extra positions in the CSC to be filled by officers. In 1959 a further proposal to appoint service officers as lecturers was supported by the argument that college lecturers in civilian universities normally had a limited stay in their first positions,[65] but shortage of suitable personnel still made a substantial increase of the military element on the faculty difficult.

The annual summer training with one of the services was a most important contribution to military experience and motivation. It was taken along with ROTP cadets from the universities, and allowed for healthy competition and friendships that would be useful in later career. In the early years, cadets had often complained about the poor quality of instruction and routine in their summer assignments, but the programs had improved. At the same time, service officers in charge of the summer training had occasionally reported that CSC cadets lacked motivation. Thus, in 1953, they had found fault with the first class to include ROTP cadets from RMC on the grounds that "their response to discipline and training was below that of the other groups at the Centres." Reports that RMC cadets lacked leadership qualities had caused serious concern, and Agnew had suggested the introduction of a course on ethics.[66] ROTP was believed to be a potent cause of this poor cadet motivation in RMC because it allegedly led to lack of interest in a service career beyond the fulfilment of the obligatory three years of service.

Morale and motivation were threatened by other factors. In 1958 Commodore Piers, the commandant, said he personally favoured ROTP graduates serving at the Queen's pleasure, but noted that the CSC obligation of three years' service was similar to that at West Point. He said that morale in RMC, especially in the third and fourth years, was lowered by the fact that cadets who served right through the college in Kingston were overwhelmed by the larger numbers who came in from Royal Roads and CMR in the third year. The newcomers, moreover, found it difficult to transfer their allegiance from the other two colleges.[67]

Criticism of ROTP's effect on morale kept alive a desire to reintroduce a reserve entry beyond the limited number of dominion cadetships that were open only to the sons of certain veterans. At a meeting of the JSUCC in

1956, the navy member said he had been instructed to explore feeling in the other services on the subject. He said it was better to admit a few reserve entry cadets in place of an equal number of poorly qualified ROTP recruits. Most of those navy cadets who were at the bottom of the list of acceptances had failed out by the end of the first year. In contrast, a considerable number of navy reserve entry cadets had entered the regular service in the past. The air force and army members retorted that past experience would not now apply because most prospective applications for regular commissions would have already joined through ROTP. The army member added that it was impossible for his service to reduce the number committed to it in advance for service in its active force.[68]

The following year, 1957, Sawyer tabled a paper in the RMC Faculty Council that claimed unanimous agreement that ROTP had in fact brought a decline in academic standards and military motivation. The increasing complexity of modern warfare, he said, required improvement in the qualities of leadership. He therefore recommended the reintroduction of reserve entry, up to no more than 15 per cent of the entering senior and junior matriculants, for applicants who had average or better officer-like qualities and a high academic qualification, who would pay fees, be eligible for scholarships, and be under no obligation to take a reserve commission. He said that the displaced ROTP entrants would have better prospects in the universities, that 50 per cent of an RMC reserve entry would go regular, and that at present good students unwilling to make an early commitment were lost to the services. Sawyer believed that introduction of a number of reserve entry cadets would have a general impact on the quality of the officers produced by RMC and on military leadership.[69] The PMC agreed to recommend that 10 per cent of the entry to the CSCs could be reserve entry cadets if suitable ROTP applicants were not forthcoming. However, the Chiefs of Staff Committee turned the proposal down. No reserve entry except

dominion cadets would be considered that year.[70]

By 1959 the acquisition of a charter to grant degrees and the confirmation of an almost exclusive ROTP intake into the services colleges had marked an important turning point in RMC's history. It paralleled the new requirements imposed by NATO obligations and the war in Korea. In this first decade of the new RMC, overall development had taken two seemingly divergent directions: towards a professional military school producing career officers, but also towards a fuller recognition of RMC's academic quality. One way in which the results of this seemingly hybrid institution may be judged is by examining statistics of commissioning and retention in the permanent forces. The class entering in 1953 was ninety-six strong, including thirteen repeaters. In the third year the additional intake from CMR and Royal Roads was sixty. Wastage over the four years at RMC was forty-four. The combined classes graduated ninety-six officers.[71] The RMC class entering in 1955 was 110 strong. With intake from CMR and Royal Roads, it was 163 in the third year, including three reserve entry cadets. In 1959 RMC graduated 128 officers. For the ten-year period 1953–63, 64 per cent of the services colleges graduates and 45.2 per cent of ROTP students in the universities stayed in service after their three-year period of obligatory service ended.[72] These figures demonstrated that there was a sharp difference in favour of the military colleges over ROTP in the universities, but they are inconclusive because in the first years they included cadets who had transferred into ROTP when it was instituted, and because the graduates in the later years had not yet had time to do three years' service. A more detailed appraisal of comparative costs and relative merits will be considered later in this book.

During the years when the college was seeking the power to grant degrees, the other battle – to make the college into a short-course military training establishment to commission junior officers – was being fought. A large number of serving officers thought a degree-granting college would not be "sufficiently military."* Many senior officers and ex-cadets, for instance, objected to the RMC officers' mess being called the Faculty Club and, as a result, the name was changed to Senior Staff Mess. There was also anxiety about the numbers and quality of RMC's graduates. That was reflected in the internal conflict between military and academic members of the staff for control of cadets' life and time – an enduring problem in accommodating military training and an academic education within a single institution.

Affecting this accommodation was the problem of the military command of a college with academic objectives, something quite different from a normal military command. Air Commodore Bradshaw anticipated this challenge and sought, unsuccessfully, to have his officer in charge of administration promoted to lieutenant-colonel. He said that in a university the faculty feels "it is the college" and regards university administrators as "meddlesome interlopers or, at best, necessarily evils"; at the same time, administrators are "natural empire builders." Bradshaw argued that irritation between these natural opponents was magnified at RMC because it was not settled internally but by the "bigger and more terrifying administrative giant at Ottawa." He said there was no easy solution, but recommended that his administrative officer should have a rank equal to that of a senior professor – a lieutenant-colonelcy or its equivalent in the other services – in order to strengthen him in relations with the academic departments.[73]

In that same year, 1955, the JSUCC proposed to strengthen the military command within the college by transferring the appointment of vice-commandant from Colonel Sawyer, the director of studies, to a military officer in one of the permanent forces. The Inter-Service Establishments Committee refused on the grounds that RMC had not requested the change.[74] In January 1957 Major-General George Kitching,** vice-chief of the General Staff, a former British officer, proposed the creation of a new appointment of officer commanding the cadet wing and deputy-commandant, a downgrading of the staff-adjutant to a captain, and the retention of the director of studies as vice-commandant. This compromise was also rejected because in service usage a "vice" is senior to a "deputy." It was therefore decided to leave the whole question of a military vice-commandant until Sawyer retired.[75] However, Queen's Regulations, as amended in 1957, stated, "In the absence of the Commandant from the Colleges, the powers of the Commandant shall, unless the Commandant shall otherwise direct, be assumed by the next senior officer on the staff of the military wing who is present at the Colleges."[76]

During the years when the new academic degree program was introduced in RMC, Sawyer had remained in a powerful position. He could therefore ensure that, as the college moved towards greater professionalism, the academic quality of RMC be kept up. Even so, the problem of accommodating professional education and training with general education remained. It continued to disturb RMC's development long after Sawyer was gone. What had been achieved at RMC, however, had been noticed outside Canada. When Bradshaw was posted to Air Defence Command, Air Marshal C. Roy Slemon, RCAF, deputy CINC of NORAD, instructed him to attend the opening of the new US Air Force Academy at Colorado Springs. There, the newly appointed superintendent, Lieutenant-General Hubert Harmon, questioned him at length about RMC. Bradshaw believes that this interview had an important input in the development of the American academy at that crucial stage in its development.[77]

*When the Duke of Edinburgh visited RMC a few years later he is reported to have asked, "What is this nonsense about degrees for naval officers?" (Information from a reliable source.) His advocacy of military degrees in a lecture to the Royal United Services Institute in 1988 takes a quite different approach. RUSI: Journal of the Royal United Services Institute for Defence Studies 128, 3 (Sept. 1983): 3–8.

**Maj.-Gen. George E. Kitching served in the British army 1930–8, and with the Canadian army in the Second World War, becoming chief of staff, 1st Canadian Corps, 1944–5. He was vice-chief of the Canadian General Staff 1956.

CHAPTER 4

Under the Director of the Regular Officer Training Plan

The creation of the Directorate of the Regular Officer Training Plan (DROTP) in Ottawa on 30 January 1958 was of greater long-term significance than either RMC's acquisition of degree-granting power in 1959 or the physical improvements that had begun at RMC in 1957. Something must be said of those improvements, however, because they were necessary to the success of both military and academic aspects of the college program.

By 1959 part 1 of Air Commodore D.A.R. Bradshaw's building plan (without the proposed chapel) provided the facilities envisaged in the original 1948 scheme. When the Canadian economy had sagged in 1957, Commodore D.W. Piers had been ready with detailed plans for further construction at RMC. Foremost of these facilities was the Massey Library to house the fine collection accumulated by chief librarian John Spurr* in less than a decade. The Massey Library was designed to support RMC's new postwar academic status by serving its liberal arts and social sciences as well as its science and engineering departments. In addition, there was to be a civil engineering annex, a senior staff mess, the Constantine Hockey Rink, and a garrison sergeants' mess. The new RMC was at last to be furnished with some of the physical plant it needed to carry out its mission.

The fortunes of the college were now dependent in large degree on relations with the new director of the ROTP (also called DROTP). His role was to coordinate all ROTP, interservice, and university reserve matters. He was responsible to the vice-chiefs of staff for Canadian Services Colleges' curricula, which were to meet appropriate academic standards, and he was to oversee training facilities. He was responsible to the Personnel Members Committee (PMC) for ensuring optimum use of the colleges and for coordinating ROTP selection. He was authorized to issue directives to the colleges on all matters concerning ROTP and he was to be their sole channel of communication on all matters of policy.[1]

This brought another radical departure from RMC's original status and function as a largely autonomous college to produce graduates who had the option of becoming part-time officers. Under the DROTP, RMC became almost exclusively a regular-officer-production college over which the Department of National Defence exercised close control. Through to 1963, the vital formative years of the new degree program, three successive directors, Brigadier-General R.P. Rothschild,** Air Com-

---

*John Wheelock Spurr was chief librarian of RMC from 1949 until his death in 1981. For an appreciation see Richard A. Preston, "In Memoriam – John Wheelock Spurr," *Association for Canadian Theatre History, Newsletter* 5, 2 (March 1982): 25–9.

**Brig.-Gen. R.P. ("Baron") Rothschild (no. 2297) graduated in 1936 and joined the RCA, serving during the war in Europe. He was director of the Staff College at RMC 1946 and became acting commandant. In 1947 he was military attaché in Greece and was appointed DROTP in March 1960. He was QMG 1962–5.

RSM J.E. ("Jack") Coggins, college sergeant major, 1947–58

modore J.B. Millward (a former commandant of Royal Roads),* and Commodore H.V.W. Groos,** supervised officer education in the Canadian Services Colleges and the universities. These officers had a voice in RMC development at a time when it was commanded in turn by Commodore D.W. Piers, Brigadier-General W.A.B. Anderson, and Brigadier-General G.H. Spencer.***

Of considerable importance almost simultaneously were internal personnel changes in the college that also seemed to signal the end of an era. On 31 July 1958 Regimental-Sergeant-Major J.E. Coggins had retired. "Coggie," as he was universally known, had come to RMC in 1929 from the British army as a physical training instructor. In 1941 he became the college RSM. During the war he joined the RCR, earned an MBE, and was commissioned. When the college reopened, he readily reverted to RSM in order to return to his former appointment there. Jack Coggins was a man of extraordinary qualities, a firm disciplinarian with a very human understanding. His influence on the re-establishment of the old RMC's traditional virtues cannot be overestimated. Widely admired by staff and cadets alike, his departure broke an important link with the prewar RMC. Fortunately, largely because of the soundness of the practices and attitudes on drill and discipline he had re-established or introduced, his successors achieved smooth transition.[2] When he died in 1984, Coggins was remembered in a special memorial service in Currie Hall in the college he loved.

At a more exalted level, RMC got its first naval commandant on 8 August 1957. As we have seen, Bradshaw had maintained and strengthened the college's development on the lines already laid down by Brigadier Donald R. Agnew and Colonel W.R. Sawyer. However, when he in his turn came to the end of his tour of duty in 1957, the navy was now prepared to take up the appointment which six years earlier it had declined. Since sailors have distinct views on practical training and academic education, and since the RCN had been unenthusiastic about the four-year course, what remained to be seen was

whether a naval commandant would bring something more than the cosmetic changes almost inevitable with every new commandant. Commodore D.W. Piers's command of RMC was inspirational for young cadets because at an early age he had commanded a destroyer squadron in the desperate Battle of the Atlantic.

When Piers hoisted his broad pennant on Point Frederick, he at once demonstrated his awareness of the historical significance of the event. He called for an investigation of the history of the commodores' pennants flown on the point when it was a naval base a century earlier.[3] Piers had arrived on the scene six weeks after his predecessor had left, and he felt that he had been inadequately briefed and that there was an unfortunate lack of continuity. Nevertheless, he quickly took effective control of college administration and the Cadet Wing through weekly conferences in normal military style with his staff officers.[4] Piers believed that the authorities in Ottawa had paid insufficient attention to the college. To counter this, he had something useful to offer. He and his wife, Janet, a member of a prominent Nova Scotia family, had extensive official and social connections in both Canada and Britain. They also had phenomenal

*Air Commodore James Bert Millward, a pilot with No. 426 Bomber Squadron and Flight-Commander of No. 405 Pathfinder Squadron, left the service in 1945 and took an MA at Bishop's University in 1946. Returning to the RCAF, he was appointed commandant, Royal Roads, 1949–52 and became DROTP in March 1960.

**Commodore H.V.W. Groos, a naval cadet attached to HMS *Erebus* 1930–1 and midshipman on the battle cruiser HMS *Hood* 1932, served with the RCN in the Second World War, and became director-general of support facilities 1962, and DPED 1963–4.

***Maj.-Gen. George H. Spencer (no. 2424) graduated in 1938. Commissioned in the RCE, he served during the war in Borneo and Northwest Europe. Assistant director of works and construction at Army HQ 1945–6, he was instructor (DS) Canadian Army Staff College 1946–9. After commanding RMC, he was chief engineer and commander of the Northwest Highway System and then director-general of training and recruiting. He retired as deputy-comptroller-general of the Canadian Forces. In 1978 he became colonel commandant of the Canadian Military Engineers.

memories for people and names. The Piers used these connections and attributes for RMC's advantage. As commandant, he distributed invitations to college functions to officers in key appointments in Ottawa and aroused awareness and interest in important quarters. The Queen and Prince Philip (who had visited in Agnew's time) came again in 1959. In the same year Field-Marshal Lord Montgomery and Admiral-of-the-Fleet Earl Mountbatten inspected the Cadet Wing. These visits increased public awareness of RMC and stimulated cadet morale.

The Piers's social skills were effective also within the college. Easy conversationalists, they worked hard to make warm contacts with staff, faculty, cadets, and parents. Like many administrators, however, Piers tended to look back on his own education and training as ideal, and some cadets were unhappy with his innovations – for instance, the reintroduction of the swagger stick used in his day as a cadet. It made saluting in town extremely difficult, especially when the cadet was carrying a parcel – something that would have been unthinkable in earlier, more formal times.

Piers found his efforts to establish the college's new image helped by the study of ROTP made by his naval colleague, Rear-Admiral W.M. Landymore (see chapter 3). Landymore had approved the current system of training and education.[5] After a year in the post, however, Piers was not happy with the process of formulating academic policy, and he solicited advice from the faculty and staff on various ideas for improvement. Established procedure was that the commandant, who was ultimately responsible for overall college policy, should receive the minutes of the Faculty Board and Faculty Council meetings and consult on academic matters with the director of studies. This practice was fundamentally different from the norm in military command and administration, where a commander consults his staff and then unilaterally makes decisions.

In RMC, academic matters were the responsibility of the director of studies. Piers complained to Sawyer that the Faculty Council and Faculty Board dealt with specific cases rather than broad policy. He may have felt that the director of studies and the minutes of the Faculty Council did not inform him adequately about the formulation of policy and that more detailed discussion and minutes would give him greater insight. He recommended that the Faculty Board hold regular scheduled meetings in addition to those at mid-term, Christmas, and the end of the session, which customarily only discussed grades. He said these other meetings should discuss policy.[6]

The Faculty Council did in fact already meet monthly to do more than discuss marks, but it was accustomed to deal with specific problems on the basis of its members' understanding of general principles of academic standards and practices, principles that were perhaps not always clear to the commandant. Piers now recommended the participation of the board in discussions, which was what some members of the faculty had been requesting since 1949 without success. This would generate information about academic activities that would be available to the commandant when he read and counter-signed the minutes.

In accordance with Piers's recommendation, a steering committee appointed by the director of studies and the Faculty Council drafted regulations to reorganize the Faculty Board. It proposed that a new board, composed of assistant professors and above, should meet three times a year in addition to the marks meetings.[7] The committee also recommended that the Faculty Board should be informed about any changes in academic regulations and curricula, and about conditions that might affect the cadets' academic work; it should be able to initiate discussion on these matters; and it should be able to make recommendations to the commandant through the Faculty Council or the director of studies. Since faculty members often believed they had more contact with individual cadets than members of the military staff, it suggested that the board should also be informed about,

Queen Elizabeth and Prince Philip, 1959

and be given opportunity to comment upon, proposed Cadet-Wing appointments, a regular practice of past years that had lately been omitted. The board should also be consulted on changes in the college calendar.* Finally, it should receive, discuss, and transmit to the Faculty Council the results of final examinations decided by its marks committees, which would consist of all those of any rank who had given the relevant courses.[8]

Piers's proposals about the powers and functions of the Faculty Board had paralleled its members' belief that RMC's new degree-granting powers imposed on them an even greater responsibility than before for the maintenance of academic standards. RMC's academic credibility was now required not merely so graduates could take a final year for a BA or a BSc in a university, but for them to secure entry to graduate schools on equal terms with graduates from other universities. This could be obtained only if graduate schools fully recognized RMC's degrees.

Shortly after this debate about organization began, an incident emphasized that it was imperative for the functions of the board and council to be more clearly defined. The board noted that the commandant had reversed a Faculty Council decision to separate a cadet who, in a sessional examination, had seven failures. He had not referred the matter back to the Faculty Council as was customary previously, but had offered this cadet the opportunity of repeating the year at his own expense.** Since Piers's immediate predecessor, Bradshaw, had reproved the board and council for not separating cadets who at Christmas had had the same number of failures, there was a need to establish and maintain consistency. Members of the council and the board felt that they were the guardians of consistency in academic standards.[9] Furthermore, the college was still negotiating to secure accreditation from engineering societies for its new degree programs. The Engineering Institute of Canada and the Chemical Institute of Canada had given their approval soon after the RMC Degrees Act was passed, but accreditation by provincial associations of engineering was still needed to permit graduates to practise their profession. Faculty members believed it was imperative that RMC's academic decisions be made only on academic grounds, otherwise accreditation might be endangered.[10]

Discussion of the reorganization of the Faculty Board at RMC dragged on for over two years, mainly because a Faculty Council subcommittee, set up to examine proposals to upgrade the functions of the board, ignored the question and failed to report back, perhaps deliberately.[11] The registrar then reverted to numbering the marks meetings as full sessions of the old Faculty Board. He held that the former regulations, being published in the calendar, still had authority because the calendar had been approved by the DROTP.[12]

On 20 July 1959, however, the newly appointed director of the Regular Officer Training Plan stepped in and issued a directive to all the colleges. Up to that time there had been a Faculty Council at all three, but only at RMC was there also a board. DROTP directed that both bodies should function at all three colleges. The councils should assist the director of studies to determine all matters of an educational character, compile the calendar, control the library, grant academic standing to officer cadets, recommend the granting of diplomas, award academic medals, prizes, and scholarships, and "make such recommendations to the Commandant as may be deemed expedient for promoting the interests of the College." The boards were to "make recommendations to the Commandant through the Faculty Council on honours standing and class failures," exercise academic supervision over officer cadets, make recommendations to the commandant on cadet appointments, and "make

---

*In British and Canadian usage, a university calendar lists faculty, courses, and regulations and is what Americans call a "catalogue."

**The cadet apparently decided instead to withdraw.

such recommendations to the commandant as may be deemed expedient for promoting the academic interest of the College." The council was authorized to tender advice on all matters concerning the college, while the board could speak on certain academic matters only.[13] Ultimately a registrar's memorandum, dated 13 December 1960, set out new regulations for RMC by virtue of which the Faculty Board was to meet four times a year in addition to its marks committee meetings.[14] Under Piers, the RMC faculty thus acquired a clearer right to exercise an influence on college policy and administration.

Piers also sought to remedy what he considered another weakness of college organization – this time on the military side. When RMC was re-established in 1948, the RCN and RCAF had wanted an officer of lieutenant-colonel's rank to command the Cadet Wing and be vice-commandant. General Foulkes, CGS at that time, had refused on the grounds that the course was mainly academic and the wing was only one hundred strong.[15] Agnew and Bradshaw had commanded the wing themselves, with the assistance of the staff-adjutant, a major, who was concerned with military training and discipline. By the late 1950s the cadet body was three or four times as big as before the war or again in 1948. Junior officers had been posted to command squadrons under the direction of the three officers of the rank of major, or equivalent, who simultaneously instructed in military studies as associate professors. The staff-adjutant's job, controlling a bigger and more complex organization, with three of its members equal to him in rank, had become onerous. Therefore, soon after his own appointment, Piers sought to upgrade the staff-adjutant to lieutenant-colonel and to change his title to that of officer commanding the cadet wing, as in American academies. The PMC rejected this proposal.[16] In 1963, however, the Rank Structure Committee was prepared to upgrade the staff-adjutant provided that, when the present RMC administrative officer, who had been routinely promoted lieutenant-colonel while in office, ended his tour, that

position would be downgraded to a major's slot. Since the army, which was responsible for administration, objected to this, a working group was set up to discuss the question. It recommended that both appointments be upgraded to lieutenant-colonel or equivalent.[17]

Always over-shadowing these internal readjustments and changes was what Bradshaw had called the "terrifying administrative giant in Ottawa," now taking the form of the DROTP. As we saw in chapter 1, differences in navy, army, and air force manpower policies had complicated the reopening of RMC. Thereafter, until the integration of the Canadian Forces in the late 1960s, each of the three services had a separate and different interest in the operation of the services colleges and the Regular Officer Training Plan (ROTP). Because the services colleges were each administered by a different service and were often commanded by an officer from another service, there were endemic problems of coordination and cooperation. At the same time, unlike ROTP in the Canadian Services Colleges (CSCs), ROTP programs in the universities were under single-service direction. Supervision of both CSCs and university ROTPs by the Joint Services Universities Coordinating Committee (JSUCC) had proved to be ineffective. The Joint Universities Military Advisory Committee (JUMAC), composed of the directors of training of the three services and three appointed academics, set up in 1955 to advise the JSUCC on academic problems, had also not been very successful: the universities had not responded easily to external suggestions. Furthermore, military training in a university unit was often a disappointment: at the beginning of their summer training, university cadets needed a special military orientation course from which the CSC cadets were exempt. Finally, although the absence of close external control had meant considerable freedom of action within the services colleges, and therefore may have expedited their academic development, it had sharpened confrontation with certain military interests, as for instance with the army's chief of the General Staff in the early 1950s.

In military eyes, one cause of CSC problems was that the organizational chart was untidy. The addition of Collège militaire royal (CMR) to the tri-service system had exacerbated an already complex arrangement. JSUCC met weekly but lacked immediate authority because it reported to the Personnel Members Committee (PMC), which in turn advised the Defence Council and the Chiefs of Staff Committee, august bodies that did not usually concern themselves with the details of administration. In practice it had become common for most questions relating to the services colleges to be referred to the PMC, including many that were not concerned with personnel.[18] Three years later the problem was more clearly identified. Five government agencies had a voice in the running of RMC: the college's military establishment came under the chiefs of staff and the Treasury Board; civilian establishments, standards, and salaries were governed by the deputy minister of national defence, the Civil Service Commission, and the Treasury Board; civilian faculty members were recruited by the deputy minister and the commission; and procurement of equipment, accommodation, and local administration came under the Canadian army.[19]

The Landymore Committee had recommended the creation of the Directorate of the ROTP in an effort to diminish complexity in the operation of the colleges, to coordinate them with programs in the universities, and to ensure that they fulfilled their military purpose. On 30 January 1958 the chairman of the Chiefs of Staff Committee approved the abolition of the short-lived Joint Universities Military Advisory Committee and agreed to the appointment of the director of the Regular Officer Training Plan (DROTP) with the rank of brigadier or equivalent.

The director was given considerable power and functions. Whereas JUMAC had been formed only to advise JSUCC on the duration and scope of academic training, on its relation to the military curricula, on university degree requirements, and on the proportion of time spent on physical recreation, the director was to review the curricula of the Canadian Services Colleges to see that they met service requirements. He was also to ensure the maintenance of academic standards and to give advice on military training at the CSC and in the universities. Finally, he was to be responsible to the PMC for all aspects of administration and policy in the colleges and for ROTP in the universities. The interservice JSUCC came under his wing and was housed in his Directorate offices in the Department of National Defence buildings on Elgin Street.[20]

In June 1958 the minister, noting this was a full-time job, agreed that a director's appointment should be for a three-year tour of duty. He said that DROTP would be responsible for tri-service officer production. The chairman of the Chiefs of Staff, General Foulkes, hastily qualified this by noting that DROTP was concerned only with ROTP and not with other single-service officer-production plans for short-service commissions, including commissioning from the ranks, or with reserve-officer training.[21]

This switch, from an advisory committee on academic matters to an executive directorate, facilitated DND supervision of the services colleges but did not eliminate the involvement of other agencies, for instance of the Treasury Board and the Civil Service Commission. Furthermore, although the tri-service aspect had been strengthened by the move, the colleges still produced for three different services. More significant was the fact that it was not at first clear how far DROTP would involve himself in the academic curriculum in the colleges. In the universities his mandate to maintain academic standards could only mean that he would ensure that ROTP cadets took courses that satisfied military requirements. It could not imply criticism of the content or standards in university courses that had been considered appropriate and were approved. The main question was how DROTP's concern with maintaining academic standards would be applied in the services colleges.

The first DROTP, Brigadier-General Rothschild, was an ex-cadet. Some indication of his views and intentions was given by his first act, which was to approve the application for degree-granting powers for RMC.[22] He soon showed he was aware of the value of the faculty that had been recruited for RMC and of the need to hold on to it. The director of studies had reported to him that, since the opening of RMC, faculty salaries had lagged far behind advances in the universities and in the civil service generally.[23] Specifically referring to the fact that universities would make attractive offers to CSC faculty members if conditions in the colleges were not favourable, DROTP approved the establishment of a Canadian Services Colleges Faculty Review Board to deal with salaries.[24]

But salaries were not the only factor in retaining faculty. Conditions must approximate to those in universities in conditions of service, opportunity for research, and participation in the discussion of academic policy. Yet, with the introduction of the Regular Officer Training Plan, the purpose of the college was, more than ever before, to produce professional officers. CSC academic curricula must therefore also serve that end. DROTP's relations with the faculty would obviously depend on mutual agreement about what was to be taught.

Stronger external control from Ottawa was in some ways offset by the simultaneous development, under Piers, of the internal college mechanisms for control of academic matters. In this respect, one consequence of the legislation empowering RMC to grant degrees was another confirmation of faculty participation. The Ontario act set up a Senate consisting of the president of the college (the minister of national defence, who from 1959 was also the chancellor), the commandant (as vice-chancellor and chairman of Senate), the director of studies, and the chairmen of the divisions (later to be called the deans of the faculties) of arts, sciences, engineering, and graduate studies (and later the dean of the Canadian Forces College), with the registrar as secretary. The Sen-

ate's original primary function was to approve the granting of degrees, including honorary degrees.[25] By this means, however, the commandant and also the Department of National Defence were admitted into an academic area that had previously been the concern solely of the Faculty Board and the Faculty Council. Except for honorary degrees where the Senate had the initiative and sole authority, at this stage the Senate only convened to approve the conferring of degrees already recommended by the council and the board. Nevertheless, its existence would have future significance in bringing outside influences on academic policy.

In 1959 RMC granted its first degrees in arts, and also an honorary degree to Prime Minister John Diefenbaker. Before that time, the award of a Rhodes Scholarship to Officer-Cadet Desmond Morton,* without the technicality of a BA degree, may perhaps be construed as an implication that the Rhodes trustees already equated the RMC course with programs in the Canadian universities. Morton was not the first RMC ex-cadet to win that prestigious award. Major A.W. Duguid,** a graduate of the class that, because of wartime acceleration, left in December 1939, having completed only a little more than two years at RMC, had been selected in 1946 from a group of military-service-qualified applicants for a special

The Honourable George Pearkes, VC, minister of national defence and president of RMC, presents the first honorary degree granted by the college to Prime Minister John Diefenbaker, 14 May 1959.

Cadet Squadron Leader Desmond P. Morton is congratulated by Prime Minister Diefenbaker on receiving his BA degree, May 1959. Morton was the first cadet to win a Rhodes Scholarship while still in the college.

*Capt. Desmond Paul Morton (no. 4393) entered CMR in 1954 and graduated from RMC in 1959. He attended Keble College, Oxford, and the London School of Economics. He served in the Royal Canadian Army Service Corps in the Canadian army from 1954, and then in the army Historical Section. Later he entered on an academic career and was also assistant secretary of the Ontario New Democratic Party. He is now principal of Erindale College of the University of Toronto. Dr Morton is one of Canada's most prolific historians.

**Maj. Adrian Winslow Duguid (no. 2565) was at RMC 1937–9. The son of Col. A.F. Duguid, director of the army Historical Section, he served overseas in the Second World War with the RCA, becoming brigade major. After the war he attended Khaki University and received his BA from Oxford in 1949. He was engineering manager, Computing Devices of Canada. He died in 1969.

Rhodes Scholarship. Morton was the first to receive the nomination while still at the college.[26]

Some members of the faculty were determined that RMC must now possess the attributes and functions of a university degree-granting institution more fully. Shortly after the first bachelors' degrees were granted, the question whether the college should grant graduate degrees was discussed in the Faculty Council on the grounds that graduate studies were needed, first, for serving officers, second, to provide research assistants for RMC professors, and, third, to accommodate junior faculty who would otherwise have to take leave of absence to obtain degrees elsewhere, perhaps at government expense.[27] The question was referred to the CSC Advisory Board, which saw merit in the proposal. The director of studies then recommended that RMC graduate degrees be made available for regular service officers, for graduates of RMC, and for service members of the college staff.[28] The first graduate course offered at RMC was History 500, naval history, given by Visiting Professor Brian Tunstall from the Royal Naval College, Greenwich, in cooperation with Queen's University, in the year 1965–6. Two officers who took it were Major B.D. Hunt,* who was attached to Queen's as army staff officer, and Lieutenant-Commander W.A.B. Douglas,** naval staff officer at RMC. Both Hunt and Douglas went on to complete a PhD at Queen's.

However, the recognition of RMC's academic qualities and progress symbolized by the BA and BSc degrees had aroused, or revived, the fears of some service officers and members of the RMC Club that the RMC course was too academic, or insufficiently military, to carry out its basic mission – the production of professional officers. DROTP was well aware of this feeling. He had approved the proposal for a degree in engineering, with the proviso that the revised third- and fourth-year courses have military as well as academic relevance. He added that the armed forces now, more than ever before, were faced with problems in human relations, and thought the ed-

ucation of engineers for the services must have a humanistic content to cope with such problems.

Colonel Sawyer told the Faculty Council that RMC's proposed engineering courses were similar to those at the Massachusetts Institute of Technology and the California Institute of Technology, where the curriculum for engineering students included a larger than average amount of humanities and social sciences.[29] A combination of arts and sciences had, of course, been a notable feature of CSC programs since 1948, and even earlier. That was now confirmed and extended in the new degree programs. In the senior years there were humanities and social science courses that were appropriate for a basic military education for the modern age. The Departments of History and Political Science cooperated to produce a new honours program in international relations that conformed with a recommendation by the National Conference of Canadian Universities and Colleges that undergraduate studies in that field should be broadly based, leaving specialization to the graduate school. Furthermore, one hour a week of military history, given in the Department of Military Studies in the early years of the reopened RMC, had already been taken over by the history department to improve its impact, though at first only for third-year engineers. A course in nuclear engineering technology was added for cadets taking chemical engineering.[30] The developing curriculum thus stressed military content. To counter fears that

---

*Barry Dennis Hunt (no. 4919) entered Royal Roads in 1956 and became cadet wing commander there. Graduating from RMC in 1960, he was commissioned in the 2nd RCR in 1960 and served with the British Royal Fusiliers in Germany 1962–4. He joined the history department at RMC in 1967, became head of department in 1987 and dean of arts in 1990.

**W. Alec B. Douglas was under training in HMC ships *Swansea* and *Ontario* 1951–4 and was commissioned lieutenant in HMCS *Quebec*. In 1964 he was appointed naval staff officer and associate professor of military studies at RMC, and in 1967 an historian at the Directorate of History, DND, where, in 1973, he became director.

the military function was being neglected, in January 1959 the Department of National Defence decided to grant new college colours to all three services colleges.[31]

About the same time, Douglas Fisher, Co-operative Commonwealth Federation (CCF) member for Port Arthur, asked the minister in the House of Commons whether a degree course at RMC would mean extending the length of the course, and whether RMC graduates had the opportunity to continue their contribution to Canada's military needs by service in the reserves after they had completed their required years of service in the regular forces. Major-General George Pearkes* assured him that the course would still be only four years long and that, out of 141 graduates, twenty-six had joined the reserve after completing the required three years of regular service.[32] Later, after a budget debate in which he attacked the government fiercely on other matters, Fisher asked the Budget Committee a long series of penetrating questions about the Royal Military College that revealed he had considerable knowledge of it. In effect, he questioned whether a military college was, and should be, equated with a university in conditions of teaching and free enquiry. He suggested that the director of studies and the registrar be called as witnesses before the committee. But a comprehensive set of replies returned by Commodore Piers apparently satisfied him.[33] The only immediate follow-up was that Paul Hellyer, the Liberal defence critic, asked the minister of national defence who decided who would get an RMC honorary degree. He apparently did this to draw attention to the fact that the first honorary degree had gone to Prime Minister Diefenbaker.[34]

With regard to Fisher's concern about RMC's military contribution, the record showed in 1960 that retention rates after three years of service were much the same as they had been earlier. The RCN lost 27 per cent, the army 33, and the RCAF 34. In contrast, figures available in 1961 showed that 43 per cent of former university ROTP cadets exercised the release option after three years, almost twice as many as former CSC cadets (24 per cent). The figures for RMC were, as usual, better than the overall CSC figure: RMC, 19 per cent; Royal Roads, 29 per cent; and CMR, 27 per cent.[35]

In 1963 the detailed answers prepared for Fisher in 1959 were made available to Paul Hellyer when, as Liberal minister of national defence, he spoke in the House to answer criticisms of RMC made by the Glassco** Commission on government organization published in December 1962. Glassco had said that the Canadian Services Colleges had one instructor for every 5.46 students compared with one for every ten students in Canadian universities, that the administrative staff was "widely out of line," that only 41 per cent of those who enrolled reached graduation, though 90 per cent of those who graduated and entered the services then made a military career their life work. He said that ROTP training in the universities cost less than one-third of that in the services colleges, and that a million dollars was spent on advertising to produce only seventy officers and thirty-five other graduates "for the general benefit of the Canadian economy."[36] The most immediately threatening statement (in volume 3 of the report) was that the Canadian Services College at Victoria, BC, should be closed because of the commission's concern over the cost of operating the three colleges. The commissioners felt that "some considerable amelioration would be possible if these training activities could be concentrated in a single institution." Since CMR fulfilled what the commissioners

---

*Maj.-Gen. George Randolph Pearkes, VC, served in the RNWMP and then in the First World War in the Canadian Mounted Rifles. He commanded the 1st Infantry Division 1940–2 and subsequently became GOC Pacific Command. He was minister of national defence 1957–60 and lieutenant-governor of British Columbia 1960–8.

**J. Grant Glassco, president of Brazilian Traction, Light, and Power Company from 1963, was chairman of the Royal Commission to Examine Government Operations 1960–3.

believed was an essential bilingual function in Canada, however, they thought its closing impractical.

Commodore Groos, the DROTP, reported to the chairman of the Chiefs of Staff Committee that the commission's information about the cost of operating the services colleges was substantially correct, in fact that the cost might even be higher than its figures suggested, especially in regard to Royal Roads. However, he said that the commission's figures on the cost of producing officers through ROTP in the universities were seriously wrong since they took no account of such obvious public expenditures as travel, clothing, cost of the services' administrative staffs at the universities, services' offices and other accommodations on campuses, and medical and dental service. In addition, the commission had ignored hidden federal, provincial, and municipal grants. Groos reminded the chiefs that the release rate of university ROTP graduates was much higher than that of CSC graduates. He outlined the problems that would emerge from transferring Royal Roads faculty and cadets to the other two colleges. The move would exacerbate problems of space and organization that were already serious.[37]

Glassco's figures on CSC costs differed from those that had been produced by RMC to answer Fisher. For instance, the ratio of students to staff in 1958–9 was actually 6.5 at RMC, 7.6 at Royal Roads, and 6.0 at CMR, not 5.46 overall as he said. However, the Advisory Board was told that the commission's view that the Department of National Defence was committed to an extravagant program of officer production had ignored the fact that the system was not yet mature. Nor were CSC faculty ranked higher than individuals in the universities. Finally, the commission had said that the cost of producing a career officer from the CSCs was $47,000 and from the universities, $14,300. Those figures should actually have been $55,555 and $26,300, respectively. By 1963–4 the former would be down to $49,300 and the latter up to about $38,000. This would compare favourably. It was similar to the cost in the US Air Force Academy, which

was $47,992, and in the US army and navy, where it was only slightly less.[38]

The Glassco Commission's suggestion that consideration be given to increasing the years of ROTP obligation of service had in fact already been anticipated. In 1960 the Advisory Board had been informed that an increase in the obligation to serve was under discussion.[39] Groos told the chiefs of staff that internal reviews of the size of instructional and administrative staffs had also been made recently, so a further review would not bring new economies. However, there would be no objection to such a review if one was needed to restore public confidence in the CSCs. The PMC directed that the matter be explored with the National Conference of Canadian Universities.[40]

When Douglas Fisher asked in the House of Commons whether the minister had had any representations about how the Glassco Commission had obtained its information about the colleges, he was told that the only sources Glassco had quoted about the CSCs were the director of the School of Public Information at Carleton University and the secretary-general of Jean-de-Brébeuf College in Montreal.[41] Only one commissioner had visited the colleges, and that was one short two-hour visit to RMC, where he had asked no questions.[42] Clearly the Glassco Commission, however salutary it may have been in other fields of government organization, had recommended cutting down the CSC system and eliminating Royal Roads on research that was insufficient, superficial, and misleading.*[43]

*Col. Charles P. Stacey, former director of the army Historical Section, in letters to the press, indicated that the Glassco Commission made serious errors when it dealt with his section. When he brought these errors to its attention before publication, Glassco apologized but took no action. Stacey said he had heard of similar errors in the commission's comments on other government departments, and he questioned the competence of the commissioners to undertake their task. *Ottawa Citizen*, 18 and 19 June 1963.

The introduction of the degree program may have confirmed the fears of some who believed that the CSCs were too academic and could not adequately carry out their military mission of producing professional officers; but DROTP had found it possible to approve a reduction of summer military training from twelve weeks to ten to make time for the extra spring term required by the engineering degree. Summer training had indeed improved considerably since the early days of the CSCs when cadets sometimes complained that the navy often kept them busy with tasks such as stripping old paint from the rails of ships. The improvement had in part been an admission that the services had themselves been guilty of alienating, instead of attracting, cadets, especially the reserve entry cadets whom they particularly wanted to induce to take permanent commissions.

Officer cadets on summer training, in addition to taking military courses, gained practical experience in command of units and performed other interesting duties. From the early 1950s until 1958, many army cadets spent their summer training with the Canadian Brigade Group in Europe. Although that practice was not continued on a regular basis in the 1960s, some cadets still spent their third-year summer training in Germany.[44] In 1962 and 1963 two naval cadets went for summer training with the United States navy. They reported that they gained valuable experience with equipment but that the period was too short. They said that the standard of sea-knowledge of US midshipmen was lower than that of RCN cadets at the same level, and that training on the US ships was poorly organized and not as thorough as that in the RCN. They recommended that the balance of the summer after the American cruise ended should be spent with Canadian destroyers, otherwise the exchange with the USN should be cancelled.[45] There was thus increasing confidence in the quality of Canada's summer training programs.

It was important to achieve some continuity between summer training and military activities in the college.

Several measures designed to restore some of the leadership training of prewar days, and to give cadets responsibility as early as possible, were implemented during Commodore Piers's period of command. Commencing in the spring of 1958, selected third-year cadets had been given a rank of leading cadet to prepare them for Cadet-Wing senior appointments, which was a reversion to prewar practice. Piers then reorganized the Cadet Wing in September 1959 in an attempt to make it a better

Over the wall, 1958. The recruit obstacle race joins all cadets — past, present, and future — in a shared experience and provides life-long memories.

Brigadier W.A.B. Anderson, OBE, CD, commandant, 1960–2

vehicle for military training. Because the recruit intake was small compared with that which came from the other two colleges in the third year, and to give RMC second-year cadets experience as cadet officers similar to that of their equivalents at Royal Roads and CMR, he segregated the junior years into a special squadron. This rearrangement was abandoned in 1962, however, and a cross-section of all years in each squadron was reintroduced to give more realistic experience of a command structure.[46]

When he left the college in 1960, Piers expressed concern that ROTP had weakened cadet motivation for the services. He recommended the reintroduction of reserve entry on scholarships in the proportion of 15 per cent of total intake. Among his other recommendations to increase motivation were the addition of more service members on the faculty, the provision of postgraduate courses for service officers, and the introduction of a system of sabbatical leaves for the academic staff.[47]

Brigadier-General W.A.B. Anderson succeeded Piers in 1960. His appointment was unusual in that he still had prospects of rising much higher in the service. After distinguished service in Northwest Europe and in his postwar appointments, a three-year appointment at RMC might divert him from the normal stream, and he came with an understanding that it would not have that effect. From the college's point of view, however, there was a certain advantage in having a commandant who, after his tenure at RMC, might be in a position to exercise favourable influence on its progress.

Anderson showed at once that he would exercise his experience, abilities, and persuasive powers to enhance the prestige, public image, and effectiveness of RMC to produce well-educated professional soldiers. He was proud of being a professional himself in the fullest sense of the word, and of following his father who, he asserts, was also a "professional."[48] By taking care to make personal contact with individual members of the faculty, Anderson brought the influence of his dominant personality to bear directly on RMC's development.

One small step at the outset of his appointment was a significant new initiative. Just as Piers had begun his tour of duty by requesting an investigation of the naval establishment that had previously occupied Point Frederick, so Anderson also used history as a formative influence. He arranged for the installation of a college museum in the Fort Frederick Martello tower as a means of augmenting cadet pride in the college, of strengthening military indoctrination, and of arousing public interest. As with the naval investigation called for by Commodore Piers, the present author was assigned the task of producing a museum by the next Ex-Cadet Weekend, only a few months away. This was achieved by Herculean efforts on the part of the administrative, supply, and works-service officers of the college.

The new museum in the fort displayed part of the college's invaluable collection of arms, one of the best in North America, a gift of ex-cadet Walter Douglas* who had been engaged in railway construction in Mexico. Douglas had bought the arms from President Porfirio Diaz. The museum also displayed part of the mess silver of the British 100th Prince of Wales Own Leinster Regiment (Royal Canadians) that, for safe-keeping when the regiment was disbanded, had been deposited in Canada because the regiment had originally been raised there. Another section of the museum displayed pictures and artifacts relating to the dockyard and naval base that had existed on Point Frederick at the time of the War of 1812 and for over half a century afterwards. Finally, and most important of all, the museum illustrated the history of RMC and its famous graduates. It was designed not only to inspire successive generations of cadets but also to bring the college to public notice.

Anderson recognized that the college's achievements

*Walter Douglas (no. 249), RMC 1887–90, was a chemist and mining engineer who became chairman of the board of the Southern Pacific Railway in Mexico. He died in 1946.

in military training and academics depended ultimately on the quality of instruction and internal organization and on the number of young Canadians of the right sort who could be attracted to apply. To increase awareness of what the CSCs had to offer, the Department of National Defence advertised widely in Canadian periodicals and cadets in uniform attended certain national organizations such as the Engineering Institute of Canada. In 1959 DND provided a public relations officer who organized visits by school guidance counselors to RMC.[49] In 1966 the RMC football team made a trip to the prairies, a venture repeated there and elsewhere in later years, and the RMC Club branch in Winnipeg hosted a reception to further the public relations effort.[50]

RMC's sports program had become one of its biggest drawing cards. As in most Canadian universities, Canadian football was the major sport, but the college's relatively small numbers were a disadvantage because of the large size of a football squad. In the 1963 season the RMC team had not won a single game.[51] However, although football was the most visible of the sports, what was particularly attractive to many potential recruits was the wide range of other activities. In March 1965 the RMC Club president reported that RMC had fielded eighteen intercollegiate teams and had won eight championships.

In the academic field there was a similar notable record. The club president was able to report that same year that the third Rhodes Scholarship since the opening of the college, two Athlone Fellowships for graduate work in science in the United Kingdom, and seventeen other awards had been won by RMC cadets.[52] The college's strength had now reached 525, the highest in its history. Compared with the civilian universities with which it competed on both the sports field and for academic honours, this was very small. RMC's academic and athletic record was therefore the more creditable.

Anderson stayed at the college only eighteen months. Before he left to become adjutant-general, he gave an impressive talk to staff and cadets in which he laid down an overall strategy for college development. He stressed that, although apparently divided between an academic faculty that was predominantly civilian and a military training staff, there was a single common objective – to produce officers.[53] He had been careful to retain direct personal command of the Cadet Wing, and personally presided over the weekly cadet-staff meetings.[54]

In this End-of-Tour Report, Anderson took as his theme the differences between the Canadian Services Colleges and Canadian universities that made it difficult for the CSC to carry out the mission set them by the Department of National Defence – "to educate officer cadets to degree level." He said that in a civilian university it is the Senate's responsibility to ensure the maintenance of academic standards and to apply to the responsible administrative body, a Board of Governors or Trustees, for the financial support essential for that purpose. "Those who have the responsibility for standards and those who must make the necessary financial provision discuss their problems face to face and reach decisions accordingly." At RMC the Senate had no Board of Governors to which it could report what it needed in the way of staff, equipment, and facilities. Instead, five government agencies exercised the functions of a university Board of Governors. Moreover, no provincial government attempted to exercise as much control of a university as the federal government did in respect of RMC. Hence RMC was inevitably disadvantaged in respect of maintenance of academic standards, staff-recruitment, equipment, and accommodation.

Anderson recalled that a former minister of national defence, Brooke Claxton, had suggested that RMC should be set up like a crown corporation with its own Board of Governors. He said that with the passage of time (and the acquisition of university status), this need had become yet more urgent. "There is no reason to think that the Dominion Government will be able to sustain successfully a university of its own . . . under the present system of control, of policy and of finance."

Anderson also called attention to a most serious difficulty experienced by RMC – the staffing of its all-important departments of engineering. After Lieutenant-Colonel G. Holbrook* had resigned as dean in 1961, the Division of Engineering had no chairman (the equivalent of a dean of a university faculty). Furthermore, the mechanical and electrical departments had no permanent heads. Pointing to reliance on service personnel for part of the faculty in engineering departments, Anderson said that, while these officers were competent and also valuable because of their military connections, they were transients. He feared that accreditation by the Province of Ontario, which had not yet been secured, might be withheld because of lack of a permanent engineering faculty. He added that one of the chief difficulties in recruiting civilian faculty was the college's lack of graduate studies.

The size of the military staff that was provided for the instruction and management of cadets was completely inadequate, Anderson warned. As a result, members of the academic staff had the impression that the armed forces believed officer training in the CSCs was distinctly secondary. At the same time, the cadets had little respect for military studies, the course that ought to awaken their interest in the profession of arms. This was not a reflection on the officers who taught military studies, but simply a result of the competing demands on their time.

Anderson advised that the commandant's job would take a senior officer accustomed to service affairs a year to learn. He must take a firm stand in both academic and cadet problems. In the academic field, acting in the best interests of the college as a whole, he must approve the recommendations of the director of studies and the Faculty Council, and not encroach upon their professional responsibilities. Anderson believed that a commandant's tour of duty should be not less than six years and that the appointee should be of the rank of major-general or equivalent, preferably one who was nearing the end of his service career and unlikely to become a chief of staff.

An officer suitable to head a military university could be retained as commandant for the remainder of his service career or even beyond.

In addition to these proposals on personnel and organization, Anderson said that RMC needed a new gymnasium, extra dormitory accommodation, two chapels, and permanent married quarters. Finally, the college also required more lecture rooms, laboratories, and offices for the Departments of Science and Engineering, which were presently housed much below the standards of Canadian universities. He concluded, "This is not simply a matter of luxury and comfort; it is a matter of good teaching standards and the recruitment of staff . . . A new science and engineering building is required if RMC is to remain competitive as a university."[55]

Brigadier-General George Spencer, who also had a fine war record, came to RMC in 1962 to fill out the eighteen months of Anderson's turn as commandant. Appending Anderson's end-of-term recommendations to his annual report, Spencer added his own significant conclusions and recommendations. In retrospect, Spencer said he saw his task as one of balancing relationships between the civilian and military staffs and between academic and military demands on the cadets, of coordinating relations with the other two colleges, and of redressing the high failure rate of francophones (partly by making them feel more at home in an anglophone milieu). His most urgent problem was to counter the misleading adverse image presented by the Glassco Commission.

Brigadier G.H. Spencer, OBE, CD, commandant, 1962–3

*Lt-Col. George W. Holbrook served in the British Royal Corps of Signals during the Second World War. In 1946 he was chief instructor of the Royal Canadian School of Signals, and from 1950 to 1961 he was professor of electrical engineering at RMC, becoming dean of engineering. Later he was appointed president of Nova Scotia Technical College 1961–71, and then director of communications research, Department of Communications, Ottawa.

Spencer supported the proposal for a Board of Governors but suggested that the present Advisory Board could be adapted for the purpose. He reported that, because of cumbersome civil service procedures, there were still delays in hiring civilian academic staff. He repeated Anderson's comments on the lack of permanent members in the engineering faculty, but noted that in the mechanical and electrical engineering departments only thirteen out of twenty staff were civilian. He confirmed that the complex duties of the military staff meant that military studies suffered in comparison with the academic courses undertaken by the cadets. He noted two other problems: a decline of the real value of library acquisitions owing to inflation, and the need to maintain recruit entry at sixty cadets in order to avoid any further reduction in the size of the two first years.[56]

On 6 May 1963, after a detailed examination, the Association of Professional Engineers of Ontario approved RMC's engineering course. However, in the case of electrical and mechanical engineering, that accreditation was only for one year, to be continued conditionally thereafter on the appointment of satisfactory heads of department. Accreditation from other provinces was now expected to follow.[57]

Meanwhile, in 1962 RMC graduated its first class to be awarded degrees in engineering. Output in that year gives some indication of what the services could expect from the new venture. The class had entered RMC sixty-five strong in 1958 and had been increased by three cadets who were repeating the first year at their own expense. In 1960, when the RMC group had been reduced to fifty-one and had added nine repeaters from the previous class, fifty-nine cadets came from Royal Roads and sixty from CMR, making a total of 179. Two years later 138 were commissioned, including two from a previous class who had repeated the fourth year. Thirty of the new commissions were in the navy, fifty in the army, and fifty-eight in the air force. This was roughly in the same proportion as the relative size of those forces, namely

21,593 navy, 49,381 army, and 53,768 air force.[58] Twenty-nine cadets had graduated with honours in engineering, including three with first-class standing. Thirty-four others graduated with a bachelor's degree in engineering (including four in September after writing supplemental examinations). Six gained an honours degree in science (five of them with first-class standing) and thirty-three others were named bachelors of science. Four were awarded the bachelor of arts with honours and thirty more got the general arts degree.[59]

From these statistics it appears that sixty-two were qualified to take technical commissions, twelve in the navy, twenty in the army, and thirty in the air force. The remaining seventy-six graduates were eligible for non-technical commissions, seventeen in the navy, twenty-eight in the army, and twenty-eight in the air force. The air force thus did a little better than the other two services in securing graduates from the CSCs to take up technical commissions. However, in the five years previous to 1963, eighty-three out of 142 (or 58.5 per cent of technically qualified RCAF graduates) had elected to take air-crew commissions.[60] Statistics for the other two services would probably show a somewhat smaller percentage making a similar combat-arms choice.

The army had not fared much worse than the other two services, but some elements were still unhappy about the ability of the CSC system to serve its needs. In a foreword to the 1961–3 issue of the Canadian Army Staff College's periodical, *The Snowy Owl*, the army's chief of the General Staff, Lieutenant-General G. Walsh,* had said that the army was facing a period of accelerated change due to such technical innovations as the armoured personnel carrier, the helicopter, the guided

*Lt-Gen. Geoffrey Walsh (no. 1941) was at RMC 1926–30 and then at Nova Scotia Technical College and McGill University. Commissioned in the RCE, he served overseas 1940–5, was DGMT 1953–5, QMG 1955–8, and CGS 1961–4, becoming vice-chief of the Defence Staff at integration.

anti-tank missile, and the surveillance instrument. This would necessitate a thorough examination of organization, tactics, and logistics. He argued that change must not "just be allowed to happen; it must be guided." The goal should be a blueprint for the future to give Canada an army organized, equipped, and trained to take full advantage of modern weapons, equipment, and techniques.[61]

An army workshop then prepared a report entitled *The Professional Officer*, which defined military professionalism. This report deplored the schism within the officer corps between the products of the ROTP system, who had degrees, and officers with short-service commissions, who did not. It said that, while the Canadian army did not need as large a proportion of graduate officers as was the case in the United States (army, 82 per cent, marines, 92 per cent), the present level of 35 per cent was much too low. It proposed the adoption of a Bachelor of Military Science (BMSc) degree as the qualification for Canadian army officers. While the concept of a military science degree had few, if any, supporters in RMC, the workshop's report suggested that important elements in the army had now come closer to the other services in their requirement from the CSC program.

The government had not yet decided how the services colleges would be developed in the future to produce graduate regular officers, whether as a system in which RMC would remain the sole provider of CSC degrees, or with RMC as one of three parallel degree-granting institutions. However, whatever that decision might be, it was quite clear that the immediate primary need was construction. The Massey Library had provided ample library service, and also offices and classrooms, for the arts courses. A new dormitory to house the influx of numbers from the other two colleges into the third and fourth years was well advanced. But what was still needed urgently was a science and engineering building to carry through the government's decision that RMC's degrees should include engineering. But, like many other urgent

defence needs at this time, approval of construction was now postponed by a radical decision that affected the services as a whole as well as the services colleges – to integrate the Canadian Forces under a single command and administration. RMC's future would inevitably be affected by that decision, and even more by the ensuing radical unification of the three armed forces into a single service.

Manpower and Integration Problems

The long-standing concern in the services about wastage in the Regular Officer Training Plan (ROTP), including the Canadian Services Colleges (CSC), was linked with general manpower problems. These were becoming more serious. The shortage of officers caused by the Korean War had eased by 1955, but five years later, when senior officers commissioned before or during the Second World War were reaching retirement age, it had re-emerged as a threat to the effectiveness of the military forces. The high proportion of the defence budget that now went into personnel costs had severely limited weapon development and equipment replacement in an age of rapid technological change. In 1961 Minister of National Defence Douglas Harkness* set up a Manpower Study Group chaired by his associate minister, Pierre Sevigny,** to review the forces' administrative, support, and operational services. He hoped the study group would forestall criticism expected from the Glassco Commission, and also suggest economies by way of rationalization. According to David P. Burke,*** an American scholar who studied the origins of unification and who interviewed widely, the group found that if different personnel systems were retained, the integration of the personnel staffs of the three services would be only marginally cost effective. Only overall unification could make the needed substantial economies.[1] In 1963 the Glassco Commission precipitated a discussion of the issue of integration by reporting publicly that there was consider-able duplication in the administration of three separate services.

The idea of rationalizing the administration of Canada's armed forces by integration was not new. In 1923 the three services were placed under one minister; during the Second World War, what were substantially separate air, naval service and army ministries again evolved, but in 1946 these were reintegrated. Brooke Claxton, named minister of national defence in that year, was convinced that wartime experience had shown a need for better interservice cooperation. In 1951 he had appointed General Foulkes as chairman of the Chiefs of Staff Committee, the first one not to be simultaneously chief of one of the three services. However, Foulkes had little control over the separate services. Claxton had also made a few basic legal changes to centralize control, and he amalgamated a few special services, for instance the chaplains and the medical services. His most publicized innovation had been the creation of the tri-service college system in 1948.[2]

---

*Lt-Col. Douglas Scott Harkness, minister of national defence 1960–3. He resigned over Diefenbaker's resistance to the arming of the Canadian Forces with nuclear weapons.

**Col. Pierre Sevigny, associate minister of national defence 1959–63, later professor of finance, Concordia University.

***Lt-Col. David P. Burke, USAF, assistant director of Western and Canadian studies, US Naval Post-Graduate School, Monterey, California.

Air Commodore L.J. Birchall, OBE, DFC, CD, commandant, 1963–7

When Paul Hellyer became minister in 1963, his cabinet colleagues insisted on severe cuts in the defence budget. Frustrated by the conflicting advice he received from the several service and departmental heads who had access to him, he decided on bold measures to end the confusion and, simultaneously, to reduce costs. In 1964 he personally wrote a white paper in which he announced the government's intention to integrate the separate service functions in his department, and also the various service commands, as a step towards the ultimate unification of all the Canadian Armed Forces into a single service. Hellyer's proposed integration in DND headquarters and of service commands aroused immediate concern, but few took the further threat of complete unification into a single service seriously. It was seen as only a vague possibility, and Hellyer might not remain in office long enough to implement it.[3] Meanwhile, the three separate services continued to rely on the tri-service Canadian Services Colleges as their primary source of officers. In 1963, when Brigadier-General Spencer completed Anderson's turn as commandant, the RMC command passed again to the RCAF. Once more the appointment went to an officer with a glamorous war record. In 1943 Air Commodore Leonard Birchall* had commanded the Catalina flying boat that spotted the Japanese fleet moving against Trincomalee, the British naval base in Ceylon, and he had given valuable warning time to its defenders, but at the cost of being shot down and made prisoner. It is reported that years later, speaking of the turning point in the war, Winston Churchill said it was "when that Canadian airman, who is now at the bottom of the Indian Ocean, reported the naval threat to Ceylon and India and so gave us time to prepare their defences." He was surprised and pleased to learn that Birchall was still serving in the RCAF in Ottawa.

Birchall had another war record that could also be a special source of inspiration for cadets. As the senior allied officer in certain Japanese war camps, he had been a nuisance to his guards. When he was exchanged he smuggled a documented statement of Japanese atrocities beneath the plaster of a leg cast. Birchall's personal account of his POW experiences, which he published reluctantly in the RMC Club *Newsletter*, is a fascinating story of courage and determination in adversity.[4] He was also a gifted raconteur, which helped him in his personal relations with staff, faculty, and cadets.

Hellyer's proposals for integration had implications for officer training, and so for RMC and its new commandant, but since the college was already tri-service, with a rank structure and uniform quite different from any current service, it at first seemed it would be little affected. Birchall implied as much in an addendum to his annual report in 1964 and in a statement to the Advisory Board the following year. He said it would be easier for a currently tri-service college to deal with one service than with three, especially about summer training programs. In the addendum he also argued that officer production required the closest possible liaison in standardizing academic programs between the users – the services – on the one hand and the producers – the colleges – on the other. What was required was a "single control in daily contact with the system." He apparently hoped that Hellyer's integration of headquarters and commands might promote that.

Working closely with Colonel Sawyer, Birchall said that few of the users understood his academic problem. The Manpower Study Group might decide what the services required from the colleges, but the colleges' academic functions still had to be a recognized part of the Canadian academic profession. Curricula would still have to meet provincial standards for degrees. To compete with the universities for students, Birchall argued that the

---

*Air Commodore Leonard Joseph Birchall (no. 2364) was at RMC 1933–7 as a cadet. He served with the RCAF from 1937 and during the war was posted to Scotland and Colombo. He was POW 1942–5 and CO of RCAF Goose Bay 1948–50. He was commandant, RMC, 1963–7.

services colleges must offer a first-class degree program, including honours courses. (Honours courses in Canadian universities are more narrowly specialized than American majors.) He added that RMC also needed honours courses to prepare selected officers for graduate courses. Birchall made no specific reference to the possible effects of a complete unification of the services, but, capitalizing on the program of integration already in progress, he proposed that the RMC commandant be made director of ROTP with responsibility for ROTP in the universities, and that the RMC director of studies be made his deputy director to supervise university ROTP and CSC academic requirements.[5]

Meanwhile, a Special Committee of the House of Commons on Defence set up in 1963 had studied the production of officers through the ROTP and published its conclusions in 1965. It argued it was too early to assess the effects of a degree program that had graduated only two classes of engineers, but some of the figures it produced were revealing. It noted that, while the services would require approximately 1500 new officers a year, of whom 450 should be university graduates, the Canadian Services Colleges had produced over the past five years an average of 172 yearly and the universities, 323. This made a combined total of 495, forty-five more than needed. Taking the averages of intake and wastage after completion of the required period of service over the past five years, 117 CSC graduates (or 68 per cent of a graduating class) would remain in the services after performing their three years of service. Among university ROTP graduates, 195 (or 60.5 per cent) would not exercise the release option after three years. Looking further back at statistics from the time of recruitment, out of sixty-seven cadets who started their military career at RMC, forty-four graduated and thirty-seven (or 55 per cent) stayed in the service in the fourth year after graduation. Comparative figures for Collège militaire royal (CMR) were 23 per cent and for Royal Roads, 39 per cent. From the ROTP intake in the first year of university, the retention rate for long-service careers was 34 per cent.[6]

These calculations were based on the senior matriculation level. Retention based on CMR's junior matriculation level, where the attrition rate is always high, was only 16.5 per cent. Language difficulties and the need for cultural adaptation explain the higher loss of CMR cadets who transferred to RMC's third year. Some Royal Roads cadets also found cultural adjustment difficult on transfer to RMC.

The Honourable Paul Hellyer, minister of national defence, presents degrees, honoris causa, to Dr H.G. Thode, the Right Honourable L.S. St Laurent, and Lieutenant-General E.L.M. Burns, fall convocation, 1964.

Statistics prepared for the House of Commons committee to show retention comparisons in various disciplines, and in the aircrew versus the non-flying list, gave further information. Altogether, 27 per cent of CSC engineering graduates and 52 per cent of university ROTP engineering graduates exercised the release option. In the arts and sciences, the comparative figures were 26 per cent and 40 per cent, respectively. The most serious attrition, on which the committee commented in its report, was the loss of 75 per cent of the RCN's engineering officers who had graduated from universities and who presumably had left to take up more lucrative civilian employment. Up to 30 August 1963, the number of all ROTP graduates who took commissions and then exercised the release option was 20 per cent for the CSCs and 46 per cent of university graduates. Thus, 73 per cent of CSC graduates, compared with 54 per cent of university ROTP graduates, stayed in the service for a lifetime career.[7]

On the basis of these statistics and its other findings, the House of Commons committee came to a conclusion quite different from that of the army's workshop that had produced *The Professional Officer.* It reported that it considered the results achieved at the Royal Military College to be very good, equal to or better than those of comparable institutions in the United States and Britain, and that they compared favourably with the results achieved in civilian Canadian universities. The committee admitted it was unable to make comparisons between the cost effectiveness of the CSCs and the ROTP in civilian universities, but said it believed the difference either way was not significant. In contrast with the Glassco Commission's recommendation that academic staff at the CSCs be reduced in qualifications and numbers, the committee urged the maintenance of the highest possible standards. It deplored the antiquated facilities at RMC, and agreed with a recent decision to extend the required obligatory service of ROTP cadets. Summing up, the committee said it was impressed by the high standards of academics, discipline, and physical fitness at the CSCs. It had no doubt they would produce well-trained and well-motivated young men as junior officers for the Canadian forces.[8]

In 1964, while the Commons' committee was deliberating, the RCN appointed Rear-Admiral W.M. Landymore to chair a naval committee to report on the effects of his earlier committee's report. This second Landymore Committee reported that all ROTP naval cadets now took a four-year degree course at either a CSC or a university, as the former committee had recommended, but that, since 1958, there had been a "considerable change of emphasis" about the type of course needed. Initially, while RMC and the naval training authorities had assumed that any engineering course would meet the requirement, they had also established a general science course for general list candidates. This had led to an assumption that any cadet not taking either engineering or the general science course was not acceptable by the navy. Yet the first report had specifically said that cadets who did economics and history courses at RMC could qualify as general list naval officers, provided they took sufficient mathematics, physics, and engineering subjects to

---

TABLE 1
Numbers Exercising Release Option up to 31 August 1963

Academic discipline	Category	Percentage Exercising Option				
		Navy	Army	RCAF	(Aircrew)	(Non-flying)
Engineering	CSCS	44	17	29	31	29
	University	75	55	49	53	49
Arts, science, and others	CSCS	44	29	18	24	18
	University	33	44	29	38	29

Source: Adapted from House of Commons, *Reports of the Special Committee on Defence,* 1964–5, table 2, 15

enable them to undertake common duties on the general list. Within the past year, however, general list cadets had begun to be drawn from a much wider variety of courses, and the committee believed that this revised practice met the navy's requirements adequately.[9] The RCN system carefully avoided motivation towards civilian careers and so had a high retention rate.

There were still serious accommodation problems at RMC, where many cadets were housed two to a single room. Commodore Groos, DROTP, had given his support for a new 200-room dormitory to accommodate both recruits and third-year intakes, and Brigadier-General Spencer had initiated the procedures that led to the building of Fort Champlain.[10] Next, Birchall and Sawyer got permission for a study by a conglomerate of three Toronto architect firms called University Planning and Consultation Engineering (UPACE), which produced a ten-year plan for development.[11]

But before dormitory-building began, the policy of integration gave rise to a new minister's manpower study. In 1965 it outlined what it believed the services wanted from the colleges. Its committee on officers, chaired by former commandant and adjutant-general, Major-General W.A.B. Anderson, now deputy chief of the Reserves, stated that integration had made its task easier. The committee defined the military as a profession, but said the public was not impressed by this claim to professional status and instead suspected that the so-called "military mind" meant rigidity and conservatism. In fact, however, "the military mind" was the military profession's need for decisiveness. In war, time was an enemy. Military decisions, which might involve life or death, could not be deferred until the situation became clearer; they must often be made on the basis of insufficient evidence and they must secure willing compliance from others.

The committee added that there is a difference between peacetime and wartime soldiering. In peace the emphasis is on developing contingency plans for events that may never occur. Priority must therefore be given to a broad military education in order to cultivate flexible responses. "What is required is the discipline and breadth of outlook which flow from a liberal university education." Many officers need a general course balanced between the humanities, social sciences, science, and engineering. An officer thus educated would be equipped to cope with the changes that would occur during his career.

Other officers needed a specialized education. Some should study geopolitics related to areas in which Canada had military interests, and some should take science or engineering courses related to military equipment and weapons systems, or management courses related to the planning and coordination of programs. The specific degrees these officers required were in engineering, science, and commerce, along with a "liberal university education." The type of degree should be based on the capacity and natural aptitude of the officer cadet, but it was necessary to see that enough chose a science degree to meet service needs. The committee approved the current science-humanities balance in the Canadian Services Colleges but recommended that the curriculum be progressively developed to give greater emphasis to geopolitics, management, foreign languages required by Canada for defence interests, and the behavioural sciences, because of their growing importance in personnel management.

The committee also noted that the retention rate of the CSCs was now 73 per cent and of the university ROTPs, 54 per cent. It said that the CSC environment could not be matched in the universities, and recommended that the university ROTP program should not continue in its current uneconomical form but should be reduced. The CSCs, where the services could influence the type of instruction that a cadet received, should remain the principal source of career officers and should not be restricted to an equal footing with ROTP in the universities.

The committee had not specifically mentioned undergraduate honours courses, which Birchall had said were essential. The current honours courses, specializing in some of the subjects traditionally given in Canadian universities, might seem to have been excluded by the committee's reference to a balanced general education, yet its overall approval of the current RMC curriculum had tacitly implied its overall approval of the honours courses also. In fact, since RMC honours arts cadets had more science and engineering subjects in their first two years than students in honours arts courses in a university, theirs was a more balanced program. Similarly, RMC honours science and engineering cadets had more liberal arts courses in their early years than university graduates. The committee left for further discussion the precise definition of a career philosophy for officers that it said was still needed. The nature of that philosophy and its application could obviously affect the way in which RMC would develop.[12]

In the two or three years following the announcement of integration and the subsequent completion of the manpower study, the Department of National Defence was preoccupied with the reorganization of its various divisions, branches, and commands. Considerable confusion resulted, seriously impeding the normal conduct of routine business. Nevertheless, NDHQ made significant changes in the administration and operation of cadet education.

One result of changes brought on by integration was that the Directorate of the Regular Officer Training Plan (DROTP) was terminated. The administration of the CSCs now became a responsibility of a director of officer cadets (Commodore Groos, the last DROTP), who was under the director general of recruiting and training in the Personnel Branch. In an addendum to his annual report for 1964–5, Birchall argued that the formulation of policy on the all-important training of officers should be at the highest staff level possible in the Canadian Forces Headquarters.[13] This was in effect what was done, but only

indirectly. From 1965 the Advisory Board, which had been used as a consultative body by successive DROTPs, was encouraged to show more initiative in discussion of college policy and progress. At the same time, the conferences of college commandants and directors of studies were put on a more formal and regular basis and divided into two separate but overlapping "councils" according to function. The commandant, staff officers, and certain senior headquarters personnel constituted a General Council to deal with all CSC matters except academic questions. The directors of studies and headquarters personnel made up an Educational (or Academic) Council to handle academic policy and operation.

The primary purpose of these councils was to aid the director of professional education and development (DPED) and his superiors in exercising the supervisory functions formerly exercised by DROTP – the coordination of the colleges and their curricula to ensure they were producing officers of the right kind, quality, and number for the newly integrated Canadian Forces. The chief of the Defence Staff or his representative, the chief of personnel, and the director general of recruiting and training, or others who, in the course of the following confused years, exercised those functions, were members of both councils. The councils carried much more weight than DROTP, but their decisions were subject to approval in NDHQ and so were affected by other considerations.

A first concern of the councils was to ensure that the ROTP system produced as many officers as the services needed. In 1963, when supporting the construction of Fort Champlain at RMC, Commodore Groos, DROTP, had stated that, even with this expansion of accommodation, RMC's graduating classes would not exceed 200 a year, or half of the forces' total need of graduate officers. From the statistics published by the House of Commons Committee on Defence and the Report of the Manpower Committee, however, it was clear that RMC was producing officers who were motivated to make military service their lifetime career. RMC's record from the recruit class to career service was better than that of the other two colleges, and very much better than that of the universities' ROTP. It was imperative, therefore, that its output be not merely maintained but, if possible, increased.

RMC's strength in the fall of 1965 was 525, the highest in its history. However, statistics of applications and rejections did not encourage hopes for further expansion. In 1964 the quotas needed in all three colleges had been filled from among 3967 applicants, but in 1965 only 2961 candidates for ROTP came forward, and from that smaller number it was not possible to find enough qualified recruits for the colleges. Royal Roads was three short, and RMC seven. Only CMR, which had its preparatory year to take in less well-qualified students, was at strength. The search for recruits to double the number of graduates was clearly scraping the bottom of the barrel of those young men who applied to enter. Increasing the size of the CSCs would be very difficult.

One reason for the fall-off in applications in 1965 may have been the introduction of the policy of filling the services colleges before permitting entry to the universities. But behind this was the fact that Canada did not have a strong tradition of career military service in peacetime. The growing use of armed forces for international peacekeeping and deterrence was not sufficiently appealing to young men. Furthermore, a peace movement was growing in Canada and there were acrimonious debates about the acquisition of nuclear weapons and American defence policies. It was also widely reported that Canada's military equipment was out-of-date because of the government's failure to meet the cost of replacement. Finally, there was much adverse publicity for the armed forces as a result of integration and unification and the feared loss of historic units, uniforms, traditions, and symbols. Service morale was said to be at a low ebb.[14] All these factors undoubtedly helped to limit applications for entry to the Canadian Services Colleges.

The recruitment of cadets depended on publicity –

for instance in the press and periodicals – but the Glassco Commission had criticized the amount of money spent on advertising the ROTP and the services colleges. The commission had exaggerated the annual average advertising bill and ignored the fact that the current-year costs included a one-time item for a film that should have been amortized over four years.[15] In 1966 Sawyer argued that the solution to the problem of securing enough applicants was not to cut the publicity effort, but to increase it – to "get out and recruit as effectively as all military academies and universities had to do." "Our recruiting efforts are very feeble compared with our competitors. There is a big pool of students who could take our courses in their stride, but we must go after them."[16]

To help with recruiting, the RMC Club offered its services to the minister, much as university alumni organizations do everywhere. Beginning in 1966, cadets were stationed for brief periods at recruiting centres, and the athletic department in the college asked the RMC Club to help in recruiting athletes; a football game in the west was again used as an advertising device.

Some members of the RMC Club were, however, dismayed, not only because the college was failing to produce sufficient officers, but also because in their opinion it was not getting enough recruits of the right type. The club wanted more like those who had once been RMC's mainstay. Youths from middle-class backgrounds would be less likely than others who were less well-off to be attracted primarily by the prospect of a cheap university education, but they might not be ready to commit themselves to a full-time military career when they were just leaving school. The Reserve Entry scheme, reintroduced on a limited scale to supplement ROTP in 1961, had thus far proved a disappointment. There were nine applicants in 1964, of whom only two were accepted. The following year only one Reserve Entry recruit was admitted.[17]

There were various explanations why RETP had had so little result. Selection was in competition with ROTP, the same standards applied, and the military accorded RETP low priority. Suitable applicants able to finance their education but already interested in a military career would naturally have chosen to enter through ROTP. Since the Second World War, and even before it, however, the social prestige of a militia commission, once a big card to draw recruits because it promised a prominent role in local communities, had lost that appeal. The militia's conversion to a civil defence role in 1956 had further weakened its image. Part-time militia service had become less attractive.[18]

Partly in an attempt to offset these difficulties for Reserve Entry, the RMC Club, which had given its full support to RMC after the introduction of ROTP, set up a foundation fund to raise $200,000 to provide scholarships for Reserve Entry cadets in all the colleges and support other worthwhile cadet activities.[19] The money was to be raised from ex-cadet contributions. The club also decided to give financial support (to be recovered from sales) for a much needed history of RMC. This was a Canada Centennial project. Some of these measures may explain why Reserve Entry increased somewhat, even though youth hostility to the military continued to grow. In 1966 eleven, and in 1967 eight, Reserve Entry cadets were admitted to the colleges.[20]

Reserve Entry was, however, only a minuscule part of the officer-production program compared with ROTP, the mainstay of the CSCs. Yet, as the figures showed, ROTP was still producing too few officers for the Canadian Forces. One measure that might increase the supply of junior officers was a lengthening of the period of obligatory service required in exchange for four years of subsidized university education. The Glassco Commission had recommended such an extension, and the services wanted the three years extended to five. In 1965 Hellyer announced that the required service would be four years.[21] If that extra year did not prove counterproductive by significantly eliminating some applicants whose objective was only to serve and leave, it would add

one-third to RMC's annual input of junior officers into the forces. But that would not be effective until 1973, when the next recruit class had graduated and completed its three years of service. Thereafter it could increase the number of junior officers by one-third, and also the proportion of those who decided to stay on when their obligatory service was completed.

Another way to increase the number of graduates entering the services might be by expansion of ROTP in the universities, though the percentage of university ROTP recruits who became long-service career officers was much lower than that from the services colleges. Moreover, an increase in university ROTP, if the desired recruits could be found, would not be felt until 1970. Expanding ROTP in the universities did not promise much hope of immediate relief from the shortage.

A more immediate way to increase output could be by reducing attrition in the colleges and improving motivation. One source of wastage was the unsuitability of some cadets for a military career. After integration, selection of recruits was all done by the system of local boards formerly used by the army, which put less stress on military motivation than the former air force and naval selection processes. Furthermore, after their first few months, cadets were hardly ever separated from the colleges on grounds of their lack of officer-like qualities. Yet it was important to weed misfits out and inspire military incentive at an early date. This could open up more places within a year and reduce wastage later on. The colleges therefore introduced a preparatory course during the summer months between selection and the start of the academic term. One of these had been held on an experimental basis at Royal Roads in 1967; it operated for three weeks instead of the six originally planned, but it was beneficial enough to be repeated.[22] Later, the same result was achieved by sending CMC recruits (excluding those for CMR's preparatory year) to a Basic Officer Training Course along with all other officer candidates before they reported to their university or service college.

An increase in the output of the colleges at an early date could also come if the rate of academic failure were reduced. This might be done by diluting course requirements, but Sawyer warned against tampering with passing standards or academic requirements. He regularly produced statistics that showed that RMC's academic failure rate was comparable with those in universities and in other military academies. At the same time he would show that, because of its restriction to matriculants who had qualified in both arts and science, because of closer staff-student relations and smaller classes, and because of the quality of its faculty, RMC was more successful than most undergraduate institutions in producing a superior product.

Royal Roads, which had more places to fill than RMC and drew on a less populous hinterland, was compelled to take in some recruits who were below the Ontario matriculation level. It had a higher failure rate in its two-year course, yet the record of its cadets who made the transfer to Kingston for the senior years was very good. The Glassco Commission had recommended the closure of Royal Roads, but in 1963 Hellyer had assured the new RMC commandant, Commodore W.P. Hayes,* when he was at Royal Roads, and Royal Roads principal, Dr Eric Graham,** that it would not be closed. The minister's unstated reason may have been the political wisdom of maintaining a military college in the west.[23]

In attempts to solve Royal Roads's problems during the next few years, Dr Graham pressed for various kinds of modification of the CSC program, including some dilution or rearrangement of its academic content.[24] Eventually, RMC agreed to the introduction of pass arts and pass science courses, albeit reluctantly.[25] Royal Roads then proposed a special arts entry without the science required in senior matriculation. The RMC Faculty Council doubted whether there would be enough applicants entering RMC to validate such a course, and this proved to be true the first year it was tried.[26] Later, an arts entry for RMC honours courses was to be more successful.

Another Royal Roads proposal to increase the attractiveness of the services colleges was to permit CSC graduates to enter government departments other than the forces. On behalf of the chief of the Defence Staff, Group-Captain A.H. Middleton,*** director of officer cadets, replied that the single purpose of the colleges was to train officers. Diversion of a number of cadets from military service could be "distracting."[27] Two years later the board was told that "a few bright graduates" were permitted to work temporarily in the Defence Research Board, but they had to complete their military obligation afterwards or repay the cost of their tuition.[28]

At the same time as it was concerned with these efforts to increase the output of the colleges, RMC was occupied with another closely related problem – the development of the curriculum to ensure it would produce modern professional military officers. In 1964 the Department of Economics and Political Science had cooperated with the Department of History to add an honours degree in international relations.[29] In 1966 the Educational Council approved the introduction of courses in electronic processing and engineering management.[30] Yet, even though such courses did have obvious potential military application, they did little to change the opinion of many skeptics who still argued that RMC was "not sufficiently

Commodore W.P. Hayes, CD, commandant, 1967–70

*Commodore William P. Hayes (no. 2576) was at RMC 1937–9 and then attended the Royal Naval College in Dartmouth, England. He began his service as a midshipman with the Royal Navy's Eastern Mediterranean fleet. After the war he attended the United States War College 1958–9 and was appointed commandant of Royal Roads 1963–7. Commandant of RMC 1967–70, he later headed the Canadian Forces College, Toronto.

**Dr Eric Stanley Graham, a chemist, served in the Canadian army in England and West Europe 1942–5. He was director of studies and then principal of Royal Roads 1961–84.

***Group-Capt. A.H. Middleton joined the RCAF in 1940. He became a flying instructor, vice-commandant Royal Roads 1945–6, and director of officer cadets, NDHQ, 1965–7.

military." This problem was not unique to Canada: the American military academies, often praised by Canadian officers for their allegedly greater military impact, have been subjected to the same kind of criticism. Conversely, academic course specialization in Canadian military colleges has met with approval by American military specialists.[31] In 1966 the chairman of the Educational Council said that CSC faculty members ought to become more aware of matters of current military interest in their areas of academic specialization.[32]

RMC cadets were themselves divided in their opinions about the kind of courses the college should give. Different cadet attitudes were outlined in an article by Officer-Cadet A.D. Chant* in *The Marker*, the cadet newspaper, in March 1967. He said there were three clearly defined cadet groups: the first group was interested only in a military career, and for them a degree was secondary; the second and largest group had some motivation towards the services, had come to RMC to obtain a degree, and would stay on after their obligatory service if the services seemed to provide a worthwhile career; the third group were misfits who, coming to the CSCs by accident or misadventure, found they hated service life but had an ironbound contract to stay for "three dreadful years." Chant said the problem was to convince the second group to stay in the armed forces. This required a "broadening of the opportunities in the services and a greater direction of the RMC courses."[33]

This issue of *The Marker* was called to the attention of General Jean Allard,** the CDS, by Air Marshal E.M. Reyno,*** chief of personnel (CP). Reyno had assured the editor of *The Marker* that the problem was under advisement and that, after unification had been carried through, there would be significant change. He told Allard that those changes might be the addition of more military-type courses and the cessation of civilian-type degrees.[34] Allard replied that, although there was a need for a "complete examination," he did not agree with either extreme view on the subject – with an entirely mil-

itary college or the retention of the present system. He felt that the Advisory Board, which he said was composed of "ex-cadets in civvy street," could no longer evaluate the system in the interest of the "new concept of the Armed Forces."[35] In actual fact, the number of ex-cadets on the board was never large.

A year later Sawyer warned the RMC faculty that there was still considerable concern in NDHQ about a supposed insufficiency of military content in service colleges' courses.[36] No doubt some of this criticism came from officers of the kind whom Laurence Radway† called "anti-intellectual" and "doers," those who wanted to put more stress on combat virtues. Yet the problem of adapting RMC's academic courses to military needs was real.

Reyno's deputy, Major-General Bruce F. MacDonald,†† asked for an opinion on these questions from a former commandant, Major-General George Spencer, now at NATO headquarters in Brussels. Spencer replied

---

*Andrew Douglas Chant (no. 7346), Royal Roads 1963–5 and RMC 1965–7, joined the RCN in 1967 and served in HMCS *Saguenay* until 1971. He trained in law and is now with the Toronto-Dominion Bank's Tax Department.

**Gen. Jean-Victor Allard, commissioned in the non-permanent Canadian militia 1933, served with the Royal 22nd Regiment in Europe and was promoted to brigadier. He was military attaché in Moscow 1945–8, before retiring. Recalled to service in 1953, he commanded the 25th Infantry Brigade in Korea. He was the first commander of mobile command and was promoted general in 1966 as chief of the Defence Staff.

***Air Marshal Edwin Michael Reyno served in the Battle of Britain. He became chief of personnel 1966–9 and deputy-commander of NORAD 1969–72.

†Laurence Radway, professor of political science at Dartmouth College, New Hampshire, co-authored with J.W. Masland *Soldiers and Scholars: Military Education and National Policy* (Princeton, NJ: Princeton University Press 1957).

††Maj.-Gen. Bruce F. MacDonald served overseas with the Armoured Corps in the Second World War. He was deputy chief of personnel, CFHQ, 1966, and commandant of the Canadian Army Staff College, Kingston, 1966–8.

that academic courses in military studies should come under one of the academic divisions of the college – arts, engineering, or science – and should be "of adequate substance to be respected by the civilian members of the staff as well as by the cadets"; but he added that civilianization of military studies must be avoided. The proper mix of courses would be "maximum military orientation of many subjects which are essentially civilian in nature, and strong recognition of any related civilian trends in subjects which are essentially military in nature." He added that the question "Who is boss?" was difficult. The best solution would be close cooperation and mutual understanding between CFHQ and RMC. "The college exists solely to meet the military needs of an academic institution . . . [it] requires careful handling and negotiation with a view to genuine acceptance, rather than routine compliance, with orders." He believed this might call for the establishment of a separate controlling organization with unique skills and understanding to deal with the colleges.[37]

In one instance, military history, where the degree of military content and application was not questioned, some complaints had been levelled against the nature and quality of the instruction in military studies courses, as well as against the elementary nature of the subject matter – for instance, lists of rank structures and general principles of "the attack" and "the defence." This complaint was made in reference to courses on military history that had been taught since 1950 in the Department of Military Studies. They were given by military officers who were not historians. In 1963, to strengthen the military content of the curriculum, it was decided that the one-hour-a-week course in military history on war and society, already given by history department professors to engineering students as a humanities input, should now be taken by all senior cadets as part of their military studies requirement.[38]

Soon afterwards, the Department of Military Studies was changed from instructing in the general area described as "military studies" to a series of courses in military leadership and management. These courses were to be given by service officers posted to RMC especially for the purpose, and not, as with military studies earlier, by officers of the Cadet Wing's military staff. The new series of courses, drawing heavily on the behavioural sciences as applied to the military field, included psychology in the first year, the sociology of leadership in the second, management and organizational theory in the third and fourth, and military history and the role of the Canadian Armed Forces in the fourth.[39] Major C.J. Crowe,* a graduate of the first postwar class at RMC, organized the program. In the early years he had to use civilian instructors for some of these courses because qualified service officers were not available.

There were compelling reasons for using military faculty wherever possible, not only in military courses but also in various other academic courses. Military professors might have an important influence in inclining cadets towards a military career. The Educational Council stated in 1967 that it was desirable that 20 per cent of the CSC faculty should be military officers posted for duty as professors. The problem was to find suitable individuals. For civilian faculty the basic qualification was the doctorate, but in 1967 there were only seven officers in all three Canadian services who had a PhD.[40] This contrasted sharply with the situation in the United States, where possibly 2000 officers held the degree.[41] Of course, Canadian officers without doctorates often had knowledge and experience that could be used in instruction in

---

*Lt-Col. Charles J. Crowe (no. 2872) was a member of the "New One Hundred" class and graduated in 1952. He served in Korea and then obtained a BComm from Queen's University. In 1956–9 he was an RMC squadron commander and then attended the Staff College in Camberley. After further service in Germany, Ottawa, and Cyprus, he returned to Queen's to take an MBA. Head of the RMC Department of Military Leadership and Management from 1967, on retirement he became director of the Prairies Region of the Ministry of State for Urban Affairs in 1976, and then in 1978 of its Western Region.

Open boxing tournament, 1958. Officer cadets Facey, Walsh, Preston, Lomhein, Fletcher, and de Chastelain show off their trophies.

version to graduate schools only as a drain on the production of junior officers. He said it was preferable that ROTP graduates go directly to full-time service to obtain early experience of command. Nevertheless, a limited number of new graduates were in fact being permitted to take graduate courses considered to be of value to a military career. They were required to extend their obligatory service by two years for every year of postgraduate work.[42] In the previous year twenty-eight university ROTP and sixteen RMC graduating cadets had applied for leave to go immediately to graduate schools. The figures for 1967 were sixteen and eight, respectively. Colonel Radley-Walters,** director of training, saw this as a further "decline in the loss rate," which he said had peaked in 1964–5.[43] He, too, was more concerned about the production of adequate numbers of junior officers than about the larger problem of educational levels in the officer corps as a whole.

The pressures of technological development and social and economic change, which were increasing the requirement for graduate education in civilian professions, were operative also in the armed forces, but reliance on university graduate schools, to which many officers now went for specific military needs, was especially expensive for the Canadian Forces. Officers on educational postings were on full pay and allowances. In 1967 Allard announced an important development in Canadian military education. He said that Canada must make use of

the colleges, especially in technical departments, but such officers were few in number and often they could not be released from corps duties to teach at RMC. Furthermore, they did not always satisfy outside appraisal committees.

As numbers of ROTP graduates were now applying for permission to go straight to graduate schools after graduation, this lack of officers qualified to instruct in the services colleges might in time be remedied. But in 1966 the chief of personnel, Vice-Admiral Dyer,* saw this di-

*Vice-Adm. K.L. Dyer was a midshipman on HMS *Hood* before the war. He served as a lieutenant-commander in the Battle of the Atlantic and became vice-admiral in 1965.

**Brig.-Gen. Sydney Valpy Radley-Walters graduated from Bishop's College 1940 and was commissioned in the 27th Canadian Armoured Regiment (Sherbrooke Fusiliers). He commanded the Fusiliers in the Occupation Force, was CO 8th Canadian Hussars (Princess Louise's) 1957, and then served at SHAPE. In 1966 he was appointed director-general of training at NDHQ.

its "military universities" to qualify officers in some postgraduate fields, "thus departing from the cadet aspect of the Colleges."[44] Graduate degrees at RMC for military personnel would open the way for production of more military faculty and so could increase the military input into cadet education. Development in this direction will be examined in detail in chapter 13.

All these measures, whether at the undergraduate or graduate level, were capable of furthering RMC's military purposes. They should have served to counter doubts that the Canadian Services Colleges were "not sufficiently military." Yet those doubts persisted. In 1967 the Advisory Board engaged in spirited discussion about an alleged "civilianization" of the military colleges, which was said to be obstructing their military function. This was, in effect, an enquiry into whether they were properly fulfilling their motivational purposes. At that Advisory Board meeting, Colonel Gordon Sellar,* an ex-cadet and a former RMC army staff officer, commented that when he used to visit summer training units he often had to fend off questions from commanding officers about RMC cadets' alleged lack of military know-how. General Allard retorted that the explanation was that the colleges were striving to "educate the whole man." Dr J.R. Dacey,** acting as director of studies at RMC when Sawyer was ill, declared it was untrue that RMC was "not military." He asserted that its retention rate was as good as West Point's.[45]

Some of those who worried about RMC's alleged military shortcomings regarded sports and athletics as activities that were especially appropriate for a young man's preparation for a military career. In this period there were important changes in the college's athletic program. The venerable Recreation Club, organized in 1876 to promote extracurricular interests within the college, was "abolished" in 1963. (It was revived later.) It was replaced by an Athletic Board that seemed at the time to be more appropriate for the administration of the very extensive list of intercollegiate fixtures that now supplemented a full program of intramural sports.[46] Soon

Dr J.R. Dacey, MBE, BSc, MSc, PhD, FCIC, director of studies and principal, 1967–78

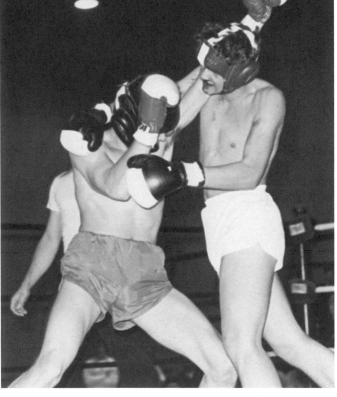

Recruit boxing, 1979: Long a part of the RMC recruit tradition, novice boxing was dropped in 1980.

*Brig.-Gen. Gordon H. Sellar (no. 2805) was at RMC 1940–2. He served with the Calgary Highlanders in the Second World War and then with the PPCLI. He was director of cadets at RMC in 1968, base commander in Europe 1969–71, and director-general of land forces 1972.

**Dr John Robert Dacey (no. HO11132), BSc 1936, MSc Dalhousie 1938, PhD McGill 1940, joined the Canadian army in 1940 and was assigned to research in chemical warfare. He retired as a major and from 1947–8 was chief superintendent, DRB Chemical Laboratories. Appointed to RMC in 1948 as professor, he became dean of science and then in 1967 director of studies and principal. He retired in 1978 but continues his researches at RMC.

The official opening of the Fort
Champlain dormitory: Governor
General Georges Vanier and
CWC Dick Cohen, 1965

afterwards the surgeon-general banned compulsory boxing in the Canadian Forces because it might cause injury. Recruit boxing, which had long been an RMC tradition, was covered by the ban. The order was a blow for many who believed that boxing was an important way to build manliness.[47]*

Many people have questioned the validity of the theory that sports are an essential key to, or a measure of, the production of officers. But periodical lamentations that RMC sports failures demonstrated the decline of the old spirit and the old virtues inevitably added fuel to inflame complaints about the college's alleged lack of effectiveness. In 1965 RMC won eight intercollegiate championships, but in 1966 it had a bad year. Again there was

talk of decline. A year later the situation was reversed: RMC fared well against civilian university competition and confounded the doomsayers.[48]

Another place where some looked to measure the success of the colleges in their task of developing military spirit and competence was at the parade square. "Steadiness on parade," beloved of reviewing officers, and the presentation of fine military spectacles are traditional visible signs of the effectiveness of military training and indoctrination. From time to time, despite the continued excellence of RMC's annual graduation parades, old-timers declared they could see nothing but deterioration in cadet bearing. However, insofar as a parade is in fact a measure of the qualities needed by an officer, the year 1967, the hundredth anniversary of Canadian Confederation, produced some solid evidence. On 26 May, over 500 cadets took part in the trooping of the colours on Parliament Hill, which was reviewed by the governor-general before 8000 spectators.[49] It was a spectacle much more brilliant than anything RMC had ever presented before.

Following the success of the Centennial Parade, the RCAF offered RMC a Sabre aircraft for display on the college grounds along with the Sherman tank and the anti-tank guns already mounted there. Many military officers believe such displays have an important influence in fostering military spirit. The college accepted the gift with pleasure as a means of providing "an RCAF atmosphere in this predominantly military area."[50]

Sports, parades, and a military atmosphere were, however, just the trimmings. What was needed was a sound educational and training program, and room in

Parading on Parliament Hill, 1976. The Cadet Wing paraded on
Parliament Hill in Canada's centennial year, 1967, and again during the
centennial year of the college, 1976.

*Recruit boxing was soon afterwards revived by General Turner because RMC still had army-qualified boxing instructors among its PT sergeants. When the last of these left in 1980, Brig.-Gen. John de Chastelain, then commandant, who had been an intercollegiate boxer, regretfully decided that recruit boxing must cease.

the college to carry it out. At a time when university expansion was in full swing with federal financial support, however, the construction that RMC needed to carry out its academic and military functions adequately was held up by a freeze. All that was built in the 1960s was the Fort Champlain dormitory. With Forts LaSalle and Haldimand, and the Stone Frigate (which had earlier been refaced with Indiana limestone to retain its buff colour), RMC was now able to house almost 600 cadets in single rooms.

The college also received authority to reclaim land across Highway 2 that had formerly belonged to it but which had been used since the Second World War by the local naval reserve unit, HMCS *Cataraqui*, and by the Kingston garrison for a Sergeants' Mess. In accordance with the ten-year plan prepared by the Toronto architectural firm UPACE, that area was now needed for recreational and instructional purposes.[51] But implementation of their recommendations was postponed while forces' integration was carried out. All the impressive progress that the college had made in recent years was hampered by obstacles located in Ottawa. Foremost among these impediments was further dislocation caused by the unification of the forces.

Unification, the Officer Development Board, and Professionalism

In December 1966 Defence Minister Paul Hellyer introduced legislation to move from the integration of Defence Headquarters to a complete unification of the Canadian Forces in a single service. There was an immediate storm of protest. Several senior officers, who held that the characteristics of specific trades in each service were different and who objected to the prospective loss of three separate services with their distinctive traditions, organizations, uniforms, and rank structures, either resigned or were forced out.[1] Despite their publicly expressed opposition, Bill C-90 passed the House of Commons on 25 April 1967.

One effect of Hellyer's unification of the forces on RMC was that it made the name Canadian Services Colleges inappropriate now that there was only one service. The colleges were soon officially renamed what they had often already been called informally before, the Canadian Military Colleges. The authorized version in French was "Les collèges militaires du Canada." The names of the individual colleges were not to be translated. Royal Military College and Le Collège militaire royal (CMR) were unchanged, but Royal Roads became Royal Roads Military College. Discussing these changes, the RMC Club executive noted that the cadet ranks and formations being used in the colleges had been borrowed from the RCAF which now, under unification, had adopted army and American nomenclature. RMC's cadet ranks therefore seemed anachronistic. Someone suggested nostalgically that a return to prewar usage – for example, battalion sergeant-major – was possible.[2] But the ranks adopted in 1948 and used by all three colleges had become hallowed by time. The postwar RMC rank structure remained intact.

Unification would eventually affect RMC in more important ways than college names and ranks because the colleges were intended to be the chief source of graduate officers for the newly unified Canadian Forces, but a local event seemed at the time to have more immediate significance. The session 1967–8, when he was on retirement leave, was to be the last year of the director of studies, Colonel Sawyer. Sawyer had been RMC's main pilot since the decision to reopen first came under discussion after the Second World War. His valuable contacts in headquarters in Ottawa dated from before and during the war. Thereafter, while commandants and staff officers had come and gone, he had been continuously at the helm, steering a steady course towards his goal of academic perfection in a military environment. His personal, almost obsessive, dedication to the well-being of the college in which he had served as both cadet and professor had earned him the soubriquet "Mr RMC." Sawyer died in February 1968 and Dr John R. Dacey, professor of chemistry, was appointed to succeed him. This led to a long-expected change in RMC's structure of command.

In addition to being director of studies, Sawyer had simultaneously been vice-commandant of the college, an arrangement that some military personnel disliked. *Queen's Regulations* stated that, unless the commandant directed otherwise, command in his absence would be assumed by the next senior military officer present at the college. Sawyer had been appointed vice-commandant shortly after the termination of his war service and had retained military status as a reserve officer, being called out to act in the absence of the commandant. It is quite common in the American military academies for academic heads to be second-in-command under the superintendent, but they are invariably regular officers. Dacey had served as a scientist in uniform during the war, but when he succeeded Sawyer he had long been a civilian. It would have been inappropriate for him to command

A going concern since 1953, the RMC Pipes and Drums are now an established part of college tradition. Here Drum Major Clark Little leads the band (and the wing) back to the college on Copper Sunday, 1971.

the college. The appointment of vice-commandant was dropped. Furthermore, in 1964 a special working group had recommended that the appointees in the Cadet Wing and the administration should hold the same rank.[3] Under Commodore W.P. Hayes, a new commandant who arrived in 1967, the successor of the former staff adjutant – now renamed director of cadets – was second-in-command. Later the officer-in-charge of administration was renamed director of administration with a lieutenant-colonel's rank.[4] The two military directors and the academic director of studies (renamed the principal and director of studies in 1972) were then co-equal heads of the three wings of the college. Thus, at a time when changes were imminent, Dacey appeared to the faculty to be in a weaker position than Sawyer to present academic views on college development.

Under this new regime RMC struggled with its perennial problem – the recruiting of cadets in sufficient numbers and of adequate quality to provide the graduate officers that the services needed. Efforts to advertise the colleges by sending uniformed cadets to recruiting centres and schools and by athletic fixtures were stepped up during 1967–8. The RMC Pipes and Drums paraded with a CMR contingent in the Place Ville Marie in Montreal, the RMC football team played three games in Calgary, Edmonton, and Windsor before a total of 13,000 spectators, the hockey and basketball teams took part in the Winter Carnival at Mount Allison University, and 500 students from Hamilton, Toronto, and Ottawa schools visited the college.[5]

Strenuous effort of this kind was becoming more necessary than before. The commandant told the Advisory Board in November 1967 that he expected during the next few years there would be increasing difficulty in securing recruits because the relative financial advantage of ROTP as a means of getting a university education had diminished. Some competing Canadian universities had reduced their entrance requirements, for instance by dropping English as a subject needed for entry to engineering. The military colleges, in contrast, still required English for entry to all their courses.[6] There was also talk of eliminating grade 13 in Ontario and in those schools in British Columbia that gave it. That would inevitably compel the military colleges to change their entrance requirements.[7]

To remove some obstacles to recruitment, a few steps were taken immediately. The director of recruiting reported that prior selection for air crew, formerly required of all air force candidates (many of whom had only one desire in life, to fly), had been dropped. Since there were still twice as many applicants for the air force as were needed, unsuccessful candidates were being offered entry to either land or sea elements, and many were accepting.[8] The establishment of new personnel selection units made the processing of ROTP candidates possible throughout the year. Finally, because many Nova Scotia applicants were weak, especially in the sciences, and because that province had no equivalent of Ontario's grade 13, some of its applicants who could not pass an RMC qualifying examination were being enrolled in CMR's preparatory year.[9]

These changes would not do enough to remedy the chronic shortage of officers, but the first step towards increasing the total number of cadets was to increase the pool of applicants from which cadets were selected. To give new vigour to the recruiting program, RMC suggested the appointment of a full-time admissions officer at each college.[10] Then, a year later, the Advisory Board learned that the situation in a vital area had become more desperate than hitherto realized. Although no professional engineers now entered the forces by direct entry, and few came from the university ROTPs, only 100 engineers had graduated from RMC in 1968, and in the new senior class there were just seventy-five.[11]

CMC recruiting problems were not caused merely by the difficulty of reaching enough high school graduates or by the colleges' determined maintenance of high academic standards. At that time student unrest was threatening to reduce interest in military careers still further. The opposition of some American youth to all things military, which had increased with the extension of the draft for Vietnam to undergraduates, had spilled over into Canada. Brigadier Ben Cunningham, an ex-cadet, wartime field commander, and former RMC commandant, spoke about this current student unrest when he addressed RMC's convocation in 1968. He said that the explanation for youth's attitudes was not just the generation gap or their specific grievances. Student frustrations and university deficiencies did provide ammunition, but youth was being skilfully exploited by dangerous demagogues. He declared that violence was not a solution for youth's problems. Cunningham then praised RMC for its "systematic training for leadership and man management." He said that, after intelligence and strength of character, the greatest attribute of leadership is the power of exact and unambiguous language in mathematics and science as well as in the humanities. He thus endorsed RMC's curriculum with its insistence on academic excellence, as well as the college's standards of disciplined behaviour.[12] But Cunningham did not have a remedy for the shortfall in recruiting. The root of the problem there was that too small a proportion of the population was willing to accept responsibility for defence, and at present that proportion was declining.

RMC could do little or nothing about youth alienation outside its walls, but it could, and did, take measures to ensure that, once admitted, cadets were handled appropriately for the circumstances of the new age. Several years earlier, members of the faculty led by Dr G.F.G. Stanley, dean of arts, had strenuously opposed a military proposal to restrict leave privileges of cadets who were weak academically. They had argued that such measures would seriously decrease the cadet's sense of personal responsibility.[13] When Hayes arrived in 1967, he and

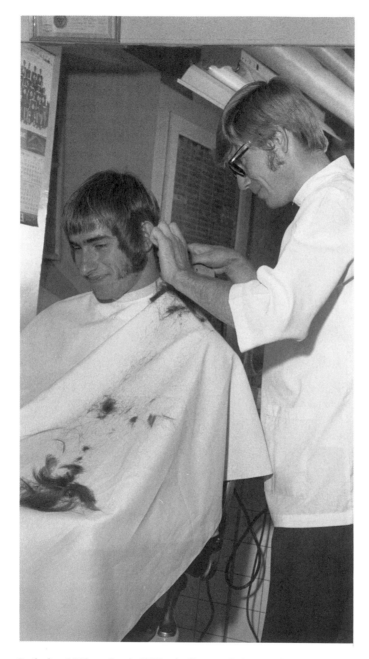

In the late 1960s and early 1970s, the first RMC haircut made a real difference.

Lieutenant-Colonel Alan Pickering,* the first director of cadets, introduced various changes in the Cadet Wing that were deliberately designed to foster cadet responsibility and maturity.

First, in order to give more opportunity for experience in command by the rotation of cadet appointments from one term to the next, Hayes planned to alter the wing organization from five squadrons to seven, to take effect in 1968–9. He also put the recruits in each squadron together in a separate flight.[14] To stress that command should be based on personal leadership rather than authority, he ordered that in the first two weeks of the session, cadet officers should not use their power to punish new recruits. The seniors responded to this very well.[15]

What was more innovative was that Hayes also introduced a number of new privileges for the senior class; they could wear civilian clothes off duty, possess and operate cars during term, have bar privileges, and purchase liquor under certain conditions. The former Sergeants' Mess across Highway 2 now became an RMC Cadet Mess for all years. This decreased internal hierarchical segregation. Second- and third-year cadets were permitted to wear blazers and flannels off the station (a privilege that had been given to fourth-year cadets by Commodore Piers), and their leave could extend from the end of classes until morning parade.[16] The right to wear civilian clothes in Kingston was especially pleasing to senior cadets.

*Maj.-Gen. Alan L. Pickering (no. 2908) graduated from RMC in 1953 and from Queen's University (BSC) in 1954. He became a test pilot on the RCAF Yukon aircraft and chief engineer on the Gemini Project with USAF-NASA Space Systems Division in California. He returned to RMC as air staff officer in 1967. In 1973 he went on the senior course at the US Naval War College. He was chief of intelligence and security and then, as a civilian, was with the Communications Security Establishment in Ottawa.

The most radical change of all was a relaxing of the regulation that cadets must remain single until they graduated. It was well known that in recent years a small number had broken this rule, but they had been deliberately overlooked until after graduation. Hayes was disturbed, for there were circumstances, he said, in which it would be dishonourable for a cadet not to marry. He therefore approached DND to ask the CMC General Council for a ruling that the commandant could approve the marriage of cadets in their senior year. He asserted that otherwise he would recommend the release of those cadets who were married, at least one of whom was an outstanding young man. He got his way.[17]

A few ex-cadets saw these measures as signs of decadence, but others believed that in adapting RMC's traditional discipline to make it conform more closely with modern realities, they served to facilitate the preservation of the college's traditions and service. Most agreed that change was both inevitable and necessary. This was a positive revolution in RMC's military training system and not, as many hostile critics said, a surrender to "liberals." These changes were introduced by a commandant from the navy, usually regarded as the most hidebound of services.

About the same time, as the director-general of individual training programs (DGITP), Commodore R.H. Leir,* told the Advisory Board in 1968, the task of coordinating military and academic excellence at RMC had become more difficult.[18] As a result of technical development, military training was more complex. Summer training, conducted much as before in separate land, sea, and air elements (except in the first year, when land and air training was undertaken jointly), was too short to be fully effective. Furthermore the summer program, now nine weeks long, imposed a heavy burden on the training staff.[19] One remedy that had been proposed was to move to a year-round operation in three semesters, abandoning the traditional adherence to university terms. This course was not adopted, partly because field training

Casey's grave: From their arrival, recruits are expected to recite, on demand of the seniors, a variety of RMC facts, figures, and trivia. The names of the "Old Eighteen," the inscriptions on the Memorial Arch, and the inscription on Casey's grave are favourites.

would be impossible in winter in Canada, and partly because a tri-semester system would have disrupted traditional university patterns for combining teaching and research. A more fruitful approach to solving the forces' officer shortage, especially of engineers, must be more effective recruiting; and it was appreciated that RMC's insistence on university standards, practices, and terms was an essential element in attracting academically qualified recruits as well as faculty.[20]

---

*In 1968 the appointments of director of training and director of officer cadets were abolished and replaced by a director of military and common training under a director-general of individual training programs. R.H. Leir was the sole appointee before there was further reorganization, and the next year he became director-general of personnel production. A midshipman in 1944, Leir had been Captain (N) of the *Venture* training school. He was promoted commodore in 1967 and became a rear-admiral in 1972.

By this time RMC's academic quality was widely established and well known. During the 1960s, RMC's innovations of the early 1950s, with its many specializations, had been vindicated when the American academies departed from their traditional lock-step curricula and began to offer a wide variety of options.[21] In 1967 RMC's introduction of graduate courses, traditionally the hallmark of university status, was recognized when the Committee of University Presidents of Ontario invited RMC's commandant to join its numbers.[22] Commodore Hayes and Dr John Dacey were the first to attend. As a result, they were made more aware of contemporary lavish provincial spending on university education at a time when CMC budgets were frozen. Meanwhile, claims of inadequate funding notwithstanding, RMC's own faculty had enthusiastically advertised the college's quality and innovations. One wrote in a British military journal that RMC was "universally acknowledged to be the West's most enlightened military college, offering a balanced military education in engineering, science, and the humanities, extensive research facilities, a high-class professoriat, and a graduate studies division."[23]

But RMC still needed to innovate further to maintain the academic reputation on which its recruiting heavily depended, to keep abreast of its peers, and to adapt more closely to military requirements. It therefore modernized its curriculum in the area long regarded as all-important in military academies – mathematics. Dr N.K. Pope,* head of the RMC mathematics department, briefed the RMC Club executive on the installation of a mathematics laboratory which, he said, was "probably the most sophisticated . . . [of any] in a university in Canada." Sixteen electronic calculators offered flexibility for statistical and scientific calculations. Four underdesk electronic packages, separate from the keyboard units, served four keyboards each. There was a card program for repetitive or particularly complex problems. When used in conjunction with the card programs, the keyboard became a miniature desktop computer. RMC could

use the keyboard and card programs to let students devise their own programs. The new mathematics laboratory cut down the workload on the RMC computer, but its most important use would be for students.

At that time, recruits who came in from the high schools were unfamiliar with even the most rudimentary electronic calculators. Dr S.D. Jog** of RMC's mathematics department devised a six-page manual on the correct use of the keyboard. Telephone-sized calculators and card programs saved students much time, and electronic data-processing facilities helped their work in other subjects as well. Furthermore, cadets could now gain experience with equipment they could use after graduation with their units.[24] RMC thus saw early that electronic processing had enormous potential for military purposes, and it moved into the computer age ahead of many of its civilian contemporaries in Canada.

Other changes in the academic programs were even more closely related to military needs. The graduate school had twenty-seven students by 1968, twenty-two in engineering and five in war studies.[25] RMC's graduate war studies program was impressive. At the same time a course in the undergraduate curriculum on the history of warfare, which had stressed its relation to political, economic, social, and technological development, was now replaced by one on modern strategic thought, hitherto regarded as more suitable for graduate students. The new RMC undergraduate course, engineering and management, which had obvious military application, was proving popular. Military leadership and man-

---

*Dr Noel Kynaston Pope graduated from the University of Canterbury in New Zealand in mathematics and physics. After taking a PhD at Edinburgh, he was a fellow at the National Research Council in 1949, working at Chalk River. He was appointed at RMC in mathematics in 1962 and became head of department in 1967, retiring in 1982.

**Dr Shridar Dattraya Jog was appointed at RMC as an associate professor in mathematics in 1967. He was head of the Department of Mathematics and Computer Science 1981–5.

agement, which emphasized sociological and psychological factors and which had replaced the old military studies, was taken more seriously than its predecessor.

These changes were not introduced without some friction between academic and military interests. When the new Department of Military Leadership and Management wanted more time, it was given an extra hour taken from drill. The Military Wing admitted it would not have similarly released an hour from drill for any purely academic subject that needed extra time. Some members of the Faculty Board were concerned that, if drill standards declined, an additional hour of drill might be reintroduced at the expense of study time. A former ex-cadet naval officer on the faculty and future principal, Dr John Plant,* said he regarded the military leadership and management courses not as military, but as academic.[26] The point was well made. The coincidence of military and general education at RMC was high.

At the root of this debate about use of time was a realization that the cadet's workload was too heavy. It was agreed by Ottawa and RMC that something must be done about it, taking care not to lose the standard of excellence that now prevailed. A 15 per cent reduction was made in the workload of the second year by reducing its content of science courses.[27] A year later steps were taken to eliminate the spring term which had been introduced before degrees in engineering were first offered.[28]

Meanwhile, Hayes had initiated a time-utilization study to establish precisely "the *actual* time now spent by cadets on various endeavours." He arranged for a representative sample of cadets to hand in a detailed record of what they did each day on a twenty-four-hour, seven-days-a-week basis. His Time Utilization Study Group came to the conclusion that recruit-time involvement in kit and room cleaning and Cadet Wing administration was excessive, that an extra hour of sleep in the morning would be advantageous for them, and that "senior cadets must not only set an example [in standards of personal appearance], but must also use considerable discretion

in the . . . standards expected from junior cadets," which must be "realistic." The group also found that RMC cadets took more part in intervarsity sports, and spent less time in intermurals, than did cadets who had come from Royal Roads and CMR, either because those colleges had put more stress on "round-robin inter-mural activity," or because cadets from those institutions had a fear of academic difficulties at RMC.[29] Hayes and his immediate successor, Brigadier-General W.K. Lye,** were able to use these findings as a basis for inaugurating further change.

Some time before this, however, behind the closed doors of the Advisory Board meeting, the Canadian Armed Forces' director of training had offered what he called a "subjective judgment" of the product of the services colleges. He reminded his audience of the well-known wartime exploits of RMC's prewar graduates and of the many prewar ex-cadets who had reached high rank in the forces after the war. He added that some of the postwar crop had now attained the rank of colonel. "This, to me, would seem an impressive record which we should keep in mind." But he warned that it was necessary to ask whether the colleges were graduating the kind of officers that would be most useful to the services, to discuss further the need for degrees, and to seek for a definite philosophy of military service and the most suitable form of organization to provide the necessary education.[30]

Brigadier-General W.K. Lye, MBE, CD, commandant, 1970–3

*Dr John B. Plant (no. 3948) entered the college in 1953 and was commissioned in the RCN in 1955. He attended the Royal Naval Engineering College at Manadon, Plymouth, and MIT, where he received a doctorate. He became lecturer in engineering at RMC in 1965, dean in 1972, and principal in 1984.

**Brig.-Gen. William Kirby Lye (no. 2530) was at RMC 1936–9, leaving for active service in the Royal Canadian Engineers. He graduated BSc, civil engineering, from Queen's University, was on the directing staff of the National Defence College, and deputy commander Canadian Forces Middle East 1957–8. After commands in the Maritimes and a chief of staff appointment in Montreal, he was director-general operations at CFHQ in 1970. Commandant of RMC 1970–3, he subsequently became director, Physical Plant Department, University of Toronto.

It was now widely believed that Canada's Armed Forces needed a definition of purpose in the form of a distinctive philosophy of military professionalism. This was partly a consequence of the country's military development. Down to the Second World War the bulk of Canada's forces were part-time reserves, and Canada had inevitably accepted British organization, concepts, and objectives. After that war, although Canada had established larger regular forces and had set up a National Defence College and staff colleges to give intellectual guidance to their approach to their profession, it had not yet developed its own concepts of a national strategy, largely because its objectives for Western and continental defence were now largely borrowed from the United States. Furthermore, within the pattern of NATO defence the three services each had different, and unrelated, tasks and, as a result, philosophies. The government had accepted a large number of international peacekeeping tasks that had added a dimension of a different kind but did not meet the desire for something that was distinctively Canadian. Finally, unification was not a completely satisfying answer because so many disliked it. How far a peculiarly Canadian approach to military organization and development was possible was still a moot point, but the desire for one was real.

The "admirals' revolt" against unification had led a staff college professor to point to the need for a *contemporary* and *Canadian* professional ethic that would apply to the now unified forces – a definition of responsibilities, competence, conduct, and loyalties that would sustain members of the Canadian Forces under all conditions of service. The RCAF Staff College's James Jackson reminded his readers in an article in *Saturday Night* that traditionally the "custody," that is to say the definition and dissemination, of such an ethic was the function of military colleges and staff colleges.[31] Jackson said that it followed that a distinctive Canadian military philosophy should take account of the country's unique national characteristics as well as its international status, aspirations, and obligations. Such a philosophy would become the basis for a Canadian strategy and for the Canadian military education to prepare for it. Some writers thought it might also be a model for other countries.[32]

General Allard, the second chief of the Defence Staff of the newly unified Canadian Forces, realized it was necessary to define what an officer corps that had formerly belonged to three distinct services held in common so the three could be welded into a single force. He asked Hellyer to consent to the creation of an Officer Development Board to investigate the way in which all Canadian officers should be educated and trained throughout their careers.[33]

In 1968 an unusually frank article by a serving officer in the yearbook of the Canadian Land Forces Command and Staff College, *Snowy Owl*, talked of frustrations caused by lack of a definition of military professionalism, by decline of budgets, by integration, by increasing neutralism in the country, by the conflict between dependence on the United States and a desire to retain a Canadian identity, and by the futility of the role that a middle power like Canada could play in a bipolar world dominated by the "balance of terror." The author, Major G.A. Zypchen, RCE, a student at the staff college, said the military should be capable of translating the general philosophies imposed by civil governments into workable rules and policies. He argued they could no longer do that because they were working in a vacuum as a separate entity divorced from political realities. He said a Canadian officer's political and professional education to meet this need now began too late in his career, and concluded, "It does not appear that the Canadian military has managed to create and maintain an ethos for itself."[34] Zypchen's call for general education for officers earlier in their careers was in fact being met by the academic programs in the military colleges, but only if those programs were really appropriate for fostering the Canadian professionalism he had alleged was lacking.

Brigadier-General W.A. Milroy,* commandant of the staff college, asked a former member of the directing staff, Colonel A. Strome Galloway,** recently military attaché in Germany, to reply to Zypchen. Galloway argued there was a great deal of educational development in the Canadian Forces, but there was a danger in peacetime that it would lead to a narrow professionalism. He asserted that when professionalism was equal on both sides in time of war, things of the spirit triumphed. Time must therefore be spent to develop the Canadian officer's moral, intellectual, and spiritual forces for eventual confrontation with an enemy.[35] This answer did not fully meet the criticisms.

This exchange of views revealed the need for what the manpower group had called a "career philosophy for Canadian officers." Both officers had assumed this would come from academic education. The controversial and still unresolved question was whether an academic program that placed emphasis on freedom of enquiry could serve the purposes of inculcating the Canadian military philosophy that many sought. Lack of clarity in defining a military philosophy was related to curriculum development and procedures in the military colleges as well as the content of officer education generally.

One problem in the colleges was that, despite an elaborate system of reporting on cadet military performance, methods of assessing military qualities contrasted sharply with those used to measure academic progress. Subjective assessments of officer-like qualities had to be set against hard college grades. At the same time, complaints about high academic barriers in officer production often cloaked a desire to secure more officers by pushing through young men of limited ability, or with limited capacity for application, who appeared to have good officer potential. What was needed in the colleges as well as in the forces was consensus on the qualities that would make a Canadian professional officer. Defining professionalism was one of the tasks that Allard set the Officer Development Board (ODB).

The ODB was chaired by Major-General Roger Rowley,*** who was responsible for developing the idea. He was assisted by Lieutenant-Colonel L. Motiuk,† Lieutenant-Colonel A.D. Wallis,†† and Major J.A. Annand.††† Dr T. Hutchison,§ professor of physics and dean of grad-

*Lt-Gen. William Alexander Milroy, squadron-commander in Lord Strathcona's Horse 1941–5, was director of military training 1962–5, commander Canadian Defence Educational Establishments 1970–1, commander Mobile Command 1972, and ADM(Per) 1973–5.

**Col. Andrew Strome Ayers Galloway served in the United Kingdom, Tunisia, Sicily, Italy, and Northwest Europe during the war. He was on the directing staff of the Canadian Army Staff College, Kingston, 1951–4. He has published fictional and other books.

***Maj.-Gen. Roger Rowley, commissioned in the Cameron Highlanders of Ottawa in 1933, commanded the Stormont, Dundas, and Glengarry Highlanders in Northwest Europe. In 1944 he was DMO, in 1950 director of infantry, in 1958 commandant of the Army Staff College, and in 1962 vice-adjutant general.

†Lt-Col. Laurence Motiuk, a squadron-leader radio navigator in the RCAF with a BEd degree, prepared a comprehensive *Reading Guide for the Study of War, Defence and Strategy* as a Centennial project and published annual supplements thereafter.

††Allan D. Wallis (no. 2854) entered Royal Roads in 1948 and graduated from RMC in 1952. Commissioned as a pilot in the RCAF, he served with the International Commission for Supervision and Control team in Vietnam in 1962. In 1966 he did graduate work at the University of London and was on the staff of RMC as deputy squadron commander 1959–62. Later be became professor and registrar at Royal Roads.

†††Lt-Col. John A. Annand (no. 4154) entered Royal Roads in 1954 and graduated from RMC in 1958 and from Queen's University in 1959. He served with the Canadian Guards at Camp Petawawa and Germany and was a squadron-commander at RMC in 1963 and director of cadets at RMC in 1977. In 1983 he was senior Canadian liaison officer with the US VII Corps at Stuttgart, Germany. In 1988 he was appointed to command Canadian troops on truce-supervisory duties on the Iran-Iraq border.

§Dr Thomas S. Hutchison (no. HO13826) graduated in physics from St Andrews University in 1942 and served with the British Admiralty, torpedo testing, 1942–5. In 1947–50 he was with the UK Atomic Energy Authority. Appointed associate professor of physics in 1952, he became professor and head of the department in 1954, chairman of the science division 1959–62, dean of graduate studies 1962–72, and dean of science 1972–80. He received an honorary DSC at Queen's University in 1985.

uate studies at RMC, was seconded to ODB as a professional adviser on education. The board was commissioned to cover the whole process of officer education from recruitment to senior rank. Since its task would take at least a year, the immediate effect of its establishment was to postpone still further any consideration of RMC's physical development – in particular, the construction of the science and engineering building necessary to fulfil its prescribed role.[36]

Rowley had been commandant of the staff college but was neither an ex-cadet nor a former member of CMC staff. Allard instructed him to consult with the CSC Advisory Board about the colleges,[37] but one of Rowley's first steps was to approach the RMC Faculty Board to explain his objectives and ask for their suggestions. He promised the RMC faculty it would be consulted as the board pursued its investigations. He reported he had been instructed to ensure that, while there was to be no decrease in the environmental skills of the soldier, sailor, and airman, they must all become potentially able to command in other environments as well. With the nuclear stand-off, he said, conventional and limited warfare was being re-emphasized. Canada had only two military commitments – to United Nations peacekeeping and to the North Atlantic Treaty Organization. In view of Prime Minister Trudeau's later emphasis on Canadian sovereignty and interests, it is significant that Rowley put those commitments in that order and that he did not mention NORAD and domestic responsibilities. He listed six areas for ODB investigation, including a rationale for standing forces in Canada, the requirements of a profession of arms, retention rates, career development, and a control system for officer education and development. He noted that no Canadian had ever before "waxed philosophical" about the Canadian Armed Forces, and he announced that he would develop not merely a rationale, but also a philosophy for the profession of arms in Canada.[38]

ODB was specifically required to recommend measures to offset the premature loss of many highly trained young officers. Rowley told the Faculty Board that one of the possibilities he had suggested to the minister to remedy the shortage of officers was a lowering of the basic minimum academic qualification for an officer to grade 11. When faculty members expressed concern that this suggestion threatened a serious drop in standards, Rowley claimed he had only advanced this possibility to help secure Hellyer's agreement to the establishment of the board.[39] ODB was also instructed to consider the question of expanding Royal Roads and CMR to four-year colleges as a means of eliminating the shortage of officers.[40] The long-awaited decision on that question was thus put off for at least another year.

Rowley's *Report of the Officer Development Board*, printed in three volumes in March 1969 but not circulated generally in official circles until some time later, began with a philosophical discussion of "the emergence of military professionalism and the military ethic." This was innovative in that for the first time a distinction was made between those two concepts. It was, perhaps, an attempt to resolve two serious dilemmas that have plagued the world since the introduction of nuclear power: first, that military power, to function properly, must be credible, yet carried to its extreme form by superpowers it could lead to universal destruction; second, how can military power be exerted by international rather than national agencies to resolve this nuclear dilemma when national states are still sovereign?

Rowley made a sharp distinction between military philosophy and military ethics. Military philosophy assumes that conflict is universal and inevitable and places emphasis on constant preparedness for war. It can be marked by irresponsible enthusiasms and can appear to glorify violence. Military ethics, in contrast, while embodying the military virtues of ordered power, discipline, and restraint, introduce a moral element and a code of moral values. They also recognize the supremacy of the civil power.[41]

Rowley's argument was one that the American military sociologist, Morris Janowitz, had partly anticipated by distinguishing between "absolutist" and "pragmatic" schools of military thought.[42] But Janowitz did not use the phrases "military philosophy" and "military ethic" in this connection, or associate his differentiation with conflicting philosophies for the individual soldier, as did Rowley. Rowley and Janowitz were probably more concerned with military policy than with the intellectual bases for individual behaviour. Rowley's concept of a military ethic proposed imposing on an individual soldier a personal obligation to make moral judgments and establish patterns of conduct that could come into conflict with some of the basic principles of military authority.

Although Rowley promised a rationale as well as a philosophy and an ethic for the Canadian military profession, he did not spell out a policy on which Canadian military ethics should be based, other than to confirm the government's strategy as stated by Prime Minister St Laurent in the Gray Lecture at the University of Toronto in 1947. St Laurent had defined the principles of Canadian policy as the maintenance of national unity, belief in respect for the rule of law, and acceptance of international responsibilities. Rowley added that Canada's postwar policy had also been to promote international multilateralism in order to attain an effective voice for a middle power, thus attempting to limit somewhat the dominance of the great powers in all aspects of international affairs.

Neither did Rowley make any reference to Cold War strategy.[43] Instead, he said there was a special Canadian context within which Canadian policy must operate – this included bilingualism and biculturalism and equality of opportunity.[44] He turned mainly to domestic Canadian considerations to show what the distinctive moral features underlying a Canadian military ethic and policy should be. He appears to have had an underlying assumption that Canadian policy is more ethical or moral than that of some other countries. Here was, perhaps, an implication that the policies of the superpowers with their own special interests were not necessarily beneficial to all.

With regard to officer development, Rowley accepted as a premise that a career officer needs a degree, "whether or not in the line of his profession . . . because it has become the symbol of a profession." He said that degrees were also necessary for technical skills. The basic requirement for all officers must therefore be the baccalaureate. The officer corps must also possess skills in the field of management. Junior officers must have an understanding of the behavioural sciences, and senior officers of government. Rowley stressed that "mind training" was fostered in the educational process, whether in history or electrical engineering or any other field, but he added that the subjects studied at a military college should also have meaning for the practice of the military art. He noted that some people thought the degrees needed by officers should have "military" in their titles, but this would jeopardize academic acceptance. Finally, he said there should be a mandatory course pattern. Civilian universities would not provide that, but military colleges could; and at the same time they could also indoctrinate.[45]

From the first Rowley assumed that the Canadian Military Colleges would be the chief source of regular officers for the Canadian Forces. The findings of the Officer Development Board, except for its call for a mandatory program (if by that it meant a single common program), had in fact largely confirmed the RMC curriculum and the way in which it was developing. But Rowley's board favoured year-round operation and the creation of a "single military college."

In April 1969, just after the *ODB Report* was printed, the Canadian Press circulated an article about the wastage rate of military college graduates after they completed their obligatory service. Based on specific questions and without further elucidation, this report was published by the *Globe and Mail*, the *Montreal Star*, and the *Ottawa Citizen* under headlines described in an RMC

Club executive meeting as "misleading, inaccurate, and unfair" and not supported by the text of the article. Members of the executive wondered what would be the impact on those Canadians who only read headlines. They were informed that eight civilian members of the RMC faculty had replied to the article to challenge it, but their reply had not been printed. That reply was therefore published in the RMC Club *Newsletter*. The professors said that wastage rates in first employment in other fields, including in the civil service, were just as high as in the forces. It added that the so-called "free education" the officers who left the service had received was not lost to the country.[46]

This report of alleged wastage and cost, coming before the government had time to consider and act on the *ODB Report*, did nothing to help solve the military educational problem, the shortage of officers, or the difficulties of the CMCs, but it did warn against further increase in the cost of producing officers. The present system was already assumed by many to be more expensive than production through university ROTPs. The expensive reorganization into the new centralized institution that Rowley proposed would obviously encounter great opposition.

By November 1969 General Rowley, due for retirement, had been replaced as chairman of the Officer Development Board by Major-General Milroy, former commandant of the Canadian Army Staff College. It was Milroy who presented the case for the report's implementation to a special meeting of the CMC Advisory Board. Milroy argued that only a single, centralized military college could provide the bilingual program, flexibility, increased output, and coordination of cadet-officer and postgraduate programs that were needed. He said that the Ottawa-Hull area was the only appropriate site for a Canadian Defence Education Centre. What would be set up in Ottawa would eventually become a "University of the Canadian Forces," with federal au-

thority to grant degrees. He added that by 1 January 1970, the headquarters, now being organized, would be put in control of all officer development, and in due course it would control the military university.

Phase 1 would have included an early consolidation of all post-commissioning officer education and training in Rockcliffe, Ottawa. For the immediate future, however, undergraduate education was to continue in its present various locations. To limit the expense of the expansion the colleges so urgently needed, facilities such as Fort Frontenac, vacated by the Land Forces Staff College and the National Defence College, could be made available to them. Phase 2 would be the eventual move of the military colleges to the Ottawa-Hull area. Milroy noted that Rowley had estimated the cost of all these changes at $60 million starting in 1973.[47]

The Advisory Board of the Canadian Military Colleges unanimously approved certain parts of the *ODB Report*, notably the necessity for a unique Canadian professional military education, the continuous upgrading of officers by an extension program that included full-time studies with a view to making the baccalaureate degree the basic standard for an officer, and the desirability of having a single executive and administrative entity to carry out the Canadian Officers' Development Programme. But it foresaw some difficulty in offering degree-level education to presently serving officers, and it doubted whether there was a need to centralize the military colleges in one location with their consequent disappearance as separate entities. Some members also questioned whether a federal charter to grant degrees would be constitutional for such an academy. Other members asked about the possible reaction of the teaching and administrative staffs in the colleges if they were required to move to a new environment, and whether centralization in one location might not reduce present advantages for regional recruiting. The board concluded that the estimated cost, which it said was "after all, only a small part of the initial cost," might in the long run bring only questionable advantage when considered in relation to the potential of the present system.⁴⁸

These serious challenges to some of the basic recommendations of the *ODB Report* seem to have caused misgivings in Canadian Forces Headquarters. Some members of the Advisory Board suspected there would be an attempt to suppress, or by-pass, their opinions. Within a month, for instance, the incoming chairman of the Advisory Board, H.D. Smith from Nova Scotia, wrote to the minister, Leo Cadieux,* to express surprise and disappointment that no minutes of the special meeting that had discussed the report had yet been circulated. He had therefore taken it upon himself to "ferret out the minutes from Headquarters," make a résumé of their contents, and send copies to Cadieux, Milroy, and Professor J.L. Corneille,** the retiring board chairman. Smith's résumé concluded by saying that some of the Advisory Board felt that the present system should be kept as strong as possible and should not give way to a new idea unless that idea offered definite and overwhelming benefits. He said that "caution was expressed that the present Colleges not be allowed to deteriorate while a final decision on their future is being sought."⁴⁹

The *ODB Report*, of which so much had been expected, thus seemed to have little immediate prospect of affecting RMC in the immediate future. In his autobiography Allard blames Brigadier-General Bernard J. Guimond,*** whom he had appointed to the committee before Rowley took over, for the failure of his plan to centralize staff colleges in Ottawa. He says that Guimond exceeded his instructions by including the CMCs. Allard also alleges that the traditionalist RMC Club organized the opposition against reorganization, thus destroying the whole plan.⁵⁰ But that was not the reason for ODB's failure in this portion of its report. One obvious cause was that Rowley, who had received no specific instructions but had been told to "devise a system," assumed he could recommend the ideal system for officer education without regard to its cost. He himself believed the cost of his proposals was not unreasonable, but the government was averse to increasing its budget on the personnel account.

ODB's impact on RMC had thus been primarily to delay the college's possible contribution towards an immediate easing of the shortage of officers. What remains to be seen was whether Rowley's ambitious, sagacious, philosophical, theoretical study of military professionalism would have any long-term influence on the college's future development and its military professionalism.

*Leo Cadieux was in army public relations 1940–4. He was elected MP in 1962 and was minister of national defence 1967–70 and ambassador to France 1970–5.

**Professor J.L. Corneille of École polytechnique, Montreal, was a member of the Advisory Board for Quebec.

***Maj.-Gen. Bernard J. Guimond was commissioned in the Royal 22nd Regiment in August 1939 and served in Sicily and Italy and in the occupation of Germany. After the war he served in the Congo, Indo-China, as military attaché in Paris, and with NATO. He retired as chief of staff, Division d'instruction francophone, at Quebec City.

The Canadian Defence Educational Establishments and Canadian Military Professionalism in RMC

On 1 January 1970, while Commodore W.P. Hayes was still commandant, the Department of National Defence transformed the ad hoc Officer Development Board into a permanent agency, the Canadian Defence Educational Establishments. CDEE was expected to carry out General Rowley's proposal to centralize in the Ottawa-Hull area all officer education, including that in the Canadian Military Colleges, and then supervise it. The team of officers that had produced the Officer Development Board report was kept together to introduce the changes that ODB had recommended. Military college relocation was to be part of a major effort to strengthen professionalism in the Canadian Forces.

By that time it had become clear that something more than cost was an obstacle to fulfilment. The *ODB Report* had conformed with the defence policy enunciated by Paul Hellyer in his 1964 white paper. Hellyer had outlined Canada's objectives in foreign policy as follows: fulfilling her international responsibilities, supporting NATO and NORAD, and providing for certain aspects of security and protection within Canada, in that order. The early chapters of the report, written while Prime Minister Trudeau was developing his philosophy for the government of Canada into an outline of policies during what Desmond Morton has called a time of "transition and ambivalence," had followed the white paper in emphasizing Canada's international responsibilities.

However, in April 1969, one month after the report was printed, Trudeau reversed the order of priorities announced by the Pearson government in 1964. In a statement to the House of Commons he put protection of Canadian sovereignty by surveillance of territory and coasts first, followed in turn by cooperation with the United States in the defence of North America, then by fulfilment of NATO obligations in ways to be agreed upon with allies, and, finally, by the performance of such international obligations as Canada should from time to time accept. Trudeau's approach to foreign policy and defence, though it did not state priorities in relation to the resources available, thus called for increased attention to Canadian sovereignty and interests.[1] The authors of the *ODB Report* had failed to anticipate this change, which was being developed while they worked. Although a strong thread of Canadianism pervaded their work, its early stress on Pearsonian internationalism made their report seem out of line with Trudeau's concern for Canada and North America.[2]

Time was to show that the so-called Trudeau revolution in foreign policy was not the complete change of direction it at first seemed. Meanwhile, however, Rowley's work appeared to have been undermined. Although this thorough study of the technical details of officer production still had much relevance, the report had lost political credibility. Possibly it was this, as well as its unwelcome financial proposals, that delayed its wide distribution until early in 1970, some time after it was printed. In the long run, however, its conclusions were to influence Canadian military education profoundly, because its main provisions stressed upgrading officer academic standards to the baccalaureate level, military professionalism, and bilingualism.

ODB had not immediately relieved the disquiet about the lack of Canadian military professionalism that had been apparent in the Zypchen-Galloway exchange, but the fate of its report had shown that, in one respect at least, Zypchen had been correct. There was inadequate communication between military and political leaders. The senior officers responsible for directing the board had been out of touch with political developments and were unable to influence, or be influenced by, current political thinking. Furthermore, despite the originality of Rowley's contribution to an intellectual understanding of military professionalism, ODB had not advanced a simple formula to solve the complicated problems of professionalism and officer production. What it had proposed was a centralized military educational institution, including a Board of Governors for the military colleges, and that was not to be.

Insofar as the establishment of CDEE partially fulfilled some of Rowley's intentions, it was actually only carrying on with a trend already set. Unification of the forces had led to extensive administrative reorganization in Ottawa. By 1968 the separate Directorates of Training and Officer Cadets had been replaced by a Directorate of Military and Common Training under a director-general of individual training programs.[3] This particular arrangement was soon revised again, but it shows that, when it set up ODB, Ottawa was already moving towards centralization of the control of the military colleges and of officer training and education. CDEE thus provided what Ottawa had probably intended from the first – tighter control of military education. The question that remained was whether this decline in independent academic influence in military education would bring greater Canadian military professionalism.

In 1970, as the head of CDEE, Major-General Milroy was responsible through the chief of personnel to the chief of the Defence Staff for all officer education in the Canadian Forces. With his headquarters in the Department of National Defence in Ottawa, he had supervision of the National Defence College, the staff colleges and staff school, and the Canadian Military Colleges. At the same time he was staff adviser to Canadian Forces Headquarters on officer development. He admitted he found this dual position difficult.[4]

To complicate matters for him, his command was not a normal military one. He had problems from below as well as from above. Supervision of a military college that was a university could not be a straightforward military command. Hence, when CDEE was set up, Milroy's position was challenged by his new subordinates. In November 1969 Dean Tom Hutchison, who had worked with ODB, informed the RMC Faculty Board that a reorganization in the chain of command was imminent. The RMC Faculty Board then requested Hayes to ask CDEE to recognize its expertise and its concern by consulting it in the planning for RMC's future that was now underway. This request received a favourable hearing.[5] Accordingly, in March 1970 the Faculty Board appointed a committee chaired by Professor David Baird* to study legal and technical considerations relating to the fact that "this degree-granting institution is [now] part of a government department."[6]

From the point of view of the RMC commandant, the appointment of a commander of CDEE at first seemed to imply a setback for the military colleges. Hayes states that

*David Carr Baird, professor of Physics and chairman and dean of the Science Division. With a doctorate from St Andrews University in temperature physics, he came to RMC as instructor after completing flying training with the RAFVR 1949–52. He was head of the Department of Physics 1972–8 and dean of science from 1980. In 1987 he was made distinguished professor.

when he commanded Royal Roads he had been appointed by a letter from the chief of the Defence Staff, he had reported to the Personnel Members Committee, and he had direct access to a four-star general that gave him a protective umbrella. Under the new administrative organization under integration, as commandant of RMC he found he now reported only to CDEE, a two-star general, which made interference in college affairs by "many more levels of the hierarchy" possible. Furthermore, not merely had the RMC commandant become subordinate to CDEE instead of to the chief of the Defence Staff through the chief of personnel, but around the same time a Canadian Forces Organization Order ruled that RMC was to be a lodger unit in Canadian Forces Base Kingston. Hayes complained in 1970 that this change weakened the college's ability to maintain its function as an institution of higher education, and its ability to carry out its mission to produce highly educated officers for the Canadian Forces.[7]

Relocation of the Canadian Military Colleges in Ottawa, if carried out, would probably have negated this local problem about the base, but the amalgamation of the three colleges in a new location might possibly have opened the way for radical changes, including a complete restructuring that might have included academic curricula. At the same time it could have threatened cherished traditions that RMC had nourished for a century in Kingston as a basis for officer production and which the two newer colleges had in large measure adopted. Those traditions were natural bases on which to develop military professionalism in an academic program. Furthermore, a location for the restructured consolidated colleges close to the Department of National Defence might have made them more vulnerable to the whims of their military or political masters, and so to more civilian, administrative, and political interference. Relocation in Ottawa might thus have increased some of the dangerous aspects of centralized control.

In February 1970 Milroy requested the revival of planning for construction at the three military colleges "in their present locations."[8] In his end-of-tour report in June 1970, Hayes, summing up his ideas for RMC's further development, said he was reporting "now that a decision has been reached about ODB's recommendations." He stated that the views he presented about RMC's future reflected the opinion of the college as a whole. Under a centralized system of control of officer education, college policy would eventually be determined by a Board of Governors made up of representatives from the Defence Council and from educational institutions across Canada. This would constitute a considerable break with RMC's past autonomy and would threaten its distinctive identity. It would set up a governing mechanism designed to respond to pre-commissioning circumstances (presumably meaning the adjustment of entrance qualifications) and to post-commissioning requirements (the needs of the services). He concluded that with those concerns the board would be unlike the traditional Board of Governors of a civilian university.

Hayes therefore recommended the retention of college involvement in its own long-term government to maintain RMC's very real academic professionalism and its future viability. At the same time, he said that the RMC degree-granting power could be extended for wider use in the forces generally. The college faculty would welcome an opportunity to undertake postgraduate education and extension work in order to play a wider role in the education of officers in the Canadian Forces at large. On another theme, he argued that the position of the RMC commandant was anomalous because he was a "lodger-unit commander" on the Kingston base, and he recommended the creation of a "Canadian Forces Base, RMC" to give him authority appropriate to his needs. Finally, he recalled that an architectural master plan for development of RMC's facilities in 1966 and the minister's commitment to build a new engineering building had not been carried out. He said that an announcement that RMC would complete that new building, and also a new gym-

"Putting the 'M' back into RMC"

nasium, by the college's centennial year, 1976, would be recognition of RMC's importance for Canada.[9]

Meanwhile, the Faculty Board's committee on college government had come to the conclusion that there was possible value in a buffer between the college and CDEE. It now argued that the "organization, government, command, control, and continuity of the Royal Military College of Canada would be greatly enhanced by the establishment of a Board of Governors or Trustees comprised of senior civilian and military members from the Department of National Defence and [appointees] from the public and academic sectors, with jurisdiction over the Canadian Military Colleges."[10] This proposal came to Brigadier-General W.K. Lye, a sapper, when he was appointed commandant. Lye was consciously "military," and was much affected by concerns expressed by some senior officers and ex-cadets that RMC was losing its "military" characteristics. Like them, he wanted to "put the 'M' back into RMC." However, to familiarize himself with faculty

members and their work, he made it his first duty on appointment to visit every academic department. Lye argued against a suggested expansion of the Educational Council, except by giving the commandants voting powers on it.

The Faculty Board's Baird Committee, anxious to maintain RMC's academic credibility in the eyes of the other Canadian universities, had dealt with the internal governance of the colleges. It recommended that a commandant and principal be co-equal in their responsibility for the college's operation, that the principal, deans, and heads of departments be appointed with faculty approval for limited terms of office, and that the Faculty Board be charged with responsibility for the supervision of all academic matters.[11] Lye, however, in a letter to CDEE, retorted that "participatory democracy" of that kind would be a "retrograde step." He said the most suitable form of government for a military college was a command structure[12] – in effect, the traditional RMC system. When RMC came under severe press criticism after a tragic drowning during the recruits' obstacle race, he stood firm in the college's defence. He said this was an isolated incident that did not justify demolishing traditional military and college procedures.

When the full Faculty Board discussed the Baird Committee's report in draft, it passed a resolution expressing grave concern about the erosion of RMC's sense of identity and reputation, about its increased difficulty in attracting good students, its dilapidated and antiquated facilities and accommodation, and its inadequate establishment and manning in face of rising cadet populations and increasing academic commitments. Like Hayes and Lye, the Faculty Board doubted the ability of the new "base concept" of the Canadian Force's reorganization to meet RMC's specialized needs for maintenance and support. On 26 January 1971 Lye forwarded this resolution to CDEE.[13] Nothing came of it, but the Baird Committee claimed to have already brought one change. Henceforward the selection of the principal, deans, and heads of department was done more formally, with the consultation of the faculty.[14]

That same month RMC's proposal of a Board of Governors was brought before the Advisory Board of the Canadian Military Colleges. Milroy replied that fiscal control could not be delegated by the government and that a different title, Board of Trustees, might be more appropriate. He doubted whether a Board of Trustees could control three separate colleges, and, since the CMC system existed to serve the Canadian Forces, he thought that NDHQ would insist on exercising control. He suggested that the current Educational Council of Military College Principals and Registrars along with headquarters personnel could in practice fulfil most of the functions of a Board of Governors if the college commandants and some representatives of the Faculty Councils and Faculty Boards were added to it.[15] The Faculty Council was not convinced. There were objections to the idea of adding the commandant and other faculty members to the Educational Council in addition to the principal who, theoretically, spoke in the name of the college. Dr Dacey, who saw this proposal as a weakening of the hold that Colonel Sawyer had established for the principal, suggested that the commandant could actually exercise a greater influence by acting in concert with the commandants of the other two colleges, rather than by becoming a member of the Educational Council.[16] When CDEE drafted a proposal to establish a Board of Trustees, Lye recommended that it not be submitted to the Defence Council without further consideration.[17]

At the same Advisory Board meeting in January 1971 Milroy revealed that, following the board's advice of the previous year, the Honourable Leo Cadieux, minister of national defence, had decided to defer the relocation of the military colleges indefinitely. His principal reason must have been financial, the high cost of ODB's proposals. In retrospect it seems that ODB had given insufficient thought to the probability that a government reluctant to incur increased defence expenses would reject large additions for military education in a budget that was already overweighted for personnel costs at the expense of materiel.

The immediate consequence of the minister's ruling was that construction was now possible at RMC. Milroy said he regarded the poor condition of certain facilities to be CDEE's most pressing problem. He was considering a modernization program that would include a new science and engineering building and a new gymnasium at RMC and also construction at Royal Roads and CMR. Major maintenance programs that had been deferred while ODB was at work, and while a decision about location was pending, were now to be revived.[18]

Milroy then went on to report that a CDEE study had come to the conclusion that, without the centralization of the colleges in a single location, year-round operation of the Canadian Military Colleges was not feasible. Royal Roads, however, would be permitted to introduce the semester system to secure more flexibility. RMC would

Soccer, 1990

remain in Kingston and would adhere to its traditional university academic sessions, with military training conducted elsewhere in the summer.

Milroy also used this same meeting of the Advisory Board to draft a new official definition of the military colleges' role and objectives. He stated that their basic function was to train officer cadets and commissioned officers for careers of effective service in the Canadian Forces. The colleges' objective was to provide cadets with a university-level education in appropriate disciplines. This could be designed to meet the unique needs of the forces in qualities of leadership, the ability to communicate in both official languages and understand the principles of biculturalism, to develop a high standard of personal physical fitness, and to stimulate an awareness of the ethic of the military profession. He added that, in respect to commissioned officers, the objective was simply to provide undergraduate and graduate courses in appropriate fields.[19]

Lye commented that these RMC roles differed little from traditional RMC objectives. They legalized the course on which the college was already steering and which it wished to follow.[20] There were, nevertheless, several aspects in Milroy's statement that were distinct departures from the not-too-distant past. Bilingual training, improved inculcation of the military ethic, and undergraduate courses for commissioned officers were all new emphases that needed to be worked out in detail. Moreover, the reference to instruction in "appropriate disciplines . . . designed on a broad base to meet the unique needs of the Forces" was capable of a number of different interpretations. Milroy confirmed, however, that CDEE was satisfied that RMC's courses did in fact provide the scope and level of education that the services wanted. This finding was essential, he explained, to secure Treasury Board approval for the new engineering facilities.

Milroy also noted ODB's opinion that the colleges ought to have a higher proportion of serving officers on their faculties to increase their military influence. In the years 1970–3 about 25 per cent of the RMC faculty were either serving or recently retired officers. Milroy said he had asked the colleges to show how a level of 30 per cent could be achieved. The ultimate goal would be 50 per cent. In order to increase the number of officers qualified to serve on college faculties, as well as for other service needs, the CMCs must develop postgraduate programs. Finally, Milroy recalled that ODB had been emphatic that provision must be made to enable officers who lacked a baccalaureate degree to work towards one. He noted that RMC and Royal Roads had already started pilot extension programs in history and economics. These could be used to carry out ODB's proposal. Milroy announced he was exploring the possibility of extending to the military colleges the University Training Plan for Officers that currently applied only to the universities.[21] Other important topics which CDEE introduced at that meeting of the Advisory Board were the DND policy of bilingualism, a need for computers in the colleges, and the provision of research funds. The commander of CDEE had thus shown with clarity and firmness the direction college policy should take.

At the same meeting, the Advisory Board was asked to consider the perennial problem of cadet recruitment. Despite the fact that the number of high school students had markedly increased, the number of applications for the Regular Officer Training Plan had diminished.[22] The problem of numbers and wastage was, indeed, becoming yet more serious. Dean James Cairns* informed the Faculty Council on 3 February 1971 that "over the past few years, in spite of a large increase of numbers entering the First Year, the entry to the Second Year had increased very little indeed and the loss rate of the First Year had

---

*James Pearson Cairns, professor of economics from 1964 and dean of the Division of Arts 1969–80. He came to RMC in 1960 with a PhD from Johns Hopkins University. He was made distinguished professor in 1985.

increased dramatically." He said that among the bottom twenty in each entry class there were many who could not succeed in any kind of course.[23] Recruiting was obviously still scraping the bottom of the barrel.

Now that the minister had decided that the colleges would remain where they were, CDEE began to arrange for their expansion to increase their output without waiting for an increased supply of qualified recruits. Construction of the first unit of the science and engineering building (which was named after Colonel Sawyer) was announced in November 1971.[24] The Macdonnell Athletic Complex across Highway 2 was to follow a few months later. The total CMCs' cadet body was to be increased to 1570, of whom almost half, 730, would be at RMC.[25] Since RMC's strength had already increased from 450 in 1961 to 550 in 1971, this proposal promised yet more difficulty in finding suitable candidates from sources already thoroughly tapped. However, had the military colleges not remained where they were, relocation might have made the recruiting problem worse still and caused a reduction of academic quality.

Meanwhile, the new centralized direction from Ottawa was getting into stride. CDEE made use of both the Educational Council and the Advisory Board to discuss its policies, especially those concerning the cooperation and coordination of the colleges. Selected members of the Advisory Board, as well as the commandants, attended Educational Council meetings. In turn, the Advisory Board executive asked for CDEE board representation when they obtained interviews with the minister. The Advisory Board began to take greater initiatives and to maintain an interest in the colleges between sessions. In 1972, after exploring the possibility of setting up standing committees on topical lines, it proposed that it should instead have three regional committees to maintain closer touch. The board was thus being restructured to perform some of the functions of the Board of Governors that had not been established.[26]

CDEE now set forth the principles that it wished to apply to officer education. In September 1970, when the Canadian military began to realize that Trudeau's defence policy was here to stay and that the main recommendation of the *ODB Report* on relocation of military education institutions would not be implemented, the chief of personnel had set up a working group of officers, some of whom had served on the Officer Development Board and were now on CDEE's staff. He had instructed this group "to identify changes in the professional development of military personnel that may be necessary to enable the Canadian Forces to contribute more effectively to the achievement of national aims." The group, headed by Rear Admiral R.W. Murdoch,* was expected to carry on where ODB had left off; to supply the definitions of the Canadian military ethic and philosophy that ODB had said were needed as a basis for officer education in order to bring ODB's findings into line with the government's current policies that were now specifically Canada-oriented; and to show how those findings could be implemented without ODB's principal recommendation, the centralization of all military education in the Ottawa-Hull area.[27] Significantly, the group was expressly told to concern itself with non-military as well as military roles for the forces and to take into account pressures exerted by social changes in Canada. In short, the military profession was to become more closely related to the prevailing features of Canadian society.

Murdoch's Working Group for the Study of Military Professionalism preceded and followed its investigations by seminars of senior officers and scholars to explore the problem. Its *Draft Report* began by recalling that RCAF staff college professor James Jackson, when discussing

*Rear-Adm. Robert Waugh Murdoch, an RCN cadet in 1936 who commanded Canadian and British destroyers in the Second World War, was director of naval education in 1961 and assistant chief of the Naval Staff (Plans) in 1963. After integration he was appointed director-general of plans.

unification in *Saturday Night*, had said what was now needed was a contemporary and professional military ethic: a definition of responsibilities, competence, conduct, and loyalties to sustain members of the Canadian Forces under all conditions of service.[28] The *ODB Report* had set out in detail the "canons" of the military ethic. The Murdoch *Report* held that the "custody" of such an ethic, that is to say its definition and dissemination, was a function of military academies and staff colleges. The working group duly sought ways to relate military professionalism in Canada to the Canadian scene. It stated that a proposed study of a Canadian philosophy of military service (perhaps this was to have been an offshoot of ODB) had not got off the ground. The difficulty in such studies was in defining the forces' peacetime tasks, aims, and objectives. If loyalty and commitment were crucial to the military profession in Canada, it was first necessary to establish a Canadian identity. In this way the report, because it focused attention on the need for, and the means of achieving, peculiarly Canadian goals, was an important step in the development of Canadian military education and professionalism.[29]

By this time, unification had led to an apparent strengthening of civilian, as against military, administration in National Defence Headquarters, and so to a possible downgrading of professional military influence in policy making.[30] Members of the working group must have been aware of this, and also of the fact that many Canadian officers were uneasy about the "civilianization" of military administration and policy making. Yet the group came to the conclusion that, while fighting skills must be maintained, the Canadian Forces must also develop a philosophy of service with a Canadian military orientation and a commitment to armed-forces goals. It added that those goals should be related to national goals. What the report did not indicate was how far such national goals might be non-military, nor did it discuss how this might affect combat-readiness.[31]

Although the working group spoke positively of the need for a *Canadian* military philosophy or ethic, it used the words "ethic" and "philosophy" interchangeably, thus dropping Rowley's seminal concept of a Canadian military ethic distinguishable from a universal military philosophy. Rowley's approach might have become a logical theoretical basis for the development of a Canadian military ethic that was distinguishable from other countries' military philosophies, but only if it could be assumed that Canada had a more moral approach to the use of armed force, for instance by its application in international peacekeeping. Rowley had avoided stressing this implication of moral superiority, and the working group shied away from it completely. Emphasis on a superior Canadian morality would have disturbed those in Canada who wished to see military competence put first, including many who regarded Canada's commitment to support American deterrence as the greatest possible good that Canada could contribute in a dangerous world.

ODB and the working group were both part of a contemporary debate about defining military professionalism in the circumstances that had been created by technological and social change. In Canada, as elsewhere, the arguments in this debate were presented in military periodicals. One root of the problem was difficulty in defining what military professionalism was, beyond the age-old stress on military virtues, including a dedication to service and a knowledge of military organization, processes, strategy, and tactics. These elements now had to be combined with satisfying a need for the development and control of new technology.

A Canadian discussion of broad subjects such as this usually follows similar trends in the United States, but the definitive issue of the American *Military Review* that was devoted to summing up the American services' discussion of military professionalism was not published until 1980, and then it approached professionalism in terms of rival views on leadership and managerial techniques. That confrontation had been partly spurred by allegations that the American failure in Vietnam was due to overdependence on a managerial system.[32] In some respects, then, discussion of military professionalism had come to a head earlier in Canada than in the United States, and it also followed a somewhat different line. Management skills did not have as much attraction for many Canadian soldiers as an alternative for traditional leadership qualities as they apparently did in the United States. As early as 1969, Colonel A.P. Wills,* a staff-college director, poured scorn on current managerial jargon on the grounds that it sought to promote efficiency and economy at the expense of capacity to fight. Wills said that a distinctively Canadian military ethic must be based on integrity and leadership.[33] Thereafter, promotion of managerial theory, which was to plague American discussions of military professionalism for at least another decade, was subordinated in Canada to a revival of the older debate about the relation of academic education to the development of leadership.

On this issue there was no consensus. Not surprisingly, those officers who were more articulate, and who therefore dominated the discussion, tended to stress the academic side. One writer said, however, that while ROTP in the universities had no military input whatever, RMC was designed primarily to produce highly qualified professional civil engineers rather than officers. He added that it gave an inadequate grounding in the humanities and social sciences, but he gave no details.[34] Others called for more opportunity for graduate work, or for more opportunity for self-development.[35]

Canadian discussion of military professionalism continued the old debate of the balance between education and training. This was partly because RMC, traditionally the most important influence in Canadian officer educational development, had long stressed academic excellence. The Canadian officer corps as a whole, however, despite RMC's lead, had lagged behind the allegedly more "professional" American officer corps in requiring academic qualifications. The Canadian officer-production system must now relate academic excellence to military

needs, and at the same time it must secure enough voluntary candidates for professional military careers.

This quest for military professionalism in Canada and at RMC was affected by several circumstances peculiar to Canada. The first was a desire to ensure that Canadian defence development was not merely the continuance of historical dependence on Britain or the assumption of a new dependence on the United States. Among others were the effects of unification, and the need to secure more recruits for a new professional force from a population that had little tradition of military service in peacetime, and especially from the large French-Canadian minority.[36] These factors explain CDEE's attempt to establish a clearer relation to Canadian circumstances and objectives, including bilingualism, in its direction of the new centralized system of control of military education for the Canadian Forces.

The commander, CDEE, restated to the RMC Faculty Council ODB's contention that, as a basis for the furtherance of military professionalism, a majority of the officers in the Canadian Forces should enter commissioned service with a baccalaureate degree, that the CMCs should produce 50 per cent of all such entries, and that officers commissioned without a degree should have the opportunity to obtain one after being commissioned. It also said that all future officers, English and French, commissioned under the ROTP scheme, including through the colleges, should have at least a minimum working knowledge of the country's other official language. To relate these requirements to military needs it stressed that, "without degrading academic standards," studies at the Canadian Military Colleges should be related to the

---

*Col. A.P. Wills trained as a pilot with the RAF 1940 and served in Beaufighter operations. Transferring to the RCAF in 1952, he served at NDHQ and SHAPE and as director of studies of the Canadian Forces Staff College, Toronto, and then as director (plans), Canadian Defence Education Establishments. In 1973 he was director of education, NDHQ.

military profession, and that the military ethic should be advanced by increasing the number of serving officers on the military colleges' faculties.[37]

These trends, developed behind the closed doors of military offices, were announced to the public in August 1971. A new defence white paper, which Desmond Morton called "the clearest rationale of Liberal defence policies in a decade," officially applied Trudeau's ideology to defence planning.[38] Repeating his priorities for foreign policy, it specifically related them to national development, and also made it clear that the forces, in addition to their traditional duty to aid the civil power in emergency, would make other domestic contributions.[39]

General F.R. Sharp,* chief of the Defence Staff, indicated military acceptance of this revised objective for the armed forces in 1972 in a closed staff college lecture. He said that the Canadian Forces could undertake nation-building roles to fulfil social purposes in conjunction with their traditional military roles, though not instead of them. Furthermore, he linked this view with what he saw as the changing nature of warfare. He said that since 1945, professional military forces had been used increasingly to deter, rather than to wage, war.[40] A Canadian sociologist, concerned with obtaining increased French-Canadian participation in the forces, carried that idea much further when he described the Canadian Armed Forces as an "emerging constabulary force," referring not merely to their frequent use in United Nations and other peacekeeping duties, but apparently also to their deterrent role in the Cold War.[41]

Such references to non-combat roles would obviously not please everybody in the Canadian Forces. Colin Gray,** a British scholar resident in the United States who had worked in Canada, warned that Canadian soldiers, accustomed to the "big league" of superpower Cold War conflict through Canada's participation in NATO and NORAD, would disdain a lesser role.[42] Many officers were convinced that any decrease in Canada's contribution, although it was minimal beside that of the United States,

could be detrimental to Western unity and security. Those who believed there was still a need to emphasize combat readiness feared any turning away from military roles. Thus, Colonel C.P. Stacey, formerly the Canadian Armed Forces official historian who was a long-time supporter of academic qualifications for CMC graduates, noticed that the 1971 defence white paper seldom used the word "war" and never spoke of "fight" or "fighting." He reminded his readers that in March 1939 Prime Minister Mackenzie King had said that the day of great expeditionary forces of Canadian infantry crossing the oceans was over, only to be confounded a few months later.[43]

For the Canadian Military Colleges, these conflicting views on how military preparation could and should contribute to the national well-being and security focused on the way in which curricula would be developed. Captain (N) R.C.K. Peers,*** commandant of the Royal Roads Military College, told visitors from CDEE that the CMC curricula should be reviewed in light of Defence Minister

---

*Gen. Frederick Ralph Sharp (no. 2420) graduated from RMC in 1938 and was commissioned in the RCAF. He served overseas and commanded No. 408 Bomber Squadron and No. 6 Bomber Group in 1944. After the war he took a Diploma in Business Education at the University of Western Ontario and became director-general of management, engineering, and automation, CFHQ, Ottawa, and then commander of training command. Vice-chief of the Defence Staff in 1966, he became CDS in 1967 and then deputy commander of NORAD in 1969. In 1972 he joined P.S. Ross and Partners, management consultants.

**Colin S. Gray, a specialist in strategic studies, was born in England and taught for short periods in one British and several Canadian universities, including RMC in the spring semester, 1971. He was also a staff member of the International Institute for Strategic Studies and of the Canadian Institute of International Affairs. He wrote, among other things, *Canadian Defence Priorities: A Question of Relevance* (1971). Since 1975 he has been a staff member of the hard-line Hudson Institute in the United States.

***Capt. (N) R.C.K. Peers, a 1946 graduate of Royal Roads, trained with the Royal Navy. He served in Canadian destroyers during the Korean War. A specialist in torpedo and anti-submarine warfare, he did a tour with the USN. After commanding sea training at Maritime Command and also the Second Canadian Escort Group, he was appointed to command RRMC in 1970.

Macdonald's white paper. He recalled that both RMC and CMR had said that, in view of the CMCs' relatively small size, they offered too many options. He argued that those two sister colleges offered the options merely because their faculties sought to advance their own college's status in academic circles and increase their appeal to prospective faculty members.

In a follow-up paper, Peers listed the fields in which the forces needed expertise, and he named the degree programs he thought appropriate to those fields. He said the various engineering degree programs already offered at RMC were in "appropriate disciplines designed on a broad base to meet the unique needs of the Forces," and that RMC was the only one of the three colleges that had the plant and faculty to offer them. He added that certain RMC arts degrees were also in "appropriate disciplines," notably international studies, economics, politics, and, to a lesser degree, history and commerce. He said it was doubtful whether degrees in English and French were suitable, except insofar as those departments helped to develop the ability to communicate and also to understand the principles of biculturalism. Peers said that the excellence of the arts faculty at RMC fully justified its appropriate arts specializations. He added that CMR specialized in teaching "administration" and had the nuclei of degree courses in mathematics, physics, and general science.

Peers concluded by saying that the CMCs did not yet cover three general areas needed by the forces: first, environmental science, including such fields as oceanography, northern studies, and ecology; second, computer science, including scientific and management applications; and third, government and law, including studies of both Canada and the international scene. He suggested that Royal Roads, which had already received approval for a limited development of environmental and science-oceanography, could expand courses in those fields to include ecology, hydrology, and geotechnics. When the expected enlargement of the Canadian Mili-

tary Colleges occurred, Royal Roads could then offer degrees in those fields as well as degrees in arts. Peers thus proposed to increase the CMC options in an area he claimed was appropriate to the naval tradition that had been Royal Roads's original interest.[44] However, in the discussion at the Educational Council meeting, Peers said that the council should follow the American academic example "and reduce the numbers of courses and options to those required by the Forces." He noted that CMR had proposed to specialize in a Bachelor of Science degree with a concentration in Arctic studies,[45] but he contended this was an unnecessary duplication of Royal Roads's proposed study of the environment.[46] Thus the problem of selecting courses appropriate for military needs or national objectives was complicated by the desire of both CMR and Royal Roads to stake out specializations for themselves in four-year courses.

Meanwhile an outside source had appraised the RMC engineering degree. RMC had asked Dr Philip A. Lapp,* a Toronto consulting engineer who was making a survey of engineering education in Ontario for the Committee of Presidents of Ontario Universities, to include the college in his study. Lapp noted some distinguishing features of what he said was, by virtue of the composition of its student body, a "national college." He found the student-staff ratio to be "very low (4.7 : 1)." He said that the officer-cadet experience was unique and involved a higher workload than in the other Ontario universities. He commended its content of humanities and social sciences (25 per cent), noted that athletics and physical training were compulsory (15 per cent), and remarked that new cadets had been carefully screened, so the attrition rate was kept low. He then recommended that RMC

*Philip A. Lapp, consulting engineer, consultant to the Committee of Presidents of Ontario Universities 1968–70, president of the York University Development Fund 1985.

should develop a "liberal" engineering curriculum similar to that his study group had seen at Dartmouth College, New Hampshire, and at Harvey Mudd College, California. These colleges emphasized what they called "applied humanities" and also stressed designing, using a team approach to real industrial situations. Lapp said that RMC could design problems relating to the Canadian Forces; he believed such a program would "help in overcoming an apparent reluctance . . . [of young men throughout Canada] to embark on a military career."[47]

Faced with the problem of deciding on curricula suitable for the military colleges, the commander, CDEE, invited Monseigneur Jacques Garneau,* who had a distinguished career as a university professor at École polytechnique and was a member of the Advisory Board, to be his academic adviser and to recommend the course patterns that should be offered in the CMCs to meet the forces' requirements. In a preliminary submission to the RMC Faculty Council, Garneau said he began with three basic premises that had already been confirmed by the Advisory Board: there would not be a single centralized college but the three separate colleges would continue; the CMCs would develop as bilingual national institutions; and curriculum development must meet the general requirements of the forces.[48] Garneau also assumed that Royal Roads would become a four-year degree-granting college as soon as possible, that "academic bridges" must be maintained between the three colleges, and that the degree-granting authority for each college might operate differently, either under its own charter or in affiliation with a university in accordance with circumstances in the province in which it was situated. Garneau added that government bilingual policy meant the offering of new courses in both English and French to ensure that both English- and French-speaking cadets were competent in their country's second language, and that financial support would be made available for this duplication. He concluded by saying that the quality of all CMC courses must meet university standards, that all the specializa-

tions the forces required could not be offered in the CMCs because of cost, and that the selection of specializations to be developed should be made on the basis of cost-effectiveness.[49]

The RMC Faculty Council, discussing Garneau's premises, had argued that the establishment of a four-year course at Royal Roads would not be consistent with academic viability and that it should therefore be approached with great caution. However, RMC's director of cadets, Lieutenant-Colonel White,** pointed out that both RMC and CMR now had four-year courses that had the virtue of offering their cadets a continuous education and training in certain fields as well as adequate socialization, the former in an anglophone milieu, the latter in a francophone. He believed that the two-year course at Royal Roads was too short for a similarly effective process of socialization. His implication appeared to be that the Royal Roads course ought to be extended.[50]

In further response to Garneau's preliminary submission, the RMC Faculty Council claimed that the traditional courses of Canadian universities that were given in the arts, sciences, and engineering divisions at RMC were a sounder basis for long-term response to future military needs than were courses especially tailored to fit short-term requirements, and also that current economic conditions and government policy precluded an

---

*Monseigneur Jacques Garneau, a graduate of Université Laval, served in the artillery in the militia before the war and was chaplain in the Canadian army 1942–6. After the war he was an administrator at Laval and then with the Association of Canadian Universities and Colleges. He was a member of the CMC Advisory Board 1964–70, and academic counsellor to CDEE 1971–3, when he retired to become a parish priest.

**Lt-Col. Robert Allan "Bud" White (no. 2893) was a member of the "New One Hundred," graduating in 1952. He trained as a pilot in the RCAF and served as a test pilot. He was director of cadets and military training 1969–72.

increase in either establishment or physical plant to introduce new courses. The RMC deans then prepared statements about RMC's courses for the guidance of the academic adviser.[51]

On 7 June 1972 Lye followed these statements by a memorandum to Garneau in which he said the introduction of the CMR degree had made the "singleness of the CMC system" questionable from an academic point of view. It had decreased mobility between the colleges. He added that curricula development should meet the needs of the services, not political considerations (possibly a reference to pressures for bilingual training or for geographic interests), and that no program should be developed at CMR or Royal Roads that would diminish enrolment at RMC.[52] Dr Dacey had brought to Lye's attention Dean Hutchison's proposal to Garneau of a three-year engineering degree. Hutchison had said that RMC cadets found the fourth year "somewhat artificial," and he had also suggested that officers doing postgraduate studies could double as squadron commanders. Lye wrote to tell Garneau that such changes would "wreck the military-college system rather than revitalize it." He added that a general engineering degree like West Point's (where many graduates went on elsewhere to take university degrees later at a higher level) would not serve the needs of the Canadian Forces. Finally, he argued that the fourth year at RMC was an important experience in command.[53]

Garneau presented his finished report in August 1972. He said he hoped it would be a basis for discussion by the CMC faculties. He called RMC the "senior Military College," perhaps evidence of a tacit assumption that the other two colleges should conform with its courses. The current RMC courses had in fact stood up well to Garneau's review, and he made only minor recommendations for changes in emphasis. But he advised that RMC should consider introducing the general engineering degree suggested in Lapp's report, *Ring of Iron*. Garneau also argued that CMC arts programs should emphasize service-orientated topics rather than traditional ones; he believed that cadets should have more free time for study; and he said that when Royal Roads got its full degree course, it should be by association with the University of Victoria rather than by a provincial charter as a university in its own right. Finally, he concluded that to avoid diversity, there should be a CMC super senate.[54]

The chief of personnel development, Rear Admiral Murdoch, said he thought Garneau's report was an excellent document, but the RMC principal, Dr Dacey, told the Educational Council that Garneau's proposed CMC super senate would require a federal charter. The chairman of the council then said that a super senate was not necessary because the Educational Council already served the purpose of avoiding diversity in the colleges. RMC's commandant rejected Garneau's suggestion of a general course in engineering. Further discussion of a course of that kind centred on its length and nature. Lapp had not mentioned that it would only be a three-year degree. Dacey said that a shorter general degree would not be acceptable to the civilian engineering profession in Canada and therefore to the better students. He believed it would only attract cadets who were academically weak and so would have the effect of associating this peculiarly military degree with mediocrity.[55]

Accordingly, a new general engineering course was not introduced at RMC. However, after studying other university curricula and in line with Garneau's recommendations, the RMC Department of Mechanical Engineering reduced the number of class hours for its third- and fourth-year cadets and gave them more time for individual study.[56] Other changes in the CMCs that followed Garneau's recommendations were a four-year degree course for Royal Roads (which, however, was already being planned) and the introduction of bilingual training at RMC (see chapters 10 and 11).

Meanwhile, independently of Garneau's study but with CDEE's approval, RMC introduced other significant changes. One that was peculiarly appropriate for the modern military profession was an arrangement for RMC cadets who took nuclear engineering to work for a week at McMaster University's reactor facility. CDEE strongly supported measures like that to bring the RMC curriculum up-to-date with modern technology. Milroy told the Advisory Board, however, that although RMC had had a computer for training cadets and staff since 1963, the Treasury Board had turned down requests four years later for computers for CMR and RRMC. Then, when "remote time-sharing equipment" was approved for the other two colleges, the Advisory Board had declared it unsatisfactory. Reviewing this history in 1971, Milroy argued that "policy had clearly been overtaken by events."[57] The CMCs now lagged behind the universities in computer equipment.

Other curriculum changes cut wastage by lowering academic barriers that were now realized to be too high. CDEE had called for a lightening of the first-year load. So, following a Royal Roads initiative, it was agreed that entrance requirements, especially the number of courses in mathematics, would be reduced at that college to facilitate the entry of recruits who could not meet all the science requirements for an engineering degree.[58] A year later, because of the high rate of wastage in the RMC first year, Dean James Cairns proposed to increase the pool of accepted recruits who had good potential by reducing the mathematics and science requirements for candidates with outstanding qualifications in arts. His rationale, different from that of Royal Roads, was to ensure a sufficient number of good-quality arts applicants to maintain the viability of the arts departments.[59] But changes of that kind were not made at that time. In 1970 history and engineering graphics had already been dropped from the first year to lighten its load of examinations.[60] History for engineers was now to be taught four hours a week in one term of the second year, to suit the engineering timetable. Members of the arts departments thought this distorted the claim that RMC was giv-

ing a broad liberal education.

Curricula changes in the other two colleges, for instance CMR's degree in administration and Royal Roads's reduction of science requirements, caused some students who wished to transfer to the third year at RMC for engineering to lack some of the qualifications that had been required of RMC cadets. They thus threatened the academic bridge between the CMCs that Garneau had thought desirable. Furthermore, the development of a four-year course after the preparatory year at CMR had caused the number of students transferring to Kingston for arts to decline. Although this permitted an increase in RMC's recruit intake, and so gave the college a better numerical balance between its classes for training purposes, the numbers in some of RMC's advanced arts courses dropped to a point where they were no longer viable. This happened especially in English and French. Accordingly, those two departments, in which Peers had thought honours specialization was less appropriate for military purposes, were actually faced with the loss of their honours courses for other reasons. This was clearly a trend towards RMC losing some of the strong arts departments that had supported its claim to give broad liberal engineering degrees. There was some danger of RMC becoming an exclusively engineering college. Such a result would be a further decline in what Lye had called the "singleness" of the system.

Another internal curriculum development at RMC related specifically to the requirement that RMC's courses improve their military content. The Department of Military Leadership and Management was in fact now more academic. Its courses were mainly in psychology and sociology, so much so that CDEE agreed the qualification for its permanent head should be a doctorate in psychology. Faculty in the department carried out research on military leadership, and also did some special projects for CDEE. Partly to underplay the idea that it was teaching military leadership, which was the mission of the Cadet

Wing, its acting head, Lieutenant-Colonel Gerald J. Carpenter,* suggested that the names of the department and its courses should be changed by dropping the word "military." He offered to include instruction in military ethics in some existing courses. Principal Dacey promptly rejected any change in name. Since MLM was the only part of the RMC curriculum that clearly professed a military content, it should not lose that description by becoming a study of leadership and management in general. That would only give more ammunition to those who believed that RMC was "not sufficiently military."[61]

Efforts to preserve the college's military emphasis were endangered at this time by outside pressures to remove the condition that an RMC degree should require a satisfactory military performance during the course. This emanated from a drug scandal. In January 1972 the RMC staff came to suspect that some cadets were acquiring and using marijuana. By early May suspicions focused on twelve individuals, but to avoid disrupting sessional examination routines, action was postponed. When the college authorities organized simultaneous investigations and searches, some of the suspects were away on visits to a military base. Their apprehension there brought in the RCMP, whose application for search warrants resulted in immediate publicity. Five of the accused were dealt with summarily and were convicted of "possession." One elected to go for court-martial, and then got his case transferred to a civil court. He was found guilty and was subsequently discharged conditionally from the forces, being denied both a degree and a commission under Academic Regulation 29. Five who were not charged with possession, but were known to be users, were released from the services.[62]

One significance of this incident is that a long-standing RMC principle, the requirement that for a cadet to pass a year he must have a favourable report on his officer-like-qualities, could be reinforced. The Senate assumed that, since the primary function of the college was to educate and train cadets for an effective career in the Canadian Forces, this principle should continue to apply as a requirement in the award of an RMC academic degree.

However, about this time the LeDain Commission was considering a recommendation that the use of marijuana should be legalized. In this context, political pressure from a high level sought to obtain a judge advocate general's ruling on the extent of RMC's legal powers; it was hoped that the convicted and released cadets could be awarded their degrees on the grounds that those degrees had been achieved solely by academic effort. The Senate opposed this compromise, arguing that the document necessary for RMC academic credit in the universities, for instance for application to take higher degrees, was not the degree certificate but the transcript of marks. It obtained the approval of Canadian universities and American military academics for this stand. The commandant, presenting this case to CDEE, complained that in recent years persons "remote from the action [in Ottawa], who could not see or understand the issues, had begun to make decisions about the disposal of individual cadets." Lye said this practice was an expression of DND's lack of confidence in him and his staff.[63] In this instance he was not overruled.

Around this same time the college confirmed certain important changes in the Cadet Wing that have been called "revolutionary." In Lye's first year as commandant his director of cadets, Lieutenant-Colonel R.A. "Bud" White, reported on progress with new wing procedures first introduced by their predecessors, Commodore Hayes and Lieutenant-Colonel Pickering, which were

---

*Lt-Col. Gerald James Carpenter was commissioned in the RCAF and passed through the Flying Training School. He graduated in 1960 with a BA from the University of Ottawa, received an MA in psychology in 1968 from York University, and a PhD in 1978. He was appointed to the Department of Military Leadership and Management in 1970.

now in full operation under Lye. He said that 1970–1 marked the fourth year of what he described as the "co-ordinated systems" approach to the "major socialization of cadets." In an effort to reduce the high attrition rates caused by requests for release during the first year, cadet lifestyles were redesigned to adjust the college to the changing attitudes and expectations of contemporary youth. Superficially these changes appeared to some critics to be a "liberalization" to offset cadet discontent with

A "Cadet Hop" at USMA, 1967

traditional military discipline and RMC's practices. White said that, on the contrary, the program was a positive plan to improve the quality of input into the forces.

What had happened since 1948 was that, in order to administer the larger college of later years, Cadet Wing Instructions (CADWINS) had been formalized. They contained a more precise definition of a code of behaviour than had previously been laid down, and this had led to an increase of the "beat-the-system" philosophy dear to former generations of gentlemen-cadets, so much so that there was a potential for the negation of basic integrity. A serious gap had developed between RMC regulations and their observance, with a consequent growth of the attitude sometimes described in the Cadet Wing as "truth, duty, valour, and don't get caught." To rectify this, CADWINS and other regulations that the college was either unwilling or unable to enforce were now modified. Enforcement was to be the rule. The real goal of the program was thus not a concession to current youth attitudes, but a restoration of basic integrity. Relaxation of regulations requiring the wearing of uniform, marriage, the use of cars, and the consumption of alcohol had been introduced as deliberate and positive steps to this end.

Over the years a greater stratification among the various class years had also developed. This, too, had been a consequence of the growth of the college. Hierarchical cadet class distinctions, privileges, and authority were now potentially more disruptive and harmful than when the college was smaller and when every cadet knew every one else. To reduce such stratification, the senior-cadet lounge located in the old Faculty Club near the dining hall was to be eliminated, to be replaced by the all-years' Mess and Recreation Centre in the former Sergeants' Mess across Highway 2. Recruits were admitted to this mess on completion of the obstacle race in the early fall. As in an officers' mess, there were to be no rank distinctions within its walls. Outside the mess, cadets progressed towards junior officer status by a gradual increase of dress and other privileges and responsibilities.[64] The

Senior Class Cadet Lounge and Bar was, however, retained under another name, "Bill and Alfie's."

Cadet requests for liberalization also extended to a request for a voice in the academic program. These were met by the appointment of a cadet-wing education officer. Most members of the Faculty Council opposed his admission to membership with them, so he was appointed a member of the Faculty Board.[65] Later that year the council rejected a cadet request that classes be made optional, but it agreed that individual instructors of second-, third-, and fourth-year classes could permit optional attendance if they so desired.[66] Ironically, similar moves elsewhere for student participation in university government had worked against what seemed to be CMC interests. Student members of the Board of Governors of Université de Québec had blocked arrangements for CMR graduates to obtain its degrees.[67]

Within RMC, the most radical change at this time was in "recruiting." This year-long process of initiation, which earlier generations had regarded as an essential part of military "toughening" (but what would now be called "a process of socialization" or "hazing"), was reduced from the whole of the first session to the twenty-six days of preparatory camp and the first three weeks of the fall term. Furthermore, the recruits' preparatory camp was used to establish an understanding that recruiting did not mean harassing or hazing but was a means of developing leadership by example. While recruits continued to be subject throughout their first year to other traditional obligations, such as "running [across] the square," and to correction for faults, excessive harassment was firmly discouraged. Possibly as a result of this change in the initiation process, there was a marked reduction of voluntary withdrawal in the first year during 1970–1. This was, however, partly offset by an increase in the compulsory release of substandard performers on military grounds. The director of cadets claimed that the year had seen a swing from voluntary withdrawal by youths who had come too soon to the belief they did not want a

Recruits return from bookstores, 1978.

military career, to college-initiated discharge of those found to be unsuitable for further training.[68]

This opinion was confirmed by Lieutenant-General W.A.B. Anderson, who visited RMC as a member of the Advisory Board when the recruits' Basic Officer Training Camp had been in session for ten days. He wrote to the minister, Donald S. Macdonald:* "No one could fail

*Donald Stovel Macdonald, elected to the House of Commons in 1962, was appointed to the cabinet in 1968. He served as minister of national defence 1970–2, and later in energy, mines, and resources and finance. He resigned from the cabinet in 1977. He was chairman of the Royal Commission on the Economic and Development Prospects of Canada 1983–5.

Brigadier-General W.W. Turner, CD, commandant, 1973–7

to be impressed with the excellent rapport which these cadet officers have established with the new recruits . . . These young men are being developed in ways entirely consistent with changes in society and yet in ways which will continue to meet the needs of the armed forces of the future. The whole enterprise reflects the greatest possible credit on the Commandant and his very effective team of military and civilian staff."[69]

White also reported on other measures that were positive steps to achieve a greater military impact. In addition to the fall preparatory camp for recruits between their selection and admission, there were military guest lecturers, military tours, a closer relation between the Cadet Wing and the military staff, and a spring pep camp prior to graduation. An adventure-expedition training program had begun in the spring of 1971. It provided air transportation on service aircraft for approved cadet ventures, and quarters and rations where available. Cadets themselves provided the rest of the expenses. Under this program some cadets went on special missions to Europe, accompanied the minister of national defence on a tour of western Canada, scuba-dived in Bermuda, and took courses in mountaineering or in jumping with the Airborne Regiment. White concluded that the changes introduced in the past four years in the Cadet Wing had resulted in significant improvements in motivation, morale, and performance by ROTP and Reserve Entry cadets alike.[70]

Even more innovative than these changes within the Cadet Wing were certain other additions. In January 1971 Lye reported that the RMC graduate program, now in its seventh year, was thirty-nine officers strong. There were six in war studies, twenty-three in engineering, and two in science. The normal posting time for completion of an MA degree was eighteen months.[71]

At the same time CDEE was being phased out. It had failed to obtain Treasury Board approval for the centralization of all post-commissioning institutions at Canadian Forces Base Rockcliffe. Although it had an-

nounced a program to implement ODB's proposal to bring the CMCs to Ottawa and use them to give degrees to serving officers, it had not followed through. It had also made little progress towards the introduction of comprehensive bilingual programs in the CMCs to carry out its proposal that, "prior to productive service all future officers from the ROTP scheme should attain a specific working knowledge of the second official language." These shortfalls may have been in part responsible for CDEE's disbandment and replacement by what was intended to be a more effective system of centrally controlling military education, and so of promoting military professionalism at RMC and in the Canadian Forces generally.

Changes in the system of control in Ottawa in 1973 coincided with a routine change of the command of RMC. Lye's tour of duty there and his military career were due to end in June. The problem of rationalizing military training and education, of producing an adequate supply of officers for the Canadian Forces, while simultaneously introducing bilingualism, serving personnel, and, later, women, was now to fall to his successors, the first of whom was Brigadier-General William Turner.*

---

*Brig.-Gen. William Wigglesworth Turner (no. 2816) graduated from a war-shortened RMC course in 1942. He was commissioned in the RCHA and served during the war in the United Kingdom and Northwest Europe. In 1947 he joined the regular army and served in Korea in 1951. In 1957–9 he served with the Truce Supervisory Organization in Palestine, and in 1959 he was brigade major in Germany. In 1966 he commanded the Canadian Contingent in the Truce Supervisory Commission in Cyprus. After serving as commandant, RMC, 1973–7, he retired and became a director of the Urban Transportation Development Corporation.

The Directorate of Professional Education and Development and the Rationalization of the Canadian Military Colleges

In 1973 Defence Headquarters was still in the throes of a structural reorganization recommended in a report by J.B. Pennefather, a Montreal business executive. The Canadian Defence Educational Establishment office had already been quietly disbanded during the preceding year. A Directorate of Professional Education and Development (DPED) would take CDEE's place in supervising the administration of the Canadian Military Colleges. The new directorate would be headed by a colonel – an officer lower in rank than the RMC commandant, who normally was a brigadier-general or equivalent.

The colonel who headed DPED was answerable through a director-general of recruiting, education and training (DGRET) to the chief of personnel development (CPD) who, in effect, had replaced the chief of personnel; and the Educational Council was informed in June that its future meetings would always be chaired by an assistant deputy minister for personnel, ADM(Per), a lieutenant-general, or by a major-general deputizing for him.[1] *Queen's Regulations and Orders* empowered ADM(Per) to "command and control" the military colleges.[2] He reported directly to the chief of the Defence Staff and to the minister. ADM(Per) would thus make policy and the director of professional education and development would have responsibility for day-to-day administration. On policy matters the college commandants usually dealt with the chief of personnel development.[3]

Clearly NDHQ was now in a stronger position to direct and impose policy on the colleges, and much depended on who became ADM(Per). A member of the Advisory Board, Dr J. Hodgins,* a former RMC chemical engineering professor, commented that often in the past many senior NDHQ personnel who had influenced RMC policy "had little idea what the CMCs were all about." He held that this had been a check on RMC's development.[4] Both the Advisory Board and RMC were therefore favourably reassured when they heard in June that the first holder of the appointment was likely to be Major-General W.A. Milroy,[5] Major-General Rowley's successor as head of the Officer Development Board. He was known to be interested in, and knowledgeable about, military professional education. Milroy, who had recently been promoted lieutenant-general to command Mobile Command, was brought back as ADM (Per) on 3 August 1973.[6] He had already told Brigadier-General Lye that he proposed an organization in which the colleges "would be directly under me," with the chief of personnel development acting as his deputy.[7]

---

*Dr John Willard ("Jack") Hodgins worked in the Chemical Warfare Laboratories in Ottawa 1940–5 and was a captain in the Canadian army. He was at DRB 1947–50 and then became professor of chemical engineering RMC until 1956. Dean of engineering at McMaster University 1958–69, he became director of research and a vice-president in the Domtar Corporation, forming the John W. Hodgins Corporation in 1980. He died in 1982.

NDHQ had not informed the CMC's Advisory Board why it had not followed up the suggestion that there should be a Board of Governors. When the board expressed its disappointment at this omission, it was told that the colleges had not supported NDHQ's proposal that the governing body should be a Board of Trustees rather than of Governors, and so it had been withdrawn.[8] The real reason was probably that the establishment of DPED as the "single agency" to administer officer education (a system suggested by the Officer Development Board that the Defence Council had approved on 4 September 1969), and of a command structure under ADM(Per) with his general and educational councils, now made a Board of Governors or Trustees redundant, especially since the Advisory Board itself had expanded its functions and had branched out with regional committees.

The new system for controlling the Canadian Military Colleges presented the ADM(Per), the chief of personnel development, the director of professional education and development, and RMC with a set of formidable tasks. During the rest of the decade there was one study after another to tackle chronic problems in the college's traditional role of officer production. Prominent among these was the need to foster military professionalism. This chapter will focus on the rationalization of the CMCs for that purpose, and will carry the story through to the early 1980s.

The most serious problem relating to the CMC system concerned the size and quality of its output of graduates. In 1971 Donald S. Macdonald, then minister of national defence, had noted that the number of applicants was declining and that the Canadian public was "unaware" of the colleges. He wondered why this was so, but did not try to answer his own question.[9] Had he done so, he probably would have discovered that it was because there was no widespread tradition of regular-force military service in peacetime Canada. Furthermore, the effect of the postwar baby boom had come to a natural end, a new generation of young people had become confrontational, and anti-war sentiment was on the increase. Efforts to increase the flow of qualified officers into the forces through the colleges had to cope with those impediments.

In 1972 there had been a sharp drop in the intake of ROTP cadets into the universities, and 81 per cent of them were now in the CMCs. The CMCs now had first pick of a

---

TABLE 2
Regular Officer Training Plan, Statistical Summary, 1972–3

	Number in Plan	Graduates June 1972	Intake Sept. 1972	Graduates June 1973 (expected)
Military colleges	1,214	142	430	187
Universities	278	102	31	98
Total	1,492	244	461	285

Note: The September intake in the civilian universities increased in the course of the following years. See tables 3 and 4 later in this chapter.

Source: Appendix I to Annex A to 4508-1 (DGEP), 17 Nov. 1972, DND 4508-1, vol. I, Academic Training General

declining pool of applicants and, by 1973, there were only thirteen ROTP officer-cadet recruits in all the English-language universities in Canada. In comparison, the number of graduates from the CMCs increased from 142 in 1972 to 187 in 1973, and that number would continue to grow when the increased CMC intakes graduated (see table 2). The colleges had become the most important source of graduate officers for the services, but they still produced less than half of the total officer intake and still too few to satisfy the need for technical officers.[10] Obviously, if the CMCs were to serve the purpose for which they existed, the recruit intake must be increased and wastage further decreased.

The second major problem was that, despite this swing from the university ROTPs towards greater reliance on the CMCs, there were still complaints that the colleges were not producing the right kind of officer. When Milroy came to brief the RMC Faculty Council early in 1974 about his plans, Dr Stanley Naldrett* of the chemistry department questioned the grounds for those complaints. He said some officers appeared to think that academics were "unmilitary." Milroy replied that complaints came from "users" of the product who were looking at the combined college and training system. Perhaps too much time was spent on classification training and not enough on the development of officer-like qualities. Since all classification training (that is training for specific military trades and functions) was done in the summer away from RMC, this appears to have been a criticism of summer training rather than of the CMCs.

Milroy went on to say it was necessary to look at the system as a whole, and he argued that the CMCs should provide a framework for the whole officer corps. They could only be justified if they developed the military ethic. They should inculcate the standards of the professional officer, as well as qualify graduates to fill positions in the forces. The council discussed with Milroy various means of achieving this dual objective: for instance, by establishing a chair of military studies at RMC, and by the

previously announced intention of having military officers make up 30 per cent of the faculty to offset cadet overexposure to civilian professors.[11]

Milroy had, in fact, not addressed Naldrett's challenge, which had reference to the complexity of the relation between academics and military motivation. Some of those who complained about the alleged lack of military motivation of CMC graduates apparently believed there was too much emphasis on academic qualifications for graduation and commissioning. Yet maintenance of university standards was essential, especially, because of accreditation, for engineers. The question was whether the CMCs' simultaneous promotion of academic standards and military professionalism, which was assumed to be superior to what could be done in the universities, was adequate. If it was not, how could it be improved?

One alleged source of declining motivation was the recent so-called liberalization introduced in RMC by Commandants W.P. Hayes and W.K. Lye, which had adjusted college regulations to bring cadet life more into line with modern youth lifestyles and had cut down wastage, especially of recruits. Some believed this liberalization had further reduced cadet interest in the military as a career of public service. It was during the interval between the demise of CDEE and the full establishment of DPED's control of the CMCs that Brigadier-General William Turner, an ex-cadet of the last wartime shortened course in the old college, had become commandant. He was even more determined than his predecessors to halt what appeared to him, and to many other senior officers, to be a decline in the old military spirit and standards.

Turner's first objective was to put more emphasis on military leadership potential as a requirement in the selection of recruits. Prior to his arrival, preparation of a

*Dr Stanley Naldrett obtained his PhD at McGill University. During the war he served in Canada and London in chemical warfare research, specializing in smoke and foam. He was a postdoctoral fellow at Chicago 1946–7 and came to RMC in 1949. He retired in 1982.

list of potential recruits sent to Ottawa for selection had been done by the principal and the registrar, based primarily on academic grades. Turner set up two subcommittees to prepare the selection list, one military and the other academic. He required lists prepared separately by those two committees and based the final selection on his own opinion of who would make the best leaders and future officers; "marks were *not* the basis." He also sent the coaches of the football and hockey teams across the country searching for likely recruits, a practice that had been followed earlier by Major "Danny" McLeod, the

RMC provided a Guard of Honour at General Vanier's funeral in 1967. Here the guard is formed up at the Ottawa railway station.

hockey coach. He attempted to revive the flow of recruits from the private schools by inviting twenty headmasters to visit RMC, but he found that initiative had little effect.[12] About this time the selection of the final list of acceptances, prepared in Ottawa by a committee that included the college registrars, underwent one important change. Henceforward, instead of giving RMC its first choices, a higher level committee selected candidates for RMC and Royal Roads in turn.

Turner told the General Council that a shortage of staff officers hampered him in his efforts to promote military interest. To improve cadet motivation, he stressed field trips to Canadian Forces' bases, squadron mess dinners, and lectures by senior military personnel. The only activity to which he gave priority over these extracurricular events were out-of-town team games.[13]

Within a year the RMC Faculty Board became thoroughly alarmed. It learned there had been a "documentable increase" of military and athletic activities organized by what it called "people on short postings," meaning military officers on the RMC staff. The board thought this to be especially serious now that, under the degree program, five years' academic work in engineering had to be done in four, and second-language training had been added.[14] The board therefore created an Activities Committee to which all cadet activities outside the routine had to be submitted for approval. It also instructed the registrar to keep a list of cadets who were in academic difficulty.[15]

When Turner noticed a decline in the standard of RMC drill on a church parade and saw that drill for the senior classes was infrequent, he ordered a weekly extra hour of drill for the whole college and more frequent church parades. Since RMC's centennial was approaching, it was obviously imperative that the college should present an appearance on parade that would compare well with earlier days. But the faculty feared this assignment of additional drill might become a precedent for further unilateral decisions that would cut into study time.[16]

Turner cut down academic fieldtrips, but the military fieldtrips remained as an infringement on academic time. Recruit visits to military bases, as an aid to classification selection for careers, had first been introduced in September 1971, immediately after the class assembled, but a cadet's final decision on classification often did not need to be made until October in his second year. Meanwhile, the trips were said to have reduced a recruit's academic year by approximately two weeks, and the Faculty Board believed that RMC now conducted trips more for indoctrination than to aid classification selection. It argued that the trips unduly interrupted the academic year and also gave the recruit a negative attitude to the importance of his studies. The board therefore suggested that fieldtrips should take place in the summer when there was spare time.[17] According to the board minutes, however, when graduation was advanced one week to accommodate second-language training, a May fieldtrip "suddenly disappeared." The board learned that recruit class time had thereby been returned to the satisfactory twenty-seven week session.[18]

The problem with squadron mess dinners was that they were often staged on Thursday evenings, eliminating the normal study time on that evening and producing fatigue on Friday mornings. The dinners were not scheduled for Friday because members of teams playing away at the weekend would not be able to attend. Furthermore, cadets objected to Friday because mess dinners would cut into their weekend leave. The commandant, noting that dinners came only once each term for a cadet, ruled in favour of Thursday in order to secure complete squadron attendance, but he gave cadets who wished to study permission to leave after dinner had been served. He said that partying after dinner would be checked. He also began inviting fathers of cadets to attend mess dinners with their sons, a popular practice with both generations.[19]

These minor confrontations provoked by the commandant's efforts to strengthen military motivation illustrate the delicate balance between academic and military objectives of RMC education and training. Even though their military colleagues in the Cadet Wing were cooperative, the Faculty Board and its Activities Committee found the task of protecting study time difficult. Assignment of specific hours for study had the effect of suggesting to some cadets that they need not study outside those hours. Furthermore, athletic events were staged after the West Point Weekend, the prescribed final event of the year, without informing the Activities Committee.

The commandant did not, however, recognize the Activities Committee as a college committee. He said it was responsible only to the board and could not approach him collectively, but he would receive individual faculty members at any time. The board was able to report one small success. It secured the reversal of a decision to issue the new Canadian Forces' green uniforms during class hours,[20] but anticipated that its difficulties would increase in 1976 because of the college's centennial celebrations, and suggested that the academic year should be extended by a week to compensate.[21] In contrast, however, a wing academic cadet officer, T.Y.M. Chong,* had earlier made a quite different diagnosis of the cause of cadet failure. He said they were due to poor study habits and suggested that more pressure should be exerted on cadets by a system of cadet supervisors.[22]

If the director of professional education and development and his superiors in Ottawa were aware of this sharpening of the chronic academic-military-athletic competition for the cadet's time under Turner, that knowledge may only have increased NDHQ's determination to ensure that the system produced the kind of officers it believed the Canadian Forces required. The

---

*T.Y.M. Ming ("Tony") Chong (no. 9477) entered CMR in 1968 and graduated from RMC with first-class honours in mechanical engineering in 1973. He is now assistant engineer of the City of Port Coquitlam, BC.

Officer Development Board's report, which had laid down the criteria for a baccalaureate degree and the attainment of military professionalism, was still considered to be definitive. The board had approved RMC's curriculum in principle, but none of its guidelines for shaping pre-commissioning education and training had yet been implemented.

There were, in fact, lingering doubts whether the military colleges, as then constituted, were the best medium to put into effect ODB's philosophy for producing professional officers to serve all Canada. Defence Minister James Richardson* had reaffirmed in 1973 that Royal Roads would not be closed, but no cabinet ruling on that subject appeared until 1976 – after Richardson had resigned. Meanwhile the Advisory Board, influenced by its new regional committees, had begun to urge that there must continue to be equable representation of Canada's regions and provinces, a point of view that meant there must be a college in the west as well as in Ontario and in Quebec. This reflected a fundamental fact of Canada's existence – that national institutions, if they are to be effective, must be supported by, or made up of, fully participating regional entities.

It was now coming to be realized that the closing of Royal Roads on the grounds that it was the smallest and least cost-effective of the three colleges might become a precedent for attacks on the other colleges that could be fatal for the CMC system as a whole. It might give renewed force to old arguments that the universities could produce officers more cheaply. Preservation of Royal Roads came to be seen as essential for the preservation of the CMC system, and so of the military professionalism that they could best promote.

Given acceptance of the CMC system, the immediate problem in NDHQ's eyes was to ensure that it produced officers who were qualified to cope with problems posed for the Canadian Forces by advancing technology and who at the same time were dedicated to a military career. To bring the Officer Development Board's findings up

to date on this point, NDHQ charged Brigadier-General Duncan McAlpine,** a Black Watch Officer who was an associate assistant deputy minister for personnel in 1972–3 and who became the chief of personnel development in 1974, to review training and education. Reflecting a widespread concern in the NDHQ, McAlpine questioned the proportion of time allotted to academic study in comparison with military training. He stated it was not enough merely to stimulate an awareness of the profession, and argued that the present academic session from September to May was "unable to satisfy all of four requirements, academic, military, athletic, and bilingual." The CMCs must be reassessed to determine their validity in relation to the forces' needs, and CMC course patterns should be examined in relation to service classifications.

McAlpine concluded there should be a controlled structure of course patterns to ensure that each cadet would take at least a minimum number of mandatory military and academic subjects while he acquired a nationally recognized baccalaureate degree. In the same year, 1974, Dr J.E. Mayhood, the director general of personnel research and development, studied the colleges' academic programs. His report supported a mix of sciences and humanities, but he suggested changes in the regulations to facilitate student transfers from one college to another.[23]

In May of the following year, 1975, NDHQ received specific complaints about two ex-cadets that appeared to reinforce the argument for radical reform of the system to promote professional motivation. Colonel W.G. Svab, an officer commanding No. 202 Workshop in Montreal,

*James Armstrong Richardson, a member of a prominent Canadian business family and minister of national defence 1972–6, resigned over the question of the constitutionality of the language program.

**Lt-Gen. Duncan Alistair McAlpine fought with the Black Watch in the Second World War. He was commander of Canadian Forces in Europe 1975–6.

reported that two of his junior officers were "completely disgusted with the military life" and were unable to accept that "they could be called upon to manage violence" – that is to say, to fight. They also found mess life to be "anathema." These two officers informed him they had entered the ROTP scheme to get an education, and their anti-military feelings dated from their second year in college. They alleged that no one had instructed them on the facts of a military career, and they now wanted to break their engagement. Svab commented that the military colleges should reorient their curricula towards developing military officers rather than engineers.[24] In reply, Brigadier-General J.E. Vance,* DGRET, warned against precipitate action on the basis of specific examples of this kind, but he agreed that one possible way to reverse the situation was to tailor CMC courses more closely to military needs, including combat.[25]

Lack of motivation was said to be particularly serious among naval officers. Vice-Admiral D.S. Boyle,** commander of Maritime Command, complained about junior-officer training and motivation. For years there had been difficulty in retaining young officers for service at sea, and although numerous changes had been made since the Tisdall Report, the wastage continued. Professional officers were also concerned about a dwindling professional capability: "It is apparent to those who know anything about producing a professional for service at sea that we are not on the right track." One of his squadron commanders had sent him a report that showed the experience level was falling: "A continuation of such a trend, when ships' systems and tactics are increasingly complicated, provides warning of possible trouble ahead." He complained specifically that, in the years when it should be developing a professional sea-going officer, Maritime Command was required to add professional development courses of a general nature to meet the Officer Development Board's general professional criteria.[26]

These statements show the conflict of opinion in the services about the reform of Canadian military education. Technical competence developed by academic engineering degrees was needed more than ever, but many senior officers wanted more emphasis on the military profession, especially combat. The commander of Maritime Command believed, moreover, that a general military professional motivation was inadequate for producing sea-going officers, and in all elements there were some who believed that academic education should take second place to general or specific professional motivation and training. There were other influential people, however, who wanted more stress on non-military content. When Richardson confirmed the continued existence of Royal Roads in 1973, he said its proposed program of environmental studies would offer a secondary interest that would appeal to young officers.[27]

Behind all this lay the question of cost. In 1975 Richardson asked the Advisory Board for proof that the cost of the CMCs was justifiable so he could respond to cabinet colleagues who frequently challenged that statement. Since the board did not have the personnel or resources to undertake such an enquiry, ADM(Per) instructed the incoming chief of personnel development, Major-General Jean-Jacques Paradis,*** and the director of

*Lt-Gen. John Elwood (Jack) Vance (no. 3536) was at RMC 1952–6, graduating in honours history. Commissioned in the RCR, he was attached as personal assistant in the Office of the Adjutant-General in 1961. In 1976 he commanded the No. 4 Mechanized Brigade in Europe and then was on the staff of Mobile Command, CFB Montreal. Thereafter he became ADM(Per) and vice-chief of the Defence Staff, retiring in 1988.

**Vice-Adm. Douglas Seaman Boyle entered the RCN as a cadet in 1941, and served as a midshipman with the Royal Navy and then in HMCS *Chaudière.* He was appointed to the staff of Royal Roads in 1945 and became director of naval training in 1964, chief of personnel in 1972, and commander of Maritime Command in 1973.

***Lt-Gen. Jean-Jacques Paradis, educated at the Collège de Jean de Brébeuf, joined the Royal 22nd Regiment as an officer cadet in 1948 and was commissioned in 1950. He served in Korea and Germany, became a director at the Army Staff College in 1968, chief of personnel in 1975, and commander, Mobile Command, 1977.

professional education and development, Captain J. Côté,* to prepare the rationalization study. By May 1976 Côté had assembled the material needed for the study, and Paradis submitted his report in October.[28]

While this work was going forward, RMC was celebrating its centennial. It postponed the graduation convocation and parade to 1 June, the opening day one hundred years earlier. An extra convocation was held on 6 October at ex-cadet weekend, when honorary degrees were conferred on six former cadets recommended by the RMC Club for distinguished service in various fields.** Nearly one thousand ex-cadets attended this centennial reunion, paraded on the square, and marched to the Memorial Arch. Earlier, in February, a convocation conferred a degree on Governor General Jules Léger.*** Léger also presented new RMC colours in Ottawa on Parliament Hill. During the year Prince Philip and Prince Andrew visited the college. The Canadian Post Office issued RMC commemorative stamps, and the Artillery Regiment at Petawawa fired a 100-gun salute. RMC received the Freedoms of the City of Kingston and the Township of Pittsburgh. In return, it entertained their citizens by a special tattoo and aerial fly-pasts, and proclaimed the annual West Point hockey game to be a part of the Centennial Year's proceedings. RMC gym and combat arms teams toured Ontario cities. Officer cadets participated as guards of honour at the Olympic Games and at the Royal Winter Fair in Toronto, and groups of cadets visited a number of foreign military colleges. RMC also organized and hosted two international symposia of scholars, one on military education and the other on military history; the latter has become an annual event. Special commemorative articles about the college appeared in the press, especially in the *Canadian Defence Quarterly* and *Sentinel*. The RMC Pipes and Drums recorded a commemorative disc.

These centennial celebrations, which Turner organized with the assistance of his director of cadets, Colonel John Gardam,† were a worthy tribute to the proud his-

tory of a great institution. When Turner retired, Governor General Léger invited him to a dinner in Ottawa at which he presented to the college a specially commissioned ceramic enamelled plate portraying the RMC Memorial Arch and the Wing of Gentlemen Cadets. Other notable gifts to the college during Turner's tour of command were the bell of HMS *Vancouver* and the Point Frederick dockyard bell, which had been hanging in the tower of St Mark's Church, Barriefield, since the closing of the Royal Naval Dockyard.[29] Under Turner, RMC's military attributes had thus been given considerably greater visibility. He had also taken a strong stand against excessive overt subordination to the Kingston Base commander, which he feared might result in a weakening of the commandant's ability to control or influence policy in RMC.

---

*Capt. (N) Jacques P. Côté (Royal Roads 94) was a cadet at Royal Roads 1942–4 and served as midshipman in HMS *King George V*. He took pilot training, served at headquarters on the Naval Planning Staff 1961–2, became vice-commandant of CMR, and was selected to command HMCS *Ottawa*.

**In 1977 the RMC Club recommended the following ex-cadets for honorary degrees: James Fergus Grant (no. 1429), RMC 1918–21, a journalist; George Brinton McClellan (no. 1921), 1926–9, commissioner of the RCMP; Charles Sydney Frost (no. 2761), 1940–2; Francis Herbert Maynard (no. 1490), 1898–1901; Paul-Émile Bernatchez (no. 2874), 1929–34 (see page 159 below); and Gordon Dorward de Salaberry Wotherspoon (no. 1945), 1926–34. Another ex-cadet, physicist J. Guy Savard (no. 2351), 1933–7, had been awarded an honorary DSC at the regular convocation that year.

In 1984, the centenary of the founding of the RMC Club, the club recommended a further six ex-cadets for degrees to recognize their contribution to the club's development: Capt. J.F.J. Blanchard (no. 810), Brig.-Gen. G.E. Beament (no. 1828), Brig.-Gen. J.H. Price (no. 1119), Maj. G.E. Ward (no. 2494), Maj.-Gen. Herbert C. Pitts (no. 2897), and Maj. T.A. Somerville (no. 2544).

***Jules Léger, diplomat, governor general of Canada 1974–9.

†Col. John Gardam joined the Canadian Army Reserves in 1948 and was commissioned in Lord Strathcona's Horse. He was director of cadets at RMC 1974–5.

During that same period the nature of the college and even its very existence were at risk. Paradis's *Study* rejected the idea that the CMCs should "deliberately serve a general educational process" to produce citizens who would benefit Canada, an argument sometimes used as a supplementary justification for their operation. He agreed, however, that this could be a by-product of the system. Paradis found it impossible to make a clear-cut cost-benefit analysis to compare with officer production through civilian universities. His *Study* reviewed the performance records of CMC graduates, however, and concluded that no one should ignore the fact that 68 per cent of all graduates and 75 per cent of the 1965–70 group who were still serving had earned good or outstanding performance evaluations. Professional military officers needed a university-level education, and the military colleges served a purpose which could not be duplicated in civilian universities. They added other benefits, such as military and physical development, a science/humanities mix of subjects, second-language training, and courses offered in both languages. The extra cost of producing a CMC graduate was therefore an excellent investment in the future of the officer corps. "On this basis," Paradis concluded, "it is considered that the cost of operating the military colleges is warranted and justifiable."[30]

Paradis said that cadet selection of course patterns should be related to the forces' anticipated needs. He proposed the development of two basic degree patterns to satisfy requirements in operational non-specialist classifications. These would be academically credited BA and BSC degrees that would have military relevance and include some military orientation. Paradis then suggested the appointment of a task force of officers to make specific course recommendations for improving CMC military development.[31]

On 13 October 1976 Richardson resigned from the cabinet over constitutional and language issues. His successor, Barnett Danson,* said he had not heard any questioning of the military colleges, and personally gave the

colleges keen support.[32] With Richardson's departure, then, the *Study* was apparently not needed, at least not for the purpose for which it had been commissioned.

After Milroy, Lieutenant-General J.W. Quinn** served briefly as ADM(Per). On 18 March 1977 a new ADM(Per), Lieutenant-General James C. Smith,*** introduced himself to the CMC Advisory Board on his "first day into the mission." Smith, an airman and statistician, was the third officer to hold the post since Turner had become commandant. Although he stressed that he was the "commander" of the colleges, he regarded military education as a challenge because it was a new field for him. He began, however, with a tone of great optimism. He told the board that the government had embarked on the re-equipment of the Canadian Forces, that the future of Royal Roads was no longer in doubt, and that the need for the CMC teaching staff to undertake research had been resolved favourably. After reviewing the work of the Paradis *Study*, he announced that a "statement of substantiation" had been prepared for the minister that would substantiate "the continued operation of the Colleges." Quinn had already appointed the task force recommended by Paradis to ensure that "the Colleges were

---

*Barnett J. Danson served with the Queen's Own Rifles 1939–45. He was elected MP in 1968, was secretary to the prime minister 1970–2, minister of national defence 1976–9, and consul-general, Boston, in 1984.

**Lt-Gen. J.W. Quinn served with the Royal Canadian Artillery and the 14th Armoured Regiment (Calgary Regiment) in Sicily, Italy, and Northwest Europe, then with Lord Strathcona's Horse in Korea. He was DGRET 1973–4, commanded Canadian forces in Europe 1974–5, and was ADM(Per) until his retirement in 1977.

***Lt-Gen. James C. Smith was commissioned in the RCAF as an air gunner during the Second World War and served overseas. In 1949 he obtained a B.Comm. at Saskatchewan and then joined the Royal Canadian Ordnance Corps. He transferred back to the RCAF in 1961 as a supply officer. In 1969 he was promoted brigadier-general and became director-general, supply, for the Canadian Forces and ADM(Per) from 1977. He retired in 1980.

in fact performing the unique and indispensable functions assigned to them."[33]

At the same time, Basic Officer Training Camps (BOTC) for all recruits were established on a permanent basis, as a preliminary to their reporting to RMC. The college had wanted something like that for a long time, and the first complete entry to pass through the BOTC had done so in 1976, though many activities had not been scheduled for it. As a result, wastage in the recruit class had already declined by 5 per cent.[34]

Meanwhile, the problem of the balance of the military and academic curricula in the CMCs remained unsettled. Quinn had actually appointed two committees to follow up on the Paradis *Study*. The Task Force on Military Development was chaired by Colonel J.D. Young* and had three RMC members, Lieutenant-Colonel C.E.S. Ryley,** director of cadets, Lieutenant-Colonel G.J. Carpenter, head of the Department of Military Leadership and Management, and Dr Ron Weir,*** a retired Canadian Forces' major, now professor of chemical engineering and president of the RMC Faculty Association. The Task Force on Academic Development was chaired by Dr A.C. Leonard,† dean of engineering at RMC, and included deans from CMR and RRMC along with three representatives from NDHQ. Young, as DPED, sat with both task forces so he could coordinate their work.

Quinn had instructed the military task force "to review CMC military training, including the Military Leadership and Management curriculum and, where appropriate, to recommend adjustments to ensure that the programme prepares officer-cadets, intellectually and physically, to meet the requirements of the Canadian Forces." Young's committee decided that what was needed was a more precise definition of the aims and objectives of the CMC military development program, and it drafted a comprehensive "CMC Graduation Specification" to fill that gap. It reviewed the athletic, physical fitness, and military leadership and management curricula, and suggested in broad terms how the three colleges compared with each

other and how they could coordinate to achieve the desired transferability and concordance. Using the term "military development" to encompass all organized cadet activities that were not specifically related to academic degree programs, Young found that, except in engineering at RMC, the CMCs provided a minimum of five-and-a-half hours a week for the military aspect of cadet training. The task force concluded that, during the academic year when academic demands had priority, the current CMC military programs made effective use of the time allocated to them. It suggested only minor improvements in this part of the curriculum.

The military task force then went on to propose CMC graduation specifications. It stated that an officer cadet should acquire, among many other things, "limited skills in applying the lessons of military history to the solution of military problems and in evaluating policies, situations, and events on the national and international scene," and that he should also "become skilled in understanding

*Col. James Derrick Young (no. 3182) entered Royal Roads in 1949, graduated from RMC in 1953, and was commissioned in the RCAF. He was in the office of the deputy chief of staff at NDHQ in 1973, before becoming director of professional education and development.

**Col. Charles Estlin Sheffield Ryley (no. 3927) entered Royal Roads in 1953, graduated from RMC in 1957, and was director of cadets at RMC 1976–8. He became commanding officer 1 Battalion RCR in 1980.

***Dr Ronald Douglas Weir, a 1963 graduate of the University of New Brunswick, served in the RCE 1958–75, reaching the rank of major. He was National Research Council research fellow 1965–8, a NATO fellow 1965–6, and he obtained a doctorate at Imperial College in 1966. He was an assistant professor on a military posting at RMC 1968–75, and thereafter became a civilian professor of chemical engineering.

†Dr A.C. Leonard was an aircraftsman electrician in the RCAF 1943–6. He graduated from the University of Saskatchewan, BEng, in 1950, and was assistant chief instructor at the RCEME School 1950–2. He then served in Korea, being decorated with an MBE. He obtained an MSE degree from Michigan in 1957 and served with the Ordnance Standardization Committee in Washington 1957–9 before his appointment at RMC. He headed the Department of Mechanical Engineering 1965–9 and became dean in 1979. He died in 1983.

and evaluating projects involving military science and technology." The military task force had thus moved into the area being covered by the academic task force. It stated that every cadet should acquire a "military core of the humanities" and a "military core of science courses." In the humanities it proposed replacing the arts electives by a military core consisting of geography and strategic concepts, including military history. The task force admitted, however, that the tight schedules of the engineering courses at RMC necessitated that there could be only an abbreviated form of the humanities core for engineers. On the science side, the attainment of a level equivalent to second-year mathematics, physics, and chemistry, instead of being required for entry into the third year, should be spread over all four years.[35] This was something that Royal Roads had long sought.

The full impact of the military task force on the academic side of the CMC curriculum is obvious when its report is considered together with that of the academic task force. Quinn had instructed the academic task force to review the range of degree patterns and courses offered by the CMCs so as to reduce the number of options that related to particular Canadian Forces' classification requirements or to the general functions of the military profession. At the same time, it was to determine how cadets should be allocated to courses of study to enable the forces to meet specialist classification requirements. The task force should consider the obligation to draw equitably on all regions of Canada, and also determine whether there should be a double-entry admission scheme for the engineering and non-engineering degree programs. In other words, should RMC introduce an arts entry to ensure the recruitment of all candidates who had officer potential?[36]

The academic task force picked up McAlpine's suggestion that branch advisers should be consulted about classification requirements for CMC courses. It circulated a questionnaire requesting the advisers, first, to indicate which degree programs were "preferred," "acceptable,"

or "unacceptable" for their branch, and, second, to supply information about the average number of new officers their branch would want from each of the various degree patterns each year.[37] The advisers were busy men with other duties. They tended to take a specialized view of the practicability of CMC academic courses, of which they had only a smattering of knowledge derived from RMC calendar descriptions. They did not think of courses in terms of potential for intellectual development.

In 1977 the CMCs produced only 230 of the thousand officers required annually by all the Canadian forces.[38] The remainder were commissioned through direct entry from the ranks or in the Officer Training Plan in the universities. From the returns it received, the academic task force calculated that 52 per cent of the graduates who came in through ROTP (including those in universities) and the University Training Plan, Men (UTPM) (see chapter 9) should have been enrolled in engineering, 25 per cent in science courses, and 23 per cent in arts and other disciplines.[39] This was an upward revision of the 1969 figures, which had estimated only a 42 per cent requirement of engineers in the CMCs and 35 per cent in ROTP as a whole.[40] One reason for this increase was the growing demand for technologists.

As a result of polling the branch advisers, the academic task force approved the CMC degree programs in civil, electrical, and mechanical engineering, in engineering physics, and in engineering management. It also approved the BSc programs in mathematics, physics, science (applied), and science (general). The economics and commerce degree and the political science and economics degree should continue, because they were preferred backgrounds for the logistic classifications and desirable in several other classifications. The RMC programs in chemical engineering and chemical engineering (nuclear option), and the specialized honours and general-level degree programs in history, English, French, and international studies, however, should no longer be offered.[41] To replace these programs, there

Cadet Wing Officer John de Chastelain, RMC, 1960 (Photo courtesy of General de Chastelain)

should be a single degree program at both the honours and the general (pass) level that emphasized military and strategic studies, with other optional offerings in economics, commerce, politics, history, English, and French. This program would be a preferred background for the operational classifications.[42] In the opinion of the task force, the amount of advanced-level teaching in these programs should satisfy faculty specialists.[43]

Although members of both task forces continued working on their normal duties, they completed their study in four months. Each committee met five times, the military task force for a total of fifteen days, and the academic task force for twelve.[44] They issued their reports on 19 March 1977. At its next meeting, the RMC Faculty Council referred the academic report to a special committee for reply. A former officer and ex-cadet member of the faculty, Dr Barry Hunt of the Department of History, and another former officer, Dr Ron Weir of chemical engineering, members of the departments most affected by the proposals, were primarily responsible for a draft submitted to a special meeting of the council on 9 May. When Dean Hutchison tabled the draft, "Comment on the Report of the Academic Development Task Force (ADTF) – A Consensus of Faculty Board and Faculty Council of the Royal Military College," he commented that the drafting committee's initial reaction to the report had been one of "outrage and indignation." "Outrage was particularly strong that a small committee with poor research facilities could, in a few short weeks, produce recommendations contrary to those of the Rowley Officer Development Board which had spent two years on the problem and used much more research manpower." All the threatened programs were relevant to the military profession and all attracted young men to the CMCs. Members of the reporting committee were indignant that a balanced university-level program should be threatened with destruction on the basis of what they called a "job-analysis survey." They felt that RMC had been insulted by being treated as a trade school.

Colonel Ryley, who had served on the military task force, asserted that there was in fact a severe gap in understanding between the RMC faculty and many senior officers at NDHQ who lacked an appreciation of higher education. At the conclusion of the meeting the principal, Dr Dacey, asked Hunt and Weir to prepare a revised draft of their reply.[45]

The final version of the faculty reply was submitted to another special meeting of the Faculty Council on 16 May. It repeated and elaborated the arguments in the draft. History and English and French literature all had value because they facilitated understanding of human behaviour in wars; moreover, history was an essential basis for the strategic studies that the task force wanted to introduce. The RMC history department's reputation was demonstrated by its internationally known annual Military History Symposium. Similarly, chemical engineering contributed to an understanding of nuclear development, materials, and fuels, and gave a necessary balance to the engineering faculty.[46]

When the Advisory Board met in July 1977, Dr G.F.G. Stanley, the former head of the RMC Department of History who was now a member of the board, presented a long memorandum to protest the methods used by the academic task force and to defend the RMC curriculum, particularly that in history. He noted that two of the senior officers serving in the Canadian Forces, Brigadier-Generals J.E. Vance and John de Chastelain,* were postwar RMC history honours graduates.[47] Weir asserted that

---

*Gen. Alfred John Gardyne Drummond de Chastelain (no. 4860) was educated at Fettes College, Scotland. He was at RMC 1956–60. Commissioned in the Princess Patricia's Canadian Light Infantry, he was ADC to the CGS in 1962. Appointed to command the Canadian contingent in Cyprus, he became chief of staff, UN forces in 1976. Commandant of RMC 1977–80, he commanded the Mechanized Brigade at Lahr, Germany, 1980–2, and was deputy commander, Mobile Command, at St Hubert, Que., in 1983. He was appointed ADM(Per) in 1986 and CDS in 1989.

General A.J.G.D. de Chastelain, CMM, CD, CDS, 1989–
(Photo courtesy of General de Chastelain)

commandant. As he told the faculty tongue in cheek, he was only as good as the education he had received! He said that Turner's emphasis on the "military" aspect of the program had left the college in a strong position in that respect. He would not have to fight with the faculty over the syllabus or with the RMC Club about "putting the M back into RMC." He soon became aware of the faculty's unhappiness over the task force reports, however, and thought that the involvement of the branch advisers in reporting on classifications other than engineering was a questionable procedure. On a positive note, he said that the senior officers at NDHQ most concerned with supervising the colleges now all had had a university education. There was also a change of guard within the college. John Annand, director of cadets, and Frank Hlohovsky,* director of administration, were both graduates of the new RMC. John Eggenberger,** head of the Department of Military Leadership and Management, also had academic credentials.[49] As a result of these appointments in the supervisory echelon in NDHQ and in the college itself, the task forces' reports would now be examined by military personnel who were not already committed to their findings.

The colleges now produced 61 per cent of the total of officers commissioned through ROTP and UTPM, but only one-fifth of the total annual officer entry. The overall deficit was still about one thousand a year. Furthermore, the ADM(Per), General James C. Smith, was concerned about the large number of CMC graduates who were in non-operational classifications. He suggested that candidates should only be allowed to enroll in those course

Dr Leonard, chairman of the academic task force, did not represent the views of the RMC Faculty Association.[48]

By the time Stanley made his presentation, de Chastelain had been selected to succeed Turner as commandant of RMC. He was the first postwar ex-cadet to become

*Lt-Col. Frank A. Hlohovsky (no. 3608) entered Royal Roads in 1952, graduated from RMC in 1957, and was commissioned in RCEME. He attended the Royal Military College of Science, Shrivenham 1963–4 and was appointed director of administration, RMC, in 1977.

**Col. John Eggenberger enlisted in the RCAF in 1955 and served in Europe. He was professor and head of the Department of Military Leadership and Management, RMC, 1977–9.

options that served operational classifications. The major problem of ensuring that CMC courses adequately met Canadian Forces' requirements thus remained to be solved. Smith felt, however, that the task forces had made their point clearly to the colleges: because of their limited size and because of their special function, they could not be considered as ordinary universities; their degree options must be tailored to meet the unique requirements of the forces.[50]

Smith also agreed with the faculty's suggestion that outside academic opinion should have been consulted. To guide him in adapting CMC courses to the forces' needs, he commissioned Dr Philippe Garigue,* dean of social sciences at Université de Montréal, to report on the training and education of Canadian officers "in its entirety." Garigue was specifically asked to say whether young officers should study global strategy at the time when they were young and keen to command troops. A related factor affecting this question was that there were too few troops to give all young officers practical experience in command.[51]

Garigue submitted a short preliminary reply which stated that his final report would not evaluate particular programs, but would discuss the relation of officer education to the defence needs of Canada. He suggested that the opinions of the heads of Canadian Forces "establishments" should be obtained at a high-level conference that would investigate the problem collectively.[52] The proposed conference would have put the whole question of military education back to square one. What was wanted at this stage was not a philosophical debate, but practical advice on how the current officer-development system could be adapted to serve present needs. Garigue's study was therefore abandoned. The Advisory Board recommended in April 1979 that the task forces' proposals "be pursued to completion," recognizing some "sequential implications."[53]

In December 1979, "in order to ensure objectivity" in carrying this policy through, Smith commissioned Pro-

fessor Corneille of École polytechnique to make a study of the CMC academic establishments. Smith reported that Corneille produced "an excellent report which was accepted by all parties concerned. In brief, it confirmed that the distribution of faculty among the CMCs was equitable, that the teaching loads are satisfactory, and that shortages in support and technical staff exist and should be rectified."[54] Earlier in 1977 Smith had also consulted his academic adviser, Dr Rosario Cousineau,** about the problem of the small size of some classes in the CMCs, especially in honours French at RMC. Cousineau advised that no course in either English or French should be given for very few students. After consulting the colleges, Smith ruled in 1980 that no course should have fewer than four students. In general, this new regulation eliminated one of the more vulnerable aspects of the CMC system.[55]

In place of the Garigue study, Smith commissioned Professor D.N. Solomon, a McGill sociologist, to examine the arts faculty at RMC where, it had been suggested, the proliferation of options was uneconomic. Solomon flatly rejected the findings of the Task Force on Academic Development. After discussions with RMC faculty, he approved the courses in history and international relations and also gave strong support to English and French literature (though not to languages) as first-language requirements and also as options for cadets in arts and engineering. Solomon said that while the elimination of specialized honours and general programs would cause some deterioration of the staff, the Department of

---

*Dr Philippe Garigue, a specialist on strategy. Born in Manchester, England, he served in the British army in the Second World War and was assistant professor of political science at McGill University 1955–7, professor, Université de Montréal, 1957–80, and principal of Glendon College, York University.

**Dr Rosario Cousineau was successively professor of commerce, dean, director of extension, and professor of administration at Université de Sherbrooke 1959–73.

French should be phased out over a number of years because it was no longer viable.[56]

The Advisory Board had been reminded, however, that it would be impolitic to cancel either English or French in a Canadian national academic institution.[57] RMC therefore investigated the possibility of a combined English and French literature program, perhaps a reversion to the concept of the combined department that had existed before the Second World War. That compromise satisfied no one. Offerings by the French department were already seriously affected by the switch to the introduction of second-language training, in which some of its members had refused to teach. It now had few honours students and had very small classes. RMC therefore proposed to discontinue the honours and general programs in French and to convert the department to a service department. Two faculty positions thus released became available for a new Automatic Data Processing Centre that was expected to be in full operation by the end of 1981.[58]

Solomon's approval of the history and international relations programs was related to a decision worked out in negotiations between RMC and the director of professional education and development. RMC proposed the introduction of a new optional program to meet the desire of some officers for one they thought was more suitable for operational classifications and more relevant to Canadian Forces' requirements than programs normally available in a university. The proposed program in military and strategic studies was made up of history and political science, with options in economics, and included more military subjects than the existing program in international relations. The history department also revived the pioneer course in the history of warfare and society that had been dropped in 1965 in favour of a Canadian history course. Other new courses were the history of strategic theory, civil-military relations, and war and diplomacy since 1945.[59] RMC's arts courses could be approved on the grounds that they were of value as an intellectual preparation for the general functions of the military profession.

Chemical engineering seemed to be difficult to defend because it had no specific relation to a Canadian Forces' classification and was a highly specialized technical program. Furthermore, its graduates could easily be lured away by high-paying jobs in industry. Smith said, however, that it would not be cancelled summarily, as the task force had recommended. Instead, he referred the problem back to the faculty to show how the program could be made more relevant to Canadian Forces' requirements. Dean Leonard directed a review by Dr W. Furter,* head of chemical engineering, and his colleagues. They suggested that chemical engineering should be replaced by a degree in fuels and materials, two substances that were basic to the capacity of the Canadian Forces to do their job. Smith asked Dr R.J. Uffen, dean of applied science at Queen's University and former member of the Defence Research Board and science adviser to the Canadian cabinet, for his opinion. In collaboration with RMC faculty members, Uffen mapped out the details of a compromise whereby the chemical engineering program was transformed into a study of fuels and materials and made more clearly appropriate for the problems that military engineers might encounter. Furthermore, RMC proposed, and the Academic Council approved in October 1978, a common course for all engineering departments to be taken by second-year cadets.[60]

Finally, in September 1981, to open the door for youths with leadership potential who did not possess the mathematical qualifications hitherto required, RMC introduced a reduced science and mathematics standard

---

*William Frederick Furter (no. 3045) was at RMC 1949–53. Commissioned in the 23rd Field Regiment of the RCE (Reserves) 1953–6, he was appointed lecturer in chemical engineering at RMC in 1958, became head of the department in 1960, dean of the Canadian Forces College in 1980, and dean of science in 1984.

for entry to some of its courses. All RMC programs maintained the balance of arts and science courses, and candidates in the arts stream were required to become scientifically literate and to develop some competence in quantitative reasoning. As a result of this new arts entry, the college expected to admit approximately twenty new recruits each year.[61]

In the conclusion to his report, Solomon had noted that there had been excessive investigation of arts programs in the past decade and that the colleges should now be allowed to settle down for a time without further interference. It was not surprising, then, that when General Smith retired in 1980, the Advisory Board noted that he had brought harmony to the military colleges.[62]

TABLE 3
ROTP Retention Profiles: CMC and Civilian University Graduates as of 1 December 1978

Canadian Military Colleges				Civilian Universities			
Year	Graduated	Remaining	Per cent	Graduated	Remaining	Per cent
1969	162	88	54.3	129	74	57.3
1970	163	99	60.7	116	77	66.3
1971	170	132	77.6	91	58	63.7
1972	138	99	72.7	113	71	62.8
1973	170	141	87.7	115	96	83.4

Note: Graduates in 1976, 1977, and 1978 were still performing their obligatory service.

Source: Annex A, Minutes of CMC Advisory Board, 18 May 1979

The vexed problem of how Canada's future professional officers could be produced in sufficient numbers to anticipate technical development in the forces was thus, for the foreseeable future, to be left to the Canadian Military Colleges to tackle, freed from the harassment that so many inquiries had inflicted on them. Critics

of the CMCs had once more failed to persuade decision-makers that a reversion to use of civilian universities, along with a military training and reduced academic program in the CMCs, would be more effective. An important factor was that, although the forces wanted training to commence as early as possible in a recruit's career, most recruits came to the CMCs to get a university degree along with their training. Unless that combination was offered, the number of applicants of adequate intellectual capacity would be seriously reduced. Although university-type degrees in the CMCs might give qualifications that could enable officers to leave the service prematurely to take up civilian employment, research showed that the long-term retention rate of CMC graduates was considerably higher than that of university ROTP graduates (see tables 3 and 4). This was especially so for those officers who had done their full four years in Kingston. The CMCs could claim with some justice to be the most effective way of producing graduate officers for the Canadian Forces.

Nevertheless, two groups in the Canadian Forces were still dissatisfied with the CMC system: one among officers in the forces' technical branches and the other in Maritime Command (MARCOM). MARCOM was the successor of the RCN, and many of its senior officers were still bitter about the unification of the Canadian Forces, including tri-service training.

At the time of Paradis's *Rationalization Study*, a Maritime Command Officer Production Study (MOPS) had suggested the introduction of a split-degree program whereby cadets in the Maritime Surface and Sub-Surface (MARS) classification would go to sea for five years after their second year in a CMC. They would then, if they wished, return to a CMC to complete a degree. This was, of course, a variation of the old RCN training system. The MARS branch adviser asked that "this question be examined in depth and a formal reply be given." Accordingly, the academic task force consulted CMC faculties and military staffs. Royal Roads respondents were generally favourable, but they feared that the split-degree program would cause some cadets to transfer to other classifications in which they could go straight on for their degree. They recommended that the proposal be tried for a period. CMR respondents supported the scheme, provided the period of sea service was no longer than two years. RMC faculty and staff, having by this time experienced good results in graduate and undergraduate programs for commissioned officers (see chapters 9 and 13), favoured the split-degree program, though with some qualifications and hesitations. The task force recommended a five-year sea-service interlude for MARS cadets, but suggested there should first be further study by DGRET.[63]

In 1978 the split-degree program came before the Advisory Board but failed to win approval there because the gap between proposals for two years at sea and five years had not been resolved. Moreover, the director of professional education and development had not given a high priority to the board's discussion of this question.[64] Many former members of the RCN and others continued to urge that professional development for naval purposes was being neglected. In the summer of 1979 some members of the Advisory Board visited the Naval Officers Training Centre at Esquimalt where, after almost three years of service, MARS and MARE ROTP cadets were given their first real taste of naval and shipboard life. The visitors were briefed extensively on the problems experienced in producing officers for the sea environment. The commandant of the centre, Captain (N) F.H. Hope (MARS), alarmed them. He said that the indoctrination and commitment of Maritime Command cadets was poor and that a large number of them requested reclassification. Some of the visitors reported their concern about the current system of producing naval officers at the next board meeting. At that same session, a naval brief complained that the CMC system ignored naval needs. That, it said, had been caused by unification.[65]

Some members of the board agreed that it might be necessary to expose Maritime cadets to the sea environment earlier. Suggestions how this might be done included the deferring of second-language training, or giving it at sea. Two Royal Roads ex-cadets said the main problem was that Maritime cadets were unhappy with the navy's approach to training young men because it was "too traditional." They argued that times had changed and "so should the Navy."[66]

TABLE 4

ROTP Retention Profiles: CMC and Civilian University Graduates as of 1 December 1979

	Canadian Military Colleges			Civilian Universities		
Year	Graduated	Remaining	Per cent	Graduated	Remaining	Per cent
1970	158	90	56.9	114	74	64.9
1971	168	125	74.4	89	50	56.1
1972	135	92	68.1	113	67	59.2
1973	170	136	80.0	105	73	69.5
1974	212	177	83.4	115	85	73.9
1975	223	188	84.3	109	91	83.4

Note: Graduates in 1977, 1978, and 1979 were still performing their obligatory service.

Source: Appendix 1 to Annex A to 1150-110/C53-80, DPED, vol. 1, 7 March 1980. See also table 2 earlier in this chapter

Except that it attempted to keep up with technical developments, MARCOM proposals were, in fact, little different from the naval education program set up after the RCN ended its training ties with the Royal Navy. But it now had to cope with an entirely different set of circumstances, including the introduction of ROTP, which had brought in many young men who were primarily seeking a cheap tertiary education. Some of these cadets were influenced by prevailing youth cultures and lacked any initial dedication to naval service. Meanwhile, unification had led to the search for a Canadian form of military professionalism that would apply to the forces generally, rather than only to the navy.

To some extent, MARCOM was sending out conflicting signals: while reducing or delaying MARS academic education, the naval command required qualifications for its technical officers that emphasized academic achievement and specialization. In this it was largely in agreement with the technical branches in the other environments. Whereas MARS cadets needed only first-year mathematics and physics, MARE commissions were restricted to mechanical and electrical honours engineering graduates, with only a small requirement for other engineering and applied science or physics graduates. So, when Royal Roads altered its degree in physics and physical oceanography to a BSc general degree level, only its honours level was classed as "desirable" for MARE. A combined major was declared to be "acceptable." The general level was specifically classed as "unacceptable."

The AERE (Aero-Space engineers) classification was one of the first to state that its shortage of electrical engineers had become critical: "There must be more if we are to do our job." One cause of trouble was that the technical branches of the forces were competing with each other for the limited number of specialized graduates, especially for the electrical engineers, and each of them would only accept the best qualified. Hence many cadets who could not continue to qualify in honours electrical engineering in the CMCs had fallen into other programs rated as less desirable. Generalizing about the value of these programs had always been difficult because courses differed from university to university and in the CMCs.

It was the branch advisers who recommended and administered the "classification specifications" on which acceptability was based. If accredited degree programs fitted the stated requirements for a classification, a graduate was accepted without question; but when a degree program did not neatly fit the requirements, a course-by-course review was conducted. Brigadier-General Baker, DGRET, said that he shared the anxiety over the AERE shortage and had no objection in principle to coercion in program selection; but he added that, since so many students could not qualify, a revision of the pertinent Canadian Forces Order 412, Annex A, was necessary to instruct branch advisers, CMC military staff, and cadets more effectively about the available options.[67]

In September 1981 the RMC Faculty Council learned that some "very senior naval officers" had again expressed concern about the thin flow of graduates into the MARE classification. Two months later it was informed that a revision of Annex A to CFAO 9-12 had now set down the relative desirability of each engineering, science, and arts offering for each military classification.[68] A chart based on this revision was used by career managers at NDHQ and by squadron commanders and counsellors in the CMCs. The council suggested that branch advisers should be contacted to find how they applied it in practice.[69] The council was assured that in the course of time this revision of classification standards would influence the composition of classifications established by the branches.[70]

The RMC faculty approached the problems in academic performance, including engineering, from another standpoint. On 27 May 1978 Professor M.F.R. Bardon,* chairman of the Faculty Board's Activities Committee, had reported concern that the decline in the quality of entering students, combined with changes in military routine that gave cadets greater freedom for independent initiative in the organization of their studies, had led, not to more study, but to more social activities. He said that, as a result, there was a serious threat to the integrity of some engineering programs. He added that

*Dr M.F.R. Bardon (no. 7851) entered CMR in 1964 and graduated from RMC in mechanical engineering in 1969. Commissioned in RCEME, he rose to captain and obtained his doctorate from the University of Calgary. He was appointed associate professor of mechanical engineering at RMC in 1978.

RMC cadets believed they were more pressed than students in civilian universities because of their military obligations, but they did not utilize their time effectively.[71]

In the early 1980s MARE posting vacancies grew to more than one hundred. Maritime Command tackled the problem by recruiting ROTP entrants directly into the engineer classification instead of into a preliminary all-inclusive Naval Operations classification along with MARS recruits. It did this by streamlining its training programs to present theoretical backgrounds before the practical training to which they were related, by adopting the principle that the minimum educational level would be an engineering technologist diploma that could be taken at the Fisheries' College at St John's Newfoundland or at St Lawrence (Community) College, Kingston, and by eliminating from it the training given to ratings.

Part of the success that Maritime Command achieved with this "Get Well Program" could, however, be attributed to the national economic slump that occurred at this time. Nevertheless, the reorganization of classification and training programs undoubtedly made exploitation of the easier recruiting climate more effective. The total MARE recruiting quota for all entry plans approximately doubled in 1983–4 to 172, and 168 maritime engineers were enrolled. This increase was expected in due course to produce 100 qualified MARE officers, spread over several years depending on the entry plan under which they had been recruited.[72] However, MARCOM continued to undertake course-by-course review for CMC general science degrees, making the selection of classification difficult for many cadets.

The number of cadets enrolled at RMC had increased, perhaps partly as a result of the depressed state of the Canadian economy but to the ultimate advantage of the Canadian Forces. In 1980–1 it was 681, and 163 baccalaureate degrees were granted.[73] The following year it had increased to 716.[74] In that year, to adjust to changing technology, the Department of Mathematics was renamed Mathematics and Computer Science, a revision that was adopted enthusiastically by the faculty and was popular with cadets.[75] A Canadian Accreditation Board gave unconditional renewal for five years to all RMC's engineering courses except for computer engineering. That course being new, it was only accredited for three.[76] In 1983 the Faculty Board noted with pride the selection of ex-cadet Commander Marc Garneau* as Canada's first astronaut. It expressed confidence that the resultant publicity would serve to increase further the interest of young Canadians in the CMCs.[77]

By the early 1980s, then, RMC (with its sister colleges) was closer to meeting the military, as well as the academic, needs of the Canadian Forces than it had been ten years before. More than anything else, this was undoubtedly the fruit of willing cooperation between supervisors in Ottawa and the faculty, administrative officers, and military staff at the college. The system now had an effective rationale for producing academically qualified military professionals. Not everyone was entirely satisfied, but some of the doubts voiced by both academics and the military had been set to rest.

Cadet Wing Officer Marc Garneau, RMC, 1970

Captain (N) Marc Garneau, the first Canadian in space, 1984 (Photo courtesy of Captain (N) Garneau)

*Capt. (N) Marc Garneau (no. 8276) entered CMR in 1965 and graduated in engineering from RMC in 1970. He obtained a doctorate from the Imperial College of Science in 1973, served in the RCN, and was seconded to the National Research Council 1984–7. He was Canada's first astronaut, 5–13 Oct. 1984.

Serving Personnel at RMC

The 1970s had brought a surfeit of bureaucratic and academic dissections of organization and methods in RMC, all of them attempting to ensure the professionalism that the Canadian Forces wanted in their young officers. At the same time the college introduced bilingualism, in support of national unity, and enrolled women in the Cadet Wing (see chapters 10–12). The entry of women was one of the efforts being made to increase the output of officers; another was the admittance into the RMC course of personnel already serving in the armed forces with and without commissions. This chapter will outline the introduction of such serving personnel as students. Many ex-cadets and others thought RMC paid an unacceptable price for all these innovations, that the college risked endangering its traditional system of training. It is timely, then, to look once more at the importance attached to the preservation of those cadet traditions, practices, and spirit that many believed to be the secret of RMC's past success in officer training.

In the course of the debate about the admission of women, Brigadier-General Lye succinctly expressed a widely held opinion about the value of the Cadet Wing. He wrote: "The Cadet Wing is a unique complex society governed by rules developed through many years for application to young men between the ages of 17 and 22 to develop self-discipline, integrity, and a sense of responsibility. Complementary military, physical-education, and social programmes are designed to turn youths into men

of good character, and the demanding academic courses ensure that graduates are well-educated officers." He then repeated a statement Lieutenant-General W.A.B. Anderson had made when discussing the acceptance of more mature personnel as students in residence: "The homogeneous characteristics of the Cadet Wing, academic, military, fitness, and social, were designed to produce a balanced citizen. It is a unique organization in the Canadian scene of which we have a right to be proud. Any intention to alter the character of the Wing by moving in another group should be examined carefully because the damage might be permanent."[1] These statements show that both Lye and Anderson had great confidence in the prewar RMC Battalion of Gentlemen Cadets, and its successor the Cadet Wing, as formative agencies for officer development. RMC's proud war record was assumed to have derived in large part from the Cadet Wing's traditional practices.

In prewar days, after their gruelling first year, cadets were the training system's most fervent advocates, sometimes even defying the authority of the commandant and staff when they feared that, as a result of external pressures, college authorities were becoming soft on cadet-exerted internal discipline. After the war, many members of the RMC Club, convinced of the value of the Cadet Battalion, felt impelled to strive to preserve it. It was the club's executive and a small inner group of members that was largely responsible for the college reopening in 1948.

When that had been achieved, the club remained active and influential. Its annual ex-cadet weekends at RMC in the autumn were well attended by civilian and service members, probably on a proportionately greater scale than the alumni "homecomings" in any Canadian university. The club created a fund to provide scholarships and various other amenities that the government did not supply. The RMC Club *Newsletter*, published twice (and later four times) a year, not only distributed news of ex-cadets and information about developments at the college, but also, when Colonel H.W.C. "Buster" Stethem* was its editor, printed his well-researched historical articles that recalled RMC ex-cadet achievements, including those in the British army.

Two recent books, each written and edited by an ex-cadet and financially supported by their college peers,

show something of the extraordinary durability and effects of the loyalties, friendships, and associations founded in the old RMC Battalion. Both books are monumental in size. R. Guy Smith's** *As You Were* is a two-volume compendium of submissions written primarily by ex-cadets of all classes.[2] T.L. Brock's*** three volumes on the RMC class of 1934 combine a collection of autobiographical narratives with the author's own account of the colourful careers of other classmates.[3] Essential features of that traditional system were carefully carried over into the Cadet Wing of the new RMC. Conservative elements in the RMC Club were therefore a powerful force resisting further change.

In the 1970s there were continued attempts to revive what many regard as the prewar battalion's most important feature – the training of young men who made no initial commitment to serve in the regular forces on graduation, now called "Reserve Entry" cadets. In 1961, ten years after the introduction of ROTP, the advocates of Reserve Entry secured an NDHQ agreement to admit up to 15 per cent of each annual entry on a Reserve Entry Training Plan (RETP) basis. The rationale was that it would bring in good-quality recruits who did not need

The Cakewalk: An old RMC tradition that has ended in recent years saw the recruit class present a series of skits satirizing the staff, senior cadets, and college life. Here recruits in 1959 strut their stuff.

*Col. Hubert Walter Carson ("Buster") Stethem (no. HO2354) was at RMC 1933–7. He was commissioned in the British Royal Corps of Signals, serving in France and Africa. He transferred to the Canadian army in 1948, serving in the Congo and becoming commandant, Royal Canadian School of Signals, Kingston. Secretary-treasurer of the RMC Club from 1971, he was made honorary life member shortly before his death in 1981.

**Lt-Col. R. Guy Carington Smith (no. 1877) was at RMC 1925–9. He joined the Department of Trade and Commerce in 1930 and served in Washington, Trinidad, Tokyo, and New York. He took leave to serve with the RCA during the war, ending with the Directorate of Military Operations in Ottawa in 1944. In recognition of his preparation of *As You Were*, he was made an honorary member of the RMC Club in 1984.

***Thomas Leith Brock (no. 2141) was a cadet 1930–4. He became secretary of the Aluminum Company of Canada. As RMC Club historian, he produced *Fight the Good Fight*, a book of cadet reminiscences, and *The RMC Vintage Class of 1934*.

financial support for tertiary education, and so did not have to make up their minds about commitment to a military career at the outset, as was required by ROTP. They would come mainly from those large private schools that had previously supplied a good part of the battalion. It was frequently stated that RETP cadets would be of superior quality and more easily adaptable to traditional RMC training, and that many of them would "go regular" later in their college career.

By the 1970s, however, the 15 per cent RETP entry quota had not been met in any year. NDHQ, faced since the Korean War with a chronic shortage of officers, was not sorry. It preferred ROTP, with its initial obligation for the graduate to serve for a stated time after commissioning. Nevertheless, the RMC Club, prewar college classes that had consisted largely of "reserve" cadets, and others who were graduates of the postwar RMC worked to secure more effective application of the 1961 agreement. Some RMC Club presidents repeatedly urged their members to seek out boys suitable to be RETP cadets.

The club believed that Canadian Forces' recruiters did not publicize RETP* and it suspected that in the selection process, NDHQ deliberately overlooked those who stated a preference for Reserve Entry. Then, in 1977, the club heard that NDHQ had restricted RETP entry in that year to nine recruits, rejecting six others who were eligible. The director of professional education and development (DPED) told the club's president there were other sources of officers for the reserves, in addition to RETP. The club complained to the minister of national defence, Barnett Danson, who replied he would maintain Reserve Entry. The Advisory Board recommended that RETP applicants should be admitted to the CMCs up to the 15 per cent limit on a competitive basis with ROTP applicants, but added that RETP cadets ought to be required to give an honourable undertaking to serve in the reserve forces for four years after graduation.[4]

Later, in November, DPED reported to the Advisory Board the results of a study of RETP that showed that, if

Winter training, 1960: The fine military arts are carefully honed.

the 15 per cent entry had been achieved over the last twenty years, 190 RETP cadets would have displaced that number of ROTP cadets committed to serve in the regular forces. In 1979 DPED told the RMC Club that in the previous year no RETP applicant had been denied admission because of his choice and that in the selection process an applicant's intentions were not considered; yet in equal competition with ROTP applicants, the 15 per cent quota

*Many recruiters are said to have regarded RETP as a means by which generals' sons could get a cheap education without being obliged to join the Canadian Forces. Information from Dr Pierre Bussières.

had not been reached. The club president drew this conclusion, "We are simply not getting the message across."[5] Later the club complained that NDHQ did not attempt to enforce the reserves commitment made by RETP recruits and that it did not maintain adequate records of their whereabouts. Air Commodore L.J. Birchall, a former commandant, suggested that the club's scholarships, which were plentiful, should be granted to RETP cadets as loans rather than as outright grants, to be forgiven progressively as a graduate completed his reserve obligation.[6]

Although there can be wide disagreement about the reasons why fewer affluent families now sent their sons to the new RMC, about the motives of some of those pressing for RETP, and about the value of Reserve Entry for the Canadian Forces, the stated objective of many of RETP's advocates was to ensure the maintenance of the Cadet Wing as a training agent equal to prewar standards. Many others who were cool about RETP also believed in the wing's value in the same way. They held that the tapping of a broader cross-section of Canadian society to increase the number of officers had made the Cadet Wing's formative influences and its system of training in responsibility even more necessary than before. They were also sure that, despite the college's growth and the resulting decline in intimacy, the Cadet Wing was still performing as well as the prewar battalion. Obviously a large number of people representing opinion on both sides of the RETP-ROTP issue were on the lookout for any developments that might diminish the Cadet Wing's function as a training process.

In the 1970s and early 1980s there were several developments that appeared to have some potential for harming the traditions of the Cadet Wing – for instance, the decline of Reserve Entry, ROTP, the postwar emphasis on the academic curriculum, RMC's alleged failure to promote military professionalism, and the introduction of second-language training. Certain other innovations were even more threatening in that they would introduce elements into the Cadet Wing that had previously been foreign to it and would thus destroy its homogeneity. What was most feared in this respect was the introduction of women. But what seemed equally disruptive and came to a climax earlier was a proposal to introduce mature students, in the form of serving personnel, as a step towards increasing the output of officers and raising the academic quality of the officer corps. This development brought about a series of tortuous negotiations.

The minimum academic qualification for direct entry to a commission in Canada was now a junior matriculation – that is, one year short of the requirement for entry to most Canadian universities and to RMC. A majority of Canadian officers had entered by this door, and many of them still had the minimum academic qualification, having taken only the short officer-training course to qualify for a commission. By contrast, most American officers had a baccalaureate before being commissioned. The growth of Canadian-American defence cooperation since the Second World War made this stark contrast with the United States disturbing. If the standard of the Canadian officer corps were to be raised to meet modern conditions, something more than the ROTP and RETP officer-cadet programs in the CMCs and the universities was needed. The academic qualifications of the rest of the officer corps must be improved.

Before the mid-1960s, many serving Canadian officers were already attending civilian universities or other academic institutions in Canada, the United States, or, more rarely, elsewhere. They were usually in summer or evening extension classes, or on leave, attempting to upgrade their academic or professional qualifications. When there was no definite service need for special qualifications, most did so entirely at their own expense. This practice was not restricted to the technical branches, where many officers in fact already had a degree. Officers in non-technical corps also took courses, partly to enhance their prospects in the service, but with post-retirement goals also in mind.

In 1966 DND instituted the University Training Plan, Officers (UTPO), to encourage and support this practice and to recognize that it had direct value for the forces. UTPO subsidized the attendance at a Canadian university of selected officers who were already within two years of a baccalaureate degree. UTPO recipients received full pay and allowances, and all their compulsory university costs for up to two years.[7] The UTPO program was more than an attempt to further the forces' specialist and technical requirements, and was designed to raise the average of academic standards generally. At the time when the plan was inaugurated, the Educational Council suggested that the CMCs might also be used for UTPO.[8]

Various studies at that time suggested that RMC was the only one of the three CMCs where such a program could be undertaken.[9] Since it had vacancies in its third- and fourth-year courses, the use of the college appeared likely to cost DND much less than UTPO in the universities. On 10 January 1967 when a former commandant, Brigadier-General G.H. Spencer, now director-general of recruiting, education, and training (DGRET), was chairman of the Educational Council, he told its members that UTPO was to have up to fifty officers at universities completing degrees with DND subsidization. He wondered whether some of these men could be admitted to the CMCs. Royal Roads and CMR at once replied they could each take up to six, Roads adding the qualification "if more advanced courses are given than at present." Since those two colleges had not yet fully developed their own degree programs, the council said that UTPO students attending them might qualify for degrees by an arrangement with a neighbouring university similar to that which RMC had long had for postgraduate degrees at Queen's University.[10]

Four months after Spencer raised the matter, Colonel S.V. Radley-Walters, speaking for the director-general of recruiting, education, and training, announced a revision of NDHQ policy in an attempt to meet the forces' need for officers with a university education more ade-

quately. RMC's facilities were to be expanded in accordance with the college's long-range development plan so it could accommodate enough students to graduate 215 a year by 1973. An obvious inference was that, with more space, RMC could educate serving personnel as well.

At the same time, Radley-Walters also revealed yet another way in which the production of graduate officers would be expanded. He said there would be a new program called University Training Plan, Men (UTPM), for commissioning from the ranks.[11] This was designed to make use of the universities. If men in the ranks already had university admission qualifications and credits, they could be subsidized to take further university courses to complete a degree.

Serving personnel in the ranks were in fact already eligible to work for a degree and a commission as RMC cadets. Several cadets in the CMCs' entry in 1948 were from the forces.* Two of them, G.C. Coops and D.W. Strong, wore on their RMC tunics ribbons for having served during the war. When ROTP was introduced in 1951, other ranks were specifically declared eligible for it. The number who could qualify was restricted, however, not only by the academic requirements but also by

---

*Five members of the ranks in the Canadian Forces who were admitted to the CSCs as cadets in 1948 eventually graduated. Three enrolled at RMC. Glen Coops (no. 2839) served in Korea with the RCE, in Indo-China, and with the Directorate of Intelligence and Security as a lieutenant-colonel. Donald William Strong (no. 2840) obtained a Queen's University BSc and served with the Liaison Office in Washington and with NATO; on retirement as a colonel, he became secretary of the RMC Club. William Alexander Ferguson (no. 2899) graduated in 1953, was commissioned in the RCR, and served in Korea 1953–4. He became assistant planner, Sarnia, in 1961 and was ordained as an Anglican priest in 1974. He is a chaplain of the RMC Club. Lt-Col. Andrew Claremont Moffat (no. 2981) enrolled at Royal Roads and graduated at RMC in 1952. He served with the RCAF in Korea, in the School of Artillery, in Germany, and with DPED. On retirement he became a sheep rancher. Maurice Albert Rhodes (no. 2917), Royal Roads and RMC, graduated in 1952 and was commissioned in the RCAF. A navigator, he became a recruiting officer.

age limits and the regulation that RMC cadets must be unmarried.

When the announcements of UTPM and UTPO were made, the details had still to be worked out. How far would the plans be applied in the universities, and what could be done in the CMCs? Those who went to the universities would, of course, have to meet university requirements, but age limits could be imposed by DND for service purposes. There would be no need to restrict entry to a university UTPM on grounds of marital status, but if service personnel were to go instead to the CMCs, it would be necessary to establish special entrance qualifications for them, including age limits and academic credits, and also to remove the marital restriction. What credits for advanced standing could be accepted from previous university education? What awarding authority would grant the degree, RMC under its Ontario charter, or a federal authority? Finally, how could "mature" students be coordinated with the Cadet Wing, with its hierarchical structure, strict discipline, and traditional procedures and punishments?

Other more difficult problems concerned the accommodation of the new students and their relation to the Cadet Wing. And could UTPO be members of the RMC Officers' (Senior Staff) Mess which was also the college's Faculty Club? Unmarried UTPM could, like previous recruits from the ranks, be fully absorbed into the Cadet Wing and barrack blocks, but married UTPM seemed to be an impossibility. Finally, the introduction of both officers and married "other ranks" might threaten the "unique society" of the Cadet Wing that ex-cadets so revered.

Before anything had been done to find a solution for these vexed questions, the Officer Development Board (ODB) had been established. It was set up to explore the availability of general education in the forces, both civil and military, and to facilitate the upgrading of academic qualifications so that officers could "perform effectively at all levels and in all areas of their profession without prejudice to their combat skills or efficiency." It covered ground that concerned the present CMCs and any further expansion to include serving personnel.

Presenting his views to the Chiefs of Staff Committee, to the Defence Council, and later to the Faculty Councils of all three colleges, Major-General Rowley showed that, for what he called "middle grade officers," a "sandwich program" of part-time or other study in a civilian university had real problems.[12] He hinted that he was contemplating full-time attendance at university-level military institutions, including the CMCs. The creation of the ODB therefore meant a postponement not only of RMC's building program but also of the possibility of introducing serving personnel until ODB's recommendations were known.

The Faculty Councils at CMR and Royal Roads strongly endorsed Rowley's concept of a military university "to promote officer intellectual development." At Roads, Dr Eric Graham and his colleagues presented Rowley with eight briefs on the subject.[13] The faculty at RMC was more restrained. When Rowley formally recommended the relocation of the staff and cadet colleges in a central location, General F.R. Sharp, chief of the Defence Staff, on the instruction of the minister, issued his call for the preparation of a brief to the cabinet to propose the establishment of a "University of the Canadian Forces" to give degrees under federal authority. The brief proposed the creation of a directorate to control officer development and the transfer of all post-commissioning education to Ottawa; in the meantime, the CMCs should continue temporarily in their present location.[14] As we have seen, the only part of that brief's long list of recommendations that was implemented was the creation of the Canadian Defence Educational Establishments (CDEE) to supervise all pre- and post-commissioning education.

ODB suggested that, after the relocation of the military and staff colleges in a central place, the preparatory courses for staff college, and all other courses for serving officers that would lead to a baccalaureate, could be given by a Canadian Military Colleges Division of the proposed military university.[15] When it was rumoured that the CMCs were to remain in their present locations, at least for the foreseeable future, a proposal to enroll serving personnel in them for pre-commissioning education was confidently expected. RMC's director of studies, Dr Dacey, requested briefs from the various RMC departments about the feasibility of their conducting extension work, "assuming that the existing services for the Armed Forces might be absorbed."[16] Then, six weeks after Sharp's call for the draft for submission to cabinet, the CMC Advisory Board recommended the "continuous upgrading" of serving officers by an extension program or by full-time university studies in the CMCs. This was, in effect, a proposal to extend UTPO to them. But there was no consensus on the board about the nature of such a program, or on whether it should give federal or RMC degrees.[17]

RMC's dean of arts, James Cairns, then reported to the commandant, Commodore Hayes, that two of his departments, economics and history, could offer evening classes to service personnel stationed in the vicinity. These courses would be similar to those being given to cadets. Cairns also drafted a proposal for regulations for an extension degree. He said that entering candidates should be at least twenty-one and have their senior matriculation, or be twenty-three and have had at least one year of military service after leaving school. He also proposed that fifteen sessional courses should be required for a baccalaureate.[18] That was the number normally required beyond senior matriculation for a pass degree in a Canadian university.

The RMC Senate sent this proposal to CFHQ, which then instructed the former RCAF's Extension School at Armour Heights – a school with long experience in running extension courses in conjunction with the University of Toronto's Department of Extension – to cooperate with RMC during the academic year 1970–1. They were to develop an extension program that would satisfy pre-staff-college requirements and also lead to a baccalaureate degree.[19] Nothing seems to have come of this proposed cooperation, perhaps because RMC, like the University of Toronto, would not give academic credit for Canadian Forces' College extension courses not graded by its own faculty. Furthermore, RMC was reluctant to receive the directing staff of the Toronto school in Kingston as part of its own staff, and the commandant in Toronto was equally unwilling to let them go.[20]

In April 1970 CFHQ directed RMC to proceed on its own with a pilot program of extension courses for up to twenty students working towards a degree. It was to cost no more than $1000. In June, a Canadian Forces Order announced courses in introductory economics and military history to commence in Kingston in September. CFHQ also authorized CMR and Royal Roads to offer their existing courses on an extension basis.[21] In July it set up a CMC Extramural Division Committee, with Dean Cairns as chairman, to investigate the feasibility of correspondence, day, and evening courses at all three colleges. The Extramural Committee was made up of four academic staff members from each college, with representation from the National Defence College, the staff colleges, and CFHQ.[22]

In the session 1970–1 RMC gave two evening courses and Royal Roads gave one. The next year all three colleges held evening and weekend classes for personnel from neighbouring units. Student response was enthusiastic, with 160 attending the courses in the two years.

Meanwhile, on 14 December 1970, CDEE had issued a directive that UTPO had been increased from fifty to seventy. It called for further investigation of the possibility of accepting some of the seventy as full-time students in the CMCs. The RMC Faculty Council was told there would be no extra funding for the program, that UTPO officers would be exempt from extracurricular activities, and that they would be members of the RMC Officers' Mess. At the Faculty Council where this was announced, Dr Naldrett argued that the officers should not be admitted to first-year courses, should not take the same degree as cadets who did the full RMC program, and, being students, should not be admitted to the mess. The new commandant, Brigadier-General Lye, when he signed the council's minutes, added a warning that RMC should be wary of taking on extra commitments lest it jeopardize future claims for additional staff.

The Faculty Council had reported that RMC's political science courses would be too full next year to take additional students and had recommended that UTPO students should not be admitted to engineering courses. The only degree program that could be offered was a BA (Extension), which could include mathematics, physics, and chemistry, to be approved by the Senate. The council recommended that limits be set on UTPO enrolment so the program would not harm cadet education, and that UTPO students should meet the conditions of the RMC extension program, should be twenty-five years old, and should have reached at least an army captain's rank. Candidates should offer acceptable credits from recognized universities, and should propose acceptable course programs. RMC would not provide rations or accommodation. The commandant informed CDEE that RMC could run a pilot program on these lines in 1971–2 but, to enable the results to be analysed by January 1973, it should not be repeated in 1972–3.[23] Lye said later that he learned that the commander CDEE decided not to go forward with this pilot proposal.[24]

In June 1971 the Cairns Extramural Division Committee reported its findings about a program of full-time courses for serving personnel. The General Council of the CMCs approved them in principle, made some revisions relating to credits and residence, and passed them on to the RMC Senate and Faculty Council to be implemented.[25] In October the Educational Council sent its views on academic regulations for the proposed extension degree program to the RMC Senate, the body which, under the RMC Degrees Act, had the authority to confer degrees. It told the Senate that twenty sessional credits beyond grade 12, or an equivalent, ought to be required for a degree, and that courses in the program ought to be suitable for staff-college preparation as well as for a degree. The general philosophy should be to accept for credit all courses taken earlier by Canadian Forces' officers who aspired to a university degree, even those outside the academic system. At least two courses should be taken in residence, and fees could be charged.[26]

The Faculty Council knew that the RMC Senate had asserted that a UTPO program could only be undertaken if the necessary teaching and administrative staff positions, as well as financial support, were forthcoming. The Senate had maintained that the creation of an extension program should not jeopardize existing cadet and graduate programs or RMC's academic reputation. There must be as much distinction as possible between the degree that cadets received and the proposed extension degree. To obtain that clear distinction, the Senate had suggested that a Canadian Forces Extension College should be established with its own dean who would, however, be a member of the RMC Senate and the RMC Faculty Council.

The RMC Senate now proposed the introduction of an extension BA (General) degree and also of a BA with majors in two broad interdisciplinary areas – physical sciences and social sciences. The Senate must have sole right to set admission requirements, and twenty courses must be taken. If, as seems probable, this was to be on the basis of entry with a junior matriculation, it is equivalent to what Cairns had suggested. But the RMC general degree requires twenty academic courses on the basis of a senior matriculation, plus four courses in military studies. The Senate stated that transfer credits could only be given for courses in which the college staff had competence, but that interdisciplinary courses could be accepted. The Senate recommended that there should be a "tutorial course program," similar to one at Queen's University, that would enrich course work by periodic meetings between students and specially appointed tutors, and that would make extensive use of audio-visual aids at various centres in Canada and Europe. Thus, at that time some form of correspondence course school was contemplated. The Senate's memorandum informing the Advisory Board about its proposals said that "a more contentious point . . . in print at least" was a requirement that at least five of the courses be taken at RMC. Courses taken at Royal Roads and CMR would not count towards this residence requirement because an RMC degree should not be granted without at least one year's contact between the student and the RMC faculty.[27]

In 1972 the Advisory Board learned that RMC was willing to proceed with the provision of extension courses in several subjects leading to a degree, with an assumption that academic standards would be maintained. It was planning classes in history, economics, political science, and other subjects. That program would require the establishment of a new academic division within the college, with its own dean.[28]

The board was also told that between one and two thousand officers and men in the forces were enrolled in various university courses across Canada and the United States. Because of postings, especially when the university with which they were associated did not conduct correspondence courses, many of these service students experienced difficulty in satisfying residence and other requirements. CFHQ had therefore come to the conclusion that there ought to be a scheme whereby serving members of the forces could earn degree credits at the three CMCs so that, regardless of postings, or of the isolation of their place of work, they could obtain a degree. That was why it had requested the RMC Senate to establish course patterns, appoint a dean and a registrar, and decide on the courses to be offered in 1972–3.

The board noted, however, that the question of a control organization, and also of funding, still had to be tackled.[29] As it stood, CFHQ's proposal would obviously require a considerable increase in the number of courses offered as evening extension classes and by correspondence, and CFHQ had not yet said how this expansion

would be financed.

By June the new CDS, General J.A. Dextraze,* had decided that immediate steps must be taken to increase the number of other ranks attending the CMCs on a UTPM program, and in October he expanded this to include UTPO.[30] His intentions were related to, but different from, proposals made thus far for correspondence and evening classes, and for degrees to be earned partly by mail. They raised as yet unresolved questions about the effect of bringing mature students into classes with the Cadet Wing. The CMC commandants met with the new chief of personnel, Rear-Admiral D.S. Boyle, to discuss the most effective and appropriate way to implement the CDS's directive that both of these groups of serving personnel undertake full-time academic programs leading to purely academic degrees. The meeting seems to have reached no clear agreement.[31]

The RMC commandant and principal believed that success in expanding RMC's activities into the extension field depended, at least in part, on the approval of the academic community in Canada. They brought the matter before the Council of Ontario Universities (COU) on which they had membership. They noted that about 1500 servicemen now took extension courses in Canadian universities but, as a result of postings and transfers from one university to another, many of them had difficulty in completing a degree. They said that RMC had a special responsibility for officer education, that it already gave graduate and undergraduate courses to serving personnel, and that, in response to an NDHQ request, the Senate had approved a BA degree "in extension."

*Gen. Jacques Alfred Dextraze was commissioned in the Fusiliers Mt Royal (Reserve) in 1939 and went on active service in 1940. He served in Western Europe and became commanding officer of the regiment in 1944. He retired after the war, but was recalled in 1950 to command a battalion of the R22eR in Korea. Chief of personnel 1970–2, he was CDS 1972–7.

The RMC officials told COU that they now proposed to create a subsidiary college called "The Canadian Forces' Extension College," a Division of RMC, with its own dean. It would grant degrees by virtue of the 1959 RMC Degrees Act of the Ontario legislature. Extension degrees would be conferred partly on the basis of credits earned for courses in other universities in regular, summer, or evening classes. Work in staff colleges and at the National Defence College could also be accepted "after proper assessment of their curricula." RMC proposed to offer its own credits for "regular courses conducted in the evenings and at summer schools," partly by correspondence and partly by visiting instructors and appointed tutors. A minimum of five such courses in residence would be needed, and the remainder could be transfer credits. RMC would also set up a counselling service.

Lye and Dacey sought the views of Dr John S. Macdonald, the director of COU, on the validity of this proposal, and also for an opinion from his council about the way in which the academic community would react to RMC's "entering the extension field in this limited way."[32] The request to COU made no reference to the attendance of serving personnel in cadet daytime classes, so it ignored the recommendation by COU that serving personnel should be in full-time attendance.

Macdonald's reply came early in October. The COU executive believed that RMC's proposal, if implemented, would fill an important educational need for armed forces' personnel. Macdonald said Canadian universities were anxious to enroll service personnel as part-time students but, because of the frequency of service postings, were unable to offer coherent programs suitable for individual needs. Universities could not provide adequate counselling, and it would be particularly difficult to modify residence requirements and regulations about external credits to meet the special needs of service personnel. Consequently, the executive encouraged RMC "in the strongest terms" to provide serving personnel with educational opportunities comparable with what universities made available to civilians. It was sure that the universities would cooperate.[33]

The proposed extension program would need an additional faculty member, or more, in each department to handle the correspondence teaching. A meeting of the Educational Council had been called for 30 November in Kingston to discuss the proposal. But less than a month after Lye received the reply from COU, he and the other commandants received a CFHQ telex forwarding a revised agenda: "CDS directed the following: Commencing academic year 73/74 increased numbers of UTPM candidates to be accepted by CMCs. Requirement for these candidates to be single waived. While at CMC's married UTPM to be treated as single . . . Immediate study on desirability of accepting UTPO as full time CMC students." The message added that the chief of personnel was to report to the CDS on the implementation of these instructions, and that, since success depended on the forces' ability to recruit larger numbers of suitable and qualified applicants, the CMCs were to review their admission qualifications – such as the age for UTPM – in order to increase their intake. Admission to UTPO was to be coordinated as far as possible with the proposed extension degree program. The CMCs were to present to the coming Educational Council meeting changes they would make in regulations for UTPM entry, and also their views about the desirability, the full implications, and the conditions of accepting UTPO as full-time students.[34]

The CMC commandants discussed the implications of this proposed alteration of UTPM admission standards and introduction of UTPO students into the colleges. All the commandants opposed creation of a double standard for cadets on the one hand and UTPO/UTPM on the other, and all were concerned about the attitudes that married and single students over twenty-five might have towards the unique society of a CMC wing. However, they recommended that married UTPM personnel could live out, attend classes in Canadian forces' uniform, and work towards an extension degree. In January they met again with Rear-Admiral Boyle, who informed them that the CDS had ordered that marital and age entrance qualifications should be relaxed for UTPM.[35]

The CDS's instruction about UTPO was related to a major reorganization of the Canadian Forces' career development system undertaken as a result of an NDHQ Officer Career Development Programme Report. That reorganization had sought to overcome certain officer-career pattern problems, especially rank stagnation, high unscheduled attrition, and lack of structural flexibility. The report had been completed in July 1972. Its major concept was the introduction of a three-stage career pattern, with an initial engagement of nine years of service, a second engagement period to around the age of forty, and a final extended service to the age of fifty-five. The proposal included the introduction of an academic upgrading program by which an officer who reached the rank of lieutenant-colonel could acquire a degree before retiring. The admission of UTPO into the CMCs would contribute to this plan.[36]

On 31 January 1973 Boyle informed the CMCs that the full program for UTPO/UTPM would begin the following September, with UTPO and married UTPM living out. All UTPM, married and unmarried, were to have normal officer-cadet status, notwithstanding the fact that the former did not live in the college. They would take part in the full officer-cadet program. Since the UTPO students would be officers with varying degrees of experience, the degree of their involvement would be commensurate with their rank, experience, and years of academic attendance. All the applicants for UTPO and UTPM support who were selected for a program of study would attend a CMC if that study was available there. Where there was conflict for available space in the colleges, UTPO/UTPM would be given priority over ROTP/RETP. Selection would begin in May, and it was expected that the CMCs would receive between twenty and thirty UTPO, and ten UTPM, for the coming session. A few of the UTPM candidates would have a year of university credits, but most would enter the first year. All UTPO applicants must have at least one year of university credits, and they would need a maximum of two further years to complete a degree.

The principal told the Faculty Council that the CDS had made this decision "unilaterally" (but presumably in consultation with his own staff), and that all three commandants were concerned about the effect that the change would have on cadet wings. The colleges had been strongly urged to make every attempt to accommodate the CDS's wishes. Lye again recommended admission of UTPO and UTPM on a trial basis. The commandant at Royal Roads suggested that the newcomers should get a normal CMC degree, but on a different basis.[37] A proposal was therefore to be submitted to the chief of personnel to provide a means of academic upgrading to degree status of as many UTPO/UTPM candidates as possible.[38]

When the Advisory Board learned of these developments in February, it expressed grave concern lest the admission of mature students, who were to live apart, would undermine the characteristics of the Cadet Wing; and it urged that there should be no double standard. It was told that a degree different from the RMC degree would be offered by a new Canadian Forces Military College that would be a division of RMC.[39]

The UTPO/UTPM program was, then, to be affiliated with the extension degree being planned at RMC. At this point, however, a new development cast doubt on the future of the extension programs in the CMCs. The president of the University of Manitoba, Dr Ernest Sirluck,* suggested to the minister, Mr Richardson, that his university could offer extension degrees to serving personnel based on transfer credits and correspondence

---

*Dr Ernest Sirluck, dean of the School of Graduate Studies, University of Toronto, 1962, president of the University of Manitoba 1970–6.

courses.[40] It may be significant that Mr Richardson was MP for Winnipeg.

Manitoba offered its services for an annual subsidy of $25,000, which was considered a bargain. Its proposal was said to be "uncannily" like the extension system on which RMC had been working for some years, and there was some fear that it might make the CMCs' proposal for extension teaching redundant. CMR promptly dropped its extension courses, but RMC offered to coordinate its extension work with that of Manitoba. Furthermore, its Faculty Council voted to continue evening classes in September 1973, giving teaching credits to the instructors. CFHQ decided, however, to go ahead with the Manitoba scheme as a distinct program on its own. RMC then informed headquarters that it would not be able to continue extension classes without extra funding. Accordingly, RMC's extension program continued as a parallel to, rather than as a part of, the Manitoba extension system, and its extension classes became subsidiary to its UTPO/UTPM program which emphasized the difference between RMC's officer-cadet degree and the Canadian Forces' Military College Division's degree.[41]

On 8 March 1973 Lye circulated details of RMC's proposals for further discussion of the plans for UTPO/UTPM attendance at the CMCs. Dextraze, however, believed that RMC was proceeding too slowly, and he summoned the RMC commandant and principal to his office. He complained at considerable length of what some in CFHQ had called RMC's "foot dragging." The programs were necessary and they should be "sensibly" implemented throughout. Coupling the UTPO and UTPM programs with another different development, the introduction of bilingualism, he instructed that he be personally advised quarterly about progress in both respects.[42] So, on 2 April, Lye sent to NDHQ proposals about UTPO/UTPM attendance that were similar to those he had circulated to the commandants for discussion a month earlier. Registering therein his own concern that "the disadvantages appeared to outweigh the advantages," he added that the

program was worth a trial.[43]

Early in 1973 it was assumed that the number who would apply for UTPO and UTPM would exceed the numbers who had applied in the previous year – ninety and 250, respectively. As of 13 April there were fifty UTPM applicants for RMC and sixteen each for RRMC and CMR, apart from applicants for the universities. There were forty-nine UTPO applicants for RMC and seventeen for CMR. Dr Dacey informed the Faculty Council, however, that relatively few of the UTPO applicants would come to RMC because they wanted courses not given there.[44]

The number of UTPO eventually selected for RMC in 1973 was seven, five of whom were admitted to the third year and two to the fourth. There were ten UTPM, all enrolled in the first year. The majority of the UTPO were in arts courses. They had no refresher course before entry. UTPM entrants, selected on the same basis as ROTP recruits, got refresher courses in mathematics and chemistry, which they greatly appreciated and which, incidentally, provided time for them to settle their wives and families in Kingston. Since they were in first year, the UTPM students had not yet entered the arts or sciences stream, but implications based on their interests suggested that by 1975–6 the third year in electrical engineering at RMC would be 30 per cent UTPM, a valuable development in a classification that was in short supply.[45]

During the debate about UTPO/UTPM in the CMCs, Boyle had said that, while ROTP was a method of indoctrinating the educated, the new scheme would "educate the motivated." This simple concept was not how things turned out at first. UTPO officers in the CMCs were on the whole well motivated, but most UTPMs admitted that they had wanted to go to a civilian university. By January, however, they said they believed their military motivation had been increased as a result of attending a CMC. That did not happen with UTPM students in the universities.

UTPO students were first enrolled in the Officers' Mess of Canadian Forces Base, Kingston, but they were soon admitted to the RMC Senior Staff Mess, where they became enthusiastic about their membership. They were not easily distinguishable there from officers doing graduate work. Like those officers, they gave valuable service to the college by advising and counselling their fellow cadets (with whom they had a good rapport) about classification and other service matters.[46]

CMR and Royal Roads set up separate squadrons for their UTPM students, but in RMC they were at first attached to existing cadet squadrons. UTPM were not obliged to take part in either intervarsity or intramural competition, though some did so voluntarily. RMC attached UTPO to existing squadrons as supernumeraries and, although most had family responsibilities off station, some participated in squadron social and athletic activities.[47] Later UTPO-UTPM at RMC were organized in their own "Otter" squadron, named after Canada's South African War general. Dr Walter Avis,* the first full-time dean of the Canadian Forces Military College, sought strenuously to have his people freed from some of the wing's extra-curricular activities. Membership in the all-years cadet mess across Highway 2 gave UTPM an opportunity for access to military social life if they wanted it.

Wives presented more serious difficulties. For UTPO wives the transition was relatively easy, not unlike previous postings in their husbands' service careers. But with UTPM it was different. Many married UTPM, living off-base with wives and often also children, found it hard to combine a full-time academic program with family life. UTPM wives, thrust into a four-year stint during which their husbands were completely absorbed as students, often found adjustment difficult. Some became disgruntled. At the instance of Mrs Faith Avis, the wife of the dean, they formed their own UTPM Wives' Club to meet for social and other purposes related to their husbands' careers. Eventually they were admitted to some special meetings of the Kingston Base Officers' Wives Club.

Some cadets complained that UTPO and UTPM students had an advantage in the competition for grades, because they were excused extracurricular activities.

Otter Squadron on parade, graduation, 1990

Captain (N)(R) J.B. Plant, OMM, CD, ndc, P.Eng, PhD, FEIC, director of studies and principal, 1984–

UTPO and UTPM classroom motivation was, indeed, excellent, and this was in part the reason why their academic results were very satisfactory. In the first term they achieved an average of 67.58 per cent compared with an RMC class average of 65.3 per cent. At the end of the first year, eight UTPMs were accepted to take engineering, one took science, and one an arts course. The UTPO/UTPM program was a success.[48]

Early in the following year one change had to be made. RMC had developed a three-year pass degree course for

*The first dean of the Extension Division was Capt. (N) John Plant. He was succeeded by Dr Walter S. Avis, professor of English and editor of the *Gage Canadian Dictionary*, published in 1973. Dr Avis had enlisted in the Governor-General's Horse Guards but transferred to the Royal Canadian Corps of Signals at the outbreak of war, serving overseas. He graduated from Queen's University and did his doctorate at Michigan. He was appointed to the English department at RMC in 1949.

both UTPO and UTPM. This was not what Ottawa had intended. On 30 May 1975 Lieutenant-General James C. Smith, ADM(Per), instructed that UTPM entering with advanced standing were to take the full CMC four-year program. He insisted that UTPO/UTPM selectees must choose combat operational and engineering classifications rather than logistics. However, the regulations on age limits had to be revised because it would be impossible for UTPM in operational classifications to qualify for command positions before reaching the age ceiling. Smith therefore rejected an RMC proposal to include those with advanced standing in a full CMC four-year program. A three-year pass degree was acceptable for UTPO. Both UTPM and UTPO had a choice between taking honours and pass courses.[49] For UTPO, a pass degree required fifteen credits as against twenty for an honours degree.[50] With these provisos, Smith informed the General Council on 17 October 1977 that the integration of UTPO/UTPM in the CMCs was now satisfactory.[51]

In 1978–9 the number of serving officers taking undergraduate courses had reached fifty-eight, but the UTPO portion had dropped slightly. As a result of a voluntary response by faculty, four evening courses in economics, science, English, and French were offered, but the latter was withdrawn because of the lack of students.[52] In 1980–1 only thirty UTPO and five UTPM completed their year, compared with forty-five a year earlier. Eleven Canadian Forces' Military College bachelors' degrees were awarded in that year, all but one of them to UTPM. RMC deplored the decline in numbers, not only because of the potential loss to the services, but also because it meant loss of benefit to the academic climate in the college – an indication that UTPO/UTPM had not been as deleterious as had once been feared, but was actually advantageous. Because of the decline in numbers, NDHQ revived the entry of applicants with advanced academic standing, which had been dropped in 1976 in favour of admission to the civilian universities.[53]

In 1982 the UTPO intake rose to seven, compared with three the previous year. The UTPM intake (in addition to fourteen transfers from RRMC and CMR) was thirteen, the highest level ever, and a considerable gain over two years before, when it had reached its lowest point.[54] The total enrolment of UTPM was now forty-one, and it was expected to be fifty in 1983–4. UTPO had similarly picked up again, but was rising more slowly than UTPM.

From the time of the introduction of extension classes in 1970–1, 341 course credits had been earned in them by 1981–2, most students taking courses either for personal interest or for university credit elsewhere.[55] In 1982 WO Tony MacNeill was the first NCO to get a Canadian Forces' Military College degree by extension courses. He had entered the forces in 1962 with a sixth-grade education and had taken extension courses in various universities and for six years at RMC.[56] In 1983 a UTPM student, M.J. Brown (no. M000177), who graduated in honours economics and commerce, placed in the top 10 per cent of all the students in North America who wrote the Graduation Management Admissions Test. UTPM intake in that year was twenty-four, the highest in the history of the program. Furthermore, UTPM at RMC had the highest retention rate of any university-level officer education program. But UTPO had declined again, partly because NDHQ was not sending officers whose retirement option could only give them fewer years of remaining service than would repay their UTPO service on a two-years-for-one basis.[57]

There had been no serious problems or harm caused to the unique society of the Cadet Wing, as some had feared. By 1984 UTPO and UTPM graduates were accepted for membership in the RMC Club.[58] Some ex-cadets did, indeed, still look back to their own training and education as the ideal one (a common opinion of many alumni in all fields) and still scorned the changes; but RMC authorities believed that the new college with its new members had retained the essential formative elements that were the solid foundation of the old.

Francophone Representation and Bilingualism

By 1973, when the Directorate of Professional Education and Development absorbed the defunct Canadian Defence Educational Establishments, the postwar RMC had attained a certain degree of stability and confidence. Some of its long-standing problems were being solved. In the previous decade, definition of the functions of the Faculty Council and the Faculty Board had changed to allow some input by the faculty, even if in the board it was only to respond to requests for advice on policy matters. Some college regulations had been revised to come to terms with contemporary youth lifestyles, including the use of cars, informal dress, and leave and bar privileges, with a subsequent reduction in cadet wastage. Some courses in the technological and military areas had adapted to the computer age, and there was more specific military content in the social sciences and in the problems studied in science and engineering. All these changes had been accomplished without abandoning vital and fundamental RMC traditions.

There seemed to be a better balance between academic education and professional military development. That balance was, however, very delicate. When Brigadier-General W.W. Turner, disappointed by the quality of RMC's drill on church parade, ordered an extra hour of drill each week for the whole Cadet Wing, the faculty feared his unilateral action might become a precedent for more serious pressures on study time.[1] Neither the commandant nor the faculty in this minor confrontation seemed to realize they were now facing an irresistible demand for much more time from a third source. A 1971 Canadian Defence Educational Establishments (CDEE) directive from Ottawa that every RMC cadet must become functionally bilingual before he graduated had yet to be implemented.[2]

CDEE's directive was in accord with the Trudeau government's policy to give Canada's two official languages equal status and Canada's two founding peoples equal opportunities. This policy had been drafted as a follow-up to the Laurendeau-Dunton Commission on Bilingualism and Biculturalism appointed in 1963. In the first volume of its report published in 1967, the commission had recommended wide measures to implement linguistic equality. With the rise of separatism in Quebec, the federal Liberal party's dependence on the French vote, and Trudeau's personal support, these proposals quickly became law. In July 1969 Parliament passed the Official Languages Act to extend the use of both languages in all federal government services. The act appeased some francophone grievances, but it was unpopular with many anglophones who ignored the fact that up to this time francophones had been disadvantaged. One retired naval pilot, Lieutenant-Commander J.V. Andrew,* wrote a

---

*Lt-Commander James Vernon ("Jock") Andrew, RCN, served with the Royal Navy for several years after 1947 on an exchange. In the 1960s he was a marine engineer with the RCN, attaining the rank of lieutenant-commander and retiring about the time of integration.

book alleging that the B&B policy was a Trudeau plot "to hand Canada over to the French-Canadian race."[3] That charge was absurd, but opposition to bilingualism in many parts of the country, in the Canadian Forces, and also at RMC and Royal Roads was strong.

At this point it will be useful to review the history of francophone representation at RMC down to 1973. From their inception in 1868, the Canadian Armed Forces had been predominantly anglophone in composition, their working language was English, and anglophones almost completely monopolized the technical corps of the militia, the RCN, the RCAF, and all high ranks. Around the turn of the century two British GOCs, Herbert and Hutton, had recommended that a knowledge of French would be useful for Canadian officers who might have to lead French troops, but they had been ignored.[4] The number of francophones who served in the two world wars was well below the Canadian national average, though in the early years of the First World War the rate of French-Canadian volunteering had been similar to that of Canadian-born anglophones. In both wars manpower crises led to bitter recriminations between English and French Canadians over conscription.[5] Despite this evidence that many in Canada's 30 per cent minority might again refuse to serve in national emergencies, governments after each war made no serious attempt to make military service more acceptable to francophones. Defence Minister Brooke Claxton claimed he favoured an increase in francophone representation in the forces, however, and General Charles Foulkes told the Defence Council it would be convenient for Canadian officers to know French.[6]

By that time the need for recruits for the war in Korea had compelled the government to take action. It was slowly coming to be realized and admitted that many francophones were unwilling to join the defence forces, not so much because they still regarded them as props of British imperialism or because of their feelings about the conscription crises, but because they were reluctant to serve when the forces' working language was English, when their own opportunities for promotion were slowed, and when they frequently had to live in English-speaking communities. The need for a more proportionate francophone presence in the forces thus came to be associated with the national policy of bilingualism and biculturalism, and also with the cause of Canadian unity and Canada's distinctive identity.[7]

The Official Languages Act, 1969

17-18 ELIZABETH II	17-18 ELIZABETH II
CHAPTER 54	CHAPITRE 54
An Act respecting the status of the official languages of Canada	Loi concernant le statut des langues officielles du Canada
[Assented to 9th July, 1969]	*[Sanctionnée le 9 juillet 1969]*

Her Majesty, by and with the advice and consent of the Senate and House of Commons of Canada, enacts as follows:

Sa Majesté, sur l'avis et du consentement du Sénat et de la Chambre des communes du Canada, décrète:

SHORT TITLE

TITRE ABRÉGÉ

Short title **1.** This Act may be cited as the *Official Languages Act.*

1. La présente loi peut être citée sous le titre: *Loi sur les langues officielles.* Titre abrégé

DECLARATION OF STATUS OF LANGUAGES

DÉCLARATION DU STATUT DES LANGUES

Declaration of status **2.** The English and French languages are the official languages of Canada for all purposes of the Parliament and Government of Canada, and possess and enjoy equality of status and equal rights and privileges as to their use in all the institutions of the Parliament and Government of Canada.

2. L'anglais et le français sont les langues officielles du Canada pour tout ce qui relève du Parlement et du Gouvernement du Canada; elles ont un statut, des droits et des privilèges égaux quant à leur emploi dans toutes les institutions du Parlement et du Gouvernement du Canada. Déclaration du statut des langues

STATUTORY AND OTHER INSTRUMENTS

ACTES STATUTAIRES ET AUTRES

Instruments directed to public **3.** Subject to this Act, all instruments in writing directed to or intended for the notice of the public, purporting to be made or issued by or under the authority of the Parliament or Government of Canada or any judicial, quasi-judicial or administrative body or Crown corporation established by or pursuant to an Act of the Parliament of Canada, shall be promulgated in both official languages.

3. Sous toutes réserves prévues par la présente loi, tous les actes portés ou destinés à être portés à la connaissance du public et présentés comme établis par le Parlement ou le Gouvernement du Canada, par un organisme judiciaire, quasi-judiciaire ou administratif ou une corporation de la Couronne créés en vertu d'une loi du Parlement, ou comme établis sous l'autorité de ces institutions, seront promulgués dans les deux langues officielles. Actes à l'intention du public

Legislative instruments **4.** All rules, orders, regulations, by-laws and proclamations that are required by or under the authority of any Act of the Par-

4. Les règles, ordonnances, décrets, règlements et proclamations, dont la publication au journal officiel du Canada est requise en Actes du pouvoir législatif

1213

Those who defended the language status quo in the Canadian Forces did so on the grounds that operational efficiency required the use of a single working language and that all francophone officers and men must learn English. Some argued that, although fewer francophones reached high rank, the average age of those who did was lower than that of English-speaking officers, suggesting that promotion was easier for francophones.[8]

In 1950 Jean-François Pouliot, MP for Temiscouata, a long-serving parliamentarian and a distinguished jurist, commented in the House on the lack of the use of French in the Canadian Forces. Pouliot was related to Brigadier Jean-Victor Allard, who had led the Royal 22nd Regiment with distinction in the war. Then in 1951 Léon Balcer,* MP for Allard's home riding of Trois-Rivières, suggested the creation of French-speaking crews for naval vessels and asked in the House of Commons how many cadets in the military colleges were francophones. On 5 November of that year, Allard, recently nominated to be vice-quartermaster general, sent a memorandum to the CGS, Lieutenant-General Guy Simonds, to propose the establishment of a French-language military college to attract francophones into the forces. When Simonds failed to respond, Allard managed to get the attention of Prime Minister Louis St Laurent. Four days before two important Quebec by-elections, Claxton announced that a bilingual military college would be opened in Quebec to prepare officers for all three services. This was not the French college for the army that Allard had proposed, but it was a follow-up on his initiative. Collège militaire royal de St-Jean (CMR) was thus a result of political expediency rather than a general conviction or consensus that language equality was a necessary step towards recruiting more francophones. With a proposed 60:40 French-English representation, CMR could be a step towards national unity.[9] Allard then made an agreement with Major-General Paul-Émile Bernatchez,** deputy adjutant-general, an RMC ex-cadet who was the senior francophone in the army, that they would both seize

every opportunity to move the balance in the army in favour of francophones.[10]

During the 1950s the Canadian army began to give basic military training in French to francophone recruits in the ranks before they were switched at more advanced levels to instruction in the English language. The navy and the air force followed suit more slowly. By 1965 it had been conceded that French could be used as a working language in some units in all three services, but as late as 1968 NDHQ still reiterated that the language of the forces above the unit level was English.[11]

Like the services as a whole, RMC had always been monolingual. The first entrance examinations in 1876, which were in English, had required translation from Latin, French, or other languages into English, but not into French. Not surprisingly, there was only one francophone among the first forty cadets, and he had been previously educated in English.[12] When the examination for the second entry year permitted translations into French as an alternative, no requirement for translation from English into French was imposed, presumably lest it reduce the number of anglophone recruits.[13] By 1900 there had been only twenty-three francophones among the first 500 cadets.[14] Desmond Morton found 255 RMC graduates in the *Militia Lists* over the same period, but

---

*Léon Balcer, RCNVR, served as a lieutenant during the war, 1940–5. He was elected to the House of Commons for Trois-Rivières in 1949. President of the Young Conservative Association of Canada, 1952, and of the Conservative Association of Canada, 1956, he was appointed solicitor-general in 1957.

**Maj.-Gen. Paul-Émile Bernatchez (no. 2074) was at RMC 1929–34. Commissioned in the Royal 22nd Regiment, he served overseas in Sicily, Italy, and Northwest Europe and commanded the 3rd Infantry Brigade. Deputy adjutant-general in 1946, he served in Korea and on the Joint Board of Defence. Vice-chief of the General Staff 1961–4, he became colonel of the Van Doos. Bernatchez said later that if he had known that creation of CMR's preparatory course would lead to the fragmentation of the military colleges he would not have worked for it because it might also fragment the Canadian Forces. Information from Capt. (N) P. Fortier.

he could identify only eleven as French.[15] Among the next 500 cadets from 1900 to 1914, only nineteen were francophones.[16] Most of those came from well-to-do families who had been largely assimilated and were already bilingual. Because their English was limited, and because Quebec francophone education stressed the classics rather than the sciences that were needed for RMC, other francophone candidates often found difficulty with the entrance examination questions. From 1879 to 1897 a francophone medical doctor, A.D. Duval,* who was a poor teacher and could not keep order in class, taught French grammar and reading, but did not stress oral French. It is significant that RMC's next three professors of French were natives of France.**[17]

Soon after the college reopened in 1948, the number of requests for copies of the application forms in French seemed to indicate increased French-Canadian interest in military careers,[18] but only 12 per cent of the New One Hundred class of 1948 were francophones. In 1949 the Operational Research Group of the Defence Research Board refused to take over the administration of the Canadian Services Colleges' entrance examinations with provision of alternative papers in French. It believed that, to be fair, this would have to be more than a literal translation of the questions in English, and so would cost more than the number of French applicants warranted.[19]

In the first term in 1948, in order to compensate for the extra time they had to devote to English, some francophones were excused from written work in the compulsory French course.[20] But although francophone cadets in at least one department (history) were permitted to write term examinations in French, all instruction continued to be in English and those who had had to write term examinations in French were encouraged to improve their English quickly. As a result, most francophone recruits rapidly became functionally bilingual.

Opened in 1952, CMR gave two or three years of bilingual instruction for its first decade before sending francophones and anglophones alike to do their third and fourth years in monolingual RMC. But relatively few anglophone recruits, even those who started out at CMR, had acquired a functional command of French by the time they graduated from RMC. In 1962 Dr Gerald Tougas,*** head of the French department, said he believed that only about 2 per cent of RMC's anglophone cadets were fully bilingual at graduation. The compulsory French courses were "nothing more than sops for Francophones." The registrar, Colonel T.F. Gelley, agreed, but claimed that francophones accepted this situation. Bilingualism, although highly desirable, was not a primary aim of the college. Tougas recommended more French courses and more French discussion classes,[21] but no progress occurred at that time.

In 1963 there were only two anglophones enrolled in French honours courses. There was a French table in the dining hall, attended by French-speaking professors, and some anglophone cadets went for a two-week visit to Quebec City.[22] There was also a French club, the Cercle Chabot, named after the first head of the French department.† All these measures fell far short of the effective promotion of anglophone bilingualism. Three periods of French grammar, composition, and literature a week for two (later three) years were far from sufficient to make many anglophone cadets capable of functioning in French. The idea that English-speaking Canadian officers should become competent in French was slower to

---

*Dr Arthur Duponth Duval, MD, professor of French at RMC 1879–97. See Preston, *Canada's RMC*, 69n.

**Capt. J.D. Chartrand, 1879–1905, Professor M. Lanos, 1905–18, and Georges Vattier, 1918–25.

***Dr Gerald Tougas, professor of romance studies, UBC, from 1958, professor of French, RMC, 1962–3, professor of French, UBC, from 1963.

†Lt-Col. Charles Chabot, lecturer in French at RMC before the Second World War, served in the Canadian army during the war and prepared a French-English military dictionary. Appointed head of the Department of Modern Languages at RMC in 1948, he was director of studies at CMR 1955–61. Madame Chabot wrote the college march, *Precision*.

gain acceptance in the military colleges than measures to facilitate the admission of francophones into the officer corps. The assumption was still that francophones would become bilingual but that bilingualism was not necessary for anglophones.

However, some important considerations pointed to change. The Canadian Forces desperately needed more engineers, the services colleges had become the chief source to supply them, and most of the francophone applicants for ROTP wanted to become engineers. Yet in 1956 only 7 per cent of 248 francophone enquiries about ROTP led to firm applications, a serious decline from the previous year when 19 per cent of 302 enquirers had applied. In contrast, 20–29 per cent of the francophone inquiries submitted to CMR produced firm applications. All French applicants were henceforth referred to St-Jean.[23]

Francophone wastage rates in CMR were lower than in RMC, and many CMR cadets who transferred into the third year in Kingston did less well than they had done at CMR. Annually from 1957 to 1962, between 30 and 50 per cent failed their third year. Many francophones were not yet fully bilingual, and the move into an English milieu had serious psychological effects on their attitude. The RMC Faculty Council initiated a special study to see how the college's handling of French-speaking cadets could be improved.[24] In 1963 RMC informed the Advisory Board that a francophone Catholic "padre" (chaplain)* had been appointed, there were elective courses in French in third-year politics and history, that every department had a professor who could speak French, and that the failure rate of CMR cadets in the third year had dropped to 11.45 per cent, which was roughly the same as the class average of 11.5 per cent.[25] Presumably this meant that French-Canadian cadets who had survived transfer to RMC were doing better.

Unification in the late 1960s eliminated the RCAF and RCN as enclaves of anglophone exclusiveness but diluted the overall French proportion of the officer corps from what it had been in the army. However, the appointment of General J.V. Allard as chief of the Defence Staff in 1966, the first francophone to serve at that level, meant there would now be more effort to eradicate the linguistic inequalities in the forces. In his memoirs, Allard says he stayed in the services after unification to improve opportunities for francophones in the forces. When he was vice-chief of the Defence Staff, the future of his Royal 22nd Regiment, along with other francophone units, had come into doubt, and he offered his resignation. Although the Official Languages Act had not yet been passed, Defence Minister Hellyer promised Allard he would find a solution for a suitable francophone presence in the forces. Later, when Allard was offered the appointment as chief of the Defence Staff, he took the unusual step of making his acceptance conditional on the government's acting on its promise to improve the lot of francophones. Hellyer told him this would require cabinet approval, which he confirmed within a few days.[26]

CMR cadets who transferred to RMC's third year were well aware that current official policy favoured bilingualism. In December 1968, before the passage of the Official Languages Act, three CMR cadets prepared a brief on the problems francophones faced at RMC, and also on the difficulties anglophone cadets from CMR experienced if they wanted to continue to improve their French in the English environment in Kingston. They said RMC did nothing to encourage either group.

An RMC Bilingualism Committee, set up in response to this brief, noted the limitations that RMC's anglophone environment imposed on bilingualism and recommended the introduction of the use of English and

---

*After the Second World War RMC had from time to time had a Catholic padre who was French Canadian. The first chaplain to be specifically appointed in the Bilingual and Bicultural program to serve Catholics in both language groups was Capt. J.C.W.P. Doyon, who arrived in 1975. He had a distinguished academic and athletic record. Information from Capt. (N) P. Fortier.

French on alternate weeks in cadet daily routine. When Commodore Hayes forwarded the cadets' brief to General Allard to show him the nature of the bilingual problem at RMC, he said he personally believed there should be a "most serious effort to increase the use of French." However, he added that alternate use of French and English, as practised at CMR, was unlikely in the near future.[27]

After 1965, all third-year RMC cadets capable of taking military history in French were required to do so. Those who could not follow the course in French were to take an extra compulsory course in French language and grammar as an arts option.[28] In 1966 Colonel Sawyer told members of the Commission on Bilingualism and Biculturalism visiting RMC that, in his opinion, courses taught in French should be given by francophones.[29] The commission report in 1969 recommended two similar military colleges "with analogous curricula," RMC an English-language institution and CMR a French-language degree-granting institution, both putting strong emphasis on the teaching of the second language.[30]

In August 1969, without previous warning, DND announced that CMR second-year cadets due to transfer to RMC (or to a university) in September could instead stay on for a further two years at St-Jean (another of Allard's objectives). This did not apply to engineers, but many CMR cadets who had pre-registered for engineering at RMC, especially those with poor English, now elected to stay at CMR to take science. The RMC Faculty Board, alarmed by these developments which severely decreased the flow of francophones into RMC's third year, discussed what could be done to ensure that those who did come would feel welcome.[31] Already in 1968 the Cadet Wing was experimenting with the use of French in two of its squadrons, but the decline of the francophone intake would now make it difficult to keep up that practice.

Impetus for further progress towards bilingualism then came from outside the CMC system. The passage of the Official Languages Act in 1969 required the Department of National Defence to communicate with members of the public in either official language as appropriate. The promotion of language equality in their external contacts as well as within the forces would obviously be more visible evidence of the government's determination to implement the act. While measures to attract more francophones into the forces could be rationalized as an increase of the reinforcement pool, and therefore of military value, efforts to increase anglophone bilingualism could be deplored on the grounds they would decrease military efficiency by taking up time and money better spent on military training and weapons. Nevertheless, bilingualism and biculturalism in the CMCs could be a key to furthering bilingualism throughout the forces; if RMC were to continue to produce francophone engineers, and be the only reliable source of them, it could do so only by becoming bilingual.

The problem remained that anglophone cadets could not easily be induced to learn French. By 1970, moreover, important elements in RMC had become disenchanted by efforts to move towards bilingualism. Professor François Gallays,* acting head of the French department, had found that the extra compulsory third-year French course for non-functional anglophones, imposed in 1965, disheartened poor students. His department wished to teach French literature and language in the traditional way, and he tried to have the French course made voluntary in the third year. At the same time, Dr James Cairns, dean of arts, argued that it would be more productive to make the compulsory course (which was taken in French by those capable of doing so) into a voluntary option. "The appropriate

---

*Dr François Gallays, BPaed, MA, PhD, assistant professor and acting head of the French department at RMC 1970–1, assistant professor of French literature, University of Ottawa, from 1972.

strategy" to deal with pressure from the CMCs' Educational Council to have a program of second-language training developed at RMC would be to show that the present compulsory program was ineffective and that "a more determined effort" was needed to achieve bilingualism. Until such an effort was forthcoming, however, it would be better to continue with the present system "rather than take a retrograde step."[32]

The principal, Dr Dacey, then told the Educational Council he did not think that a university was a place for instruction in conversational French. Furthermore, like Cairns, he believed that the third-year course taken in French should be an option. Major-General W.A. Milroy, commander of CDEE, retorted that "a more imaginative approach" was needed, but, since CDEE was currently studying second-language training in the forces, he would defer comment on how it should be developed in the CMCs until he had received the study's conclusion. He therefore ruled that the French program at RMC must continue unchanged for 1970–1.[33]

On 21 December 1970 the Educational Council approved a CDEE directive to "develop language training at the CMCs to permit all officer cadets to achieve level-four standards [in their second language] prior to graduation."[34] Level-four proficiency (sometimes called "functional") was defined as "an ability to understand, speak, read, and write the language with sufficient accuracy to satisfy normal work requirements."[35] Milroy assumed that one-third of an RMC class was now reaching level four by graduation, in contrast to a majority of those members of a CMR class who stayed on to complete a four- or five-year course at CMR, where French instruction stressed audio-lingual instruction and where all courses were offered in both languages. He said that the immediate objectives at RMC would be to allow francophones to take courses in French, and to have anglophones attain a high degree of proficiency in the language. RMC would need to introduce courses conducted in French, though only

"on a gradual basis as bilingual faculty became available." The faculty and staff should be given the opportunity to take language courses, and cadets should be motivated towards bilingualism by rewards, such as preference in cadet-appointments and attachment to British and French forces in the summer.

CDEE was well aware that the problem would be to find time for language study. Milroy emphasized that there must be no downgrading of the quality and viability of the RMC degree, and no interference with military classification training during the summers. Development of a bilingual atmosphere at RMC, like that at CMR, was seen as the key to the success of the program, but that would, of course, be difficult to achieve in an anglophone locality. The sum of $90,000 was allotted to "programmes of improvement" at the CMCs, and additional funds might be made available "subject to justification." The first step in 1971–2 would be to have the cadet body tested for language proficiency.[36]

Making anglophones bilingual was, however, only one facet of the government's policy. In April 1971 the Educational Council announced that another objective was that 28 per cent of CMC graduates would be francophones. This was, of course, the proportion of francophones in the population of Canada. Currently, 27 per cent of the population of the three military colleges was French, but only 20 per cent of the graduating classes. Because most francophone cadets took five years from their entrance to CMR's preparatory year, the population of the CMCs would have to be greater than 28 per cent of the colleges' total enrolment if they were to achieve the required proportion in their output. Furthermore, because the Official Languages Act stipulated that a Canadian had the right to be educated in his own language, there would also have to be an increased emphasis on offering francophone cadets courses in French, subject to a possible recognition that, for special reasons, engineering courses were an exception to the general rule.

Finally, a francophone presence must be encouraged in RMC and RRMC outside the classroom.[37]

Milroy had consulted Raymond Duplantie* of the Public Service Commission and Gerry Blackburn of Université d'Ottawa on how language proficiency could be improved in the forces. They replied that language training at RMC and RRMC was hardly more than a token and that an "all-out effort" was warranted; otherwise, the second-language effort should be abandoned. Ignoring the fact that Royal Roads cadets needed competence in French if they entered RMC's bilingual third year, they accepted the idea of RRMC remaining English but felt that RMC should be made as nearly bilingual as possible because it provided an ideal means of bringing officers of both language groups together to their mutual advantage. They believed that cadets were well disposed to the proposed bilingual effort but that goodwill was being lost because the colleges were half-hearted in its implementation.[38] Rear-Admiral Murdoch, then commander of CDEE, asked the CDEE academic adviser, Monseigneur Jacques Garneau, to report on the academic viability and the cost-effectiveness of a bilingual program at RMC that was due to receive forty French recruits in September 1972. He told Garneau that compulsory courses were to be given in French, but not necessarily the options. He said that degree patterns in certain specializations would be harder to offer in both languages, and that this might delay the process of bilingualization.[39]

Now that RMC was being asked to submit a plan to match CMR's bilingual program, Brigadier-General W.K. Lye, commandant of RMC, was faced with the fact that, while CMR had a strong representation of both language groups in the Cadet Wing, RMC was currently receiving only a handful of francophones who were usually already bilingual. Six had entered in September 1970. However, 50 per cent of CMR's first-year class of that session would presumably come into RMC's third year in 1972 or 1973. With a proposed simultaneous intake of forty extra francophone recruits, that could mean a possible RMC intake of eighty francophones in 1972 for whom education and training must be provided in French. Lye cautioned that to deal with such a vast increase of francophone cadets, there was much more to do than merely hire French-speaking staff. He also suggested that the recruits' preparatory camp should be given in both languages.[40]

Dr Dacey called a special meeting of the Faculty Council on 5 January 1972 to establish the minimum requirements needed to provide for an intake of the forty French and ninety English recruits expected in August 1972. He said every class would have to be duplicated to give instruction in both languages. The council concluded that a four-year degree program of that kind would need seventy-three and a half extra professors, preferably with doctorates, twelve and a half technicians, seven clerical staff, and appropriate extra offices, language laboratories, and other laboratories. These figures were breathtaking. Nevertheless, Dr Dacey reported that the RMC faculty was in favour of going ahead with the program "if the difficulties could be resolved."[41]

Whereas the B&B Commission had suggested a second engineering college at CMR, the Garneau Report favoured bilingual engineering at RMC to provide facilities for francophones to become engineers. Dean Hutchison claimed that RMC was in agreement with Garneau's views on all aspects of the problem. He added that a Working Group of RMC deans had also agreed with Garneau that "RMC is not located in a favourable bilingual environment, and [that] the implementation of courses in French

*Raymond C. Duplantie was director of the Language Bureau in the Public Service Commission. Gerald A. Blackburn was a former director-general of the Language Bureau. He had a B.Eng. from the Nova Scotia Technical College. On retiring from the civil service he had been appointed assistant professor in the Department of Public Administration at Université d'Ottawa. Armand Letellier, DND *Language Reform: Staffing the Bilingualism Programs, 1967–1977* (Ottawa: Socio-Military Series No. 3, Directorate of History 1987), 123.

may not [by itself] be sufficient to solve the problems of Francophone cadets." This was indeed a serious problem. Many francophone cadets found Kingston to be "ultra-Anglo" and they felt isolated there. Furthermore, public schools would be needed for the francophone children of francophone staff. Other needs were francophone cultural facilities, extensive opportunities for exposure to French-language instruction for the whole RMC staff, and increased provision for staff and cadet travel between Ontario and Quebec. The RMC deans approved Garneau's recommendation that francophone language instructors be provided for the first few years of the program. They could be non-tenured but, to persuade them to come from Quebec to work in a traditionally anglophone institution in Ontario, they would have to be paid above the level of the national average salary level.

The deans recommended that technical subjects in the senior years should still be taught in English because it was customary to use English for such courses in Quebec. They agreed with Garneau that bilingualism and biculturalism must not be introduced at the expense of academic quality, and, in view of the expected intake of CMR francophones into third-year engineering at RMC in September 1972, that it would not be practicable to admit the first batch of francophone recruits at the same time.

Hutchison then went on to estimate the cost of minimal second-language training and teaching for an intake of francophone recruits and third-year cadets in 1972. The salaries of the professors required in the first year of phase 1 of the program would cost $135,000 per annum, apart from other expenses. In phase 2 in 1973–4 they would cost $675,000; and successive annual phases thereafter would each add $195,000, $45,000, and $60,000 per annum, respectively. He declared that the cumulative annual increase in cost for teaching staff would be $1,110,000 per annum. In addition, there was the need for space. Building costs over four years would amount to $2,430,000. A draft of a simplified program presenting this information was submitted to CDEE on 24

February.[42] Surprisingly, these figures did not daunt those in Ottawa who were pressing for bilingualism. Major-General D.C. Laubman,* chief of personnel, wrote to congratulate Lye and his staff on the excellent second-language training program produced at such short notice.[43]

In April 1972 Murdoch informed the Advisory Board that he found it impossible to increase the francophone proportion of the CMCs' recruit entry immediately to 50 per cent, as had been suggested in an attempt to move more rapidly towards a francophone representation of 28 per cent in the officer corps. Because it was easier to attract francophones to CMR, he intended to alter the francophone/anglophone ratio there to 70:30 from the present 60:40 (which French quota had, however, often not been met), and also in the future to establish a capability for direct francophone input at RMC, and eventually at RRMC. He said that each of those colleges would become a "mirror image of CMR bilingually." He now proposed that French courses begin in the third year at RMC in 1972, to be followed in 1973 by first-year courses, more third-year courses, and the beginning of courses in French for the fourth year. In 1974 the second-year courses in French would begin, and the fourth-year courses would be expanded. In 1975 there would be a complete third and fourth year in French. This was, in effect, a proposal for instant bilingualism.

The Language Bureau of the Civil Service Commission had estimated that at the current ratio of language training at RMC, it would take from six to eight years for a cadet to reach an acceptable standard of competence.

*Lt-Gen. D.C. Laubman enlisted in the RCAF in 1940, served overseas, was shot down, and became a POW. When released from the RCAF he re-enlisted. In 1970 he was commander, HQ, Canadian Forces Europe, and the next year was deputy chief of personnel (military) at CFHQ, becoming chief of personnel in 1972.

Murdoch said 1200–1600 hours of formal classroom instruction were needed before graduation. The cost of French-language instruction at present was $500 per cadet. Adjusted for wastage, that came to $750. With intensive language training, the cost would rise to $4000 per cadet, and adjustment for wastage would make that $6000. In view of these figures, RRMC might have to stay monolingual for the present.[44] This was the only time that an Ottawa proposal to promote bilingualism in the CMCs showed any signs of being restricted by costs, but the government was determined to introduce it, whatever the cost.

The annual intake of francophones into the officer corps was 19.5 per cent of the whole intake, mainly through the preparatory year at CMR, but only 16 per cent of the CMCs' intake was francophone. The proportion of francophone officers in the forces was still only about 12 per cent, less than half of the percentage of francophones in the population of Canada. Pressure on DND to recruit francophone officers was therefore very great.[45] In fact, in 1970 a training command officer had reported that direct entry male and female francophones were being commissioned without any prior military training, and, he added, "it shows."[46]

In 1972 the appointment of General J.A. Dextraze, a protégé of Allard's, to be the second francophone chief of the Defence Staff suggested that NDHQ intended to push on with its efforts to increase francophone representation in the forces. Reports had begun to circulate about a proposal that the French intake into the CMCs was to be increased to 50 per cent of the first-year total intake. As a result, the Executive Committee of the RMC Club, which as usual was watching developments that affected the college, sought an interview with the minister, Edgar Benson.* They were somewhat mollified when he said that the optimum of a 50 per cent francophone intake would apply to all routes to a commission, not just to the CMCs. The intake approved for the CMCs as a whole was to be as high as 37 per cent. Approximately 30 per

cent of the RMC recruit class would be francophone by 1973, but the number of anglophone cadets would also be increased. Both RMC and CMR were to be expanded.[47]

The means by which bilingualism could be achieved at RMC had still not yet been fully worked out. In June 1972 the RMC principal had told the CMCs' General Council that five years of experience with the compulsory third-year French course at RMC had been a failure: it was "demoralizing, demotivating, and ineffective." It would be wasteful to continue a similar program when it would require extra staff. RMC therefore proposed to make the third-year course in the second language optional. To this the principal of Royal Roads, Dr Graham, replied that the compulsory course in French should continue at RMC until a better system was introduced.[48]

Meanwhile, on the grounds that the plans were not in accord with the recommendations of the B&B Commission that there should be two engineering colleges, the Treasury Board had turned down DND's first proposals for making the CMCs bilingual. In September 1972, however, it approved a B&B program for RMC, "provided there was no expansion of the student body."[49] The program was to offer degree courses in either of the official languages, to build upon the French character of CMR, to increase francophone representation to meet Canadian Forces' requirements, and to provide cultural amenities at CMC locations to support and enhance B&B programs. There was no reference to Royal Roads.[50]

In November 1972 the RMC Faculty Council had discussed the Blackburn-Duplantie report which CDEE had commissioned to obtain advice on the teaching of French in the forces. This report had confirmed that by the present system only a few cadets could reach level four by graduation. It had recommended a six-week immersion

*Edgar Benson, assistant professor of commerce, Queen's University, 1952–63, MP for Kingston and the Islands from 1962, minister of national defence, 1972.

course on graduation, or when employment in the forces required the use of the second language. Dr Dacey suggested that the amount of time currently given to French instruction at RMC should continue unchanged, that the RMC French department faculty should teach language in the manner of DND language instructors (perhaps with the help of some female instructors), that there should be a summer school in Quebec and a language school at RMC, but that credit in French should not be required for the RMC degree. When Hutchison supported these suggestions, he quoted Jacques Barzun, author of *The American University*, who had written, "the mechanics of a language . . . [bear] little relation to study of a culture."[51]

About this same time the Educational Council learned of CDEE's decisions about the CMCs' overall academic program. These decisions were based on Garneau's report, which was to be used as a guide. The four-year CMC degree program would continue, the courses in the first two years would be as "common" as possible in all three colleges, efforts would be made to find more time for cadets' private study, the arts program at RMC would be reviewed, a complete arts program at CMR was a goal for the future, arts programs at RMC would be given in both languages, and the development of a four-year program at Royal Roads was to be held in abeyance.[52] Upon hearing of these decisions, the RMC Faculty Council discussed plans for French classes for the RMC civilian staff; and the Educational Council asked DND's Personnel Branch to recommend a language-training program for cadets that would make RMC "institutionally bilingual" by 1975.[53]

The Personnel Branch informed the Advisory Board in April 1973 that (despite what Benson had stated) future emphasis in recruiting for the military colleges should in fact be to obtain 50 per cent whose mother tongue was French, because "to achieve the [required] distribution throughout the Forces over the fifteen years of our programme it is essential we have a selection of sufficient size just as soon as we can." This was a call for affirmative action. It also reported that the classrooms, offices, and research facilities "to provide instruction in two languages" that could be charged to the program would cost an estimated $2,220,000, and that RMC staff would need an additional 107 military and civilian man-years over a period of five years, to be made up of two military personnel, seventy-four civilian academics, and thirty-one support staff. Equipment chargeable to the program would cost $450,000. The RMC faculty must be given opportunities to obtain language training. A study was also being made of the introduction of intensive language training for officer cadets as an integral part of the RMC academic curriculum.[54]

Some members of the Advisory Board had initially doubted the wisdom of the move towards institutional bilingualism in RMC. But L.J. L'Heureux,* a former chairman of the Defence Research Board, pointed out that the present trends might lead to Canadian Forces' engineers becoming predominantly anglophones produced through RMC, with the majority of francophone officers having followed courses in arts and sciences at CMR. In the long run this kind of segregation would be detrimental to national unity and to the services. Rear-Admiral D.S. Boyle, chief of personnel, said that it followed that, since RMC was the only CMC engineering college, it must offer courses in French to produce francophone engineers. For that, institutional bilingualism was essential in the college. The board then endorsed the plan.[55] With this approval, the way towards implementing a program of bilingualism and biculturalism at RMC seemed at last to have been cleared.

Progress towards bilingualism at RMC was apparently going ahead in March 1973 when General Dextraze sud-

---

*Léon-Joseph L'Heureux was commissioned in the RCCS 1943–7. He joined the Canadian Armament Research and Development Establishment 1947, was scientific adviser to the General Staff 1961–3, chief superintendent, Canadian Armament Research and Development Establishment, 1963–7, vice-chairman DRB 1967, and chairman 1969–77.

denly called the RMC commandant, his designated successor, and the principal to an interview. Despite the satisfaction with RMC's second-language planning recently expressed by Major-General Laubman, Dextraze severely criticized Lye and Dacey for alleged procrastination and linked the bilingual and bicultural program at RMC with UTPO and UTPM as problems that were to be planned and implemented without further delay.[56] A week later Lye advised Dacey to appoint a member of the academic staff to draw up a master plan for the implementation of the B&B program to begin in the fall of 1975. The plan was to assume that additional classrooms, offices, and research facilities would be available by that time and that francophone recruits would arrive as envisaged. Lye added, "I do not think we can delay this much longer and I urge you to give the matter priority attention."[57] On 1 June the registrar, Dr R.E. Jones,* presented a plan in accordance with these instructions.[58] Since Lye was to retire after the coming convocation, it was left to his successor, Brigadier-General W. Turner, to put the plan into effect. After the graduation ceremonies on 18 and 19 May, Lye remained in office to arrange for a visit by Her Majesty, Queen Elizabeth, on 27 June 1973 to unveil a bronze plaque to be placed on the main entrance of the new Sawyer Building when it was completed. Among the letters that Lye received for his fine organization of this royal event was a gracious tribute from General Dextraze.[59]

---

*Dr Richard E. Jones served in the RCAF overseas 1941–6. He got his MA degree in geology at Queen's University in 1958 and was a lecturer and assistant professor at McMaster University from 1954 to 1963, serving at the same time as a flight-lieutenant in the RCAF reserve. In 1963 he was appointed assistant director of studies and registrar at RMC. In 1981 he resigned as registrar to become professor of geology. He retired in 1987.

Institutional Bilingualism

It was easier to secure approval for bilingualism at RMC in principle than to introduce it in practice. Dr Dacey, commenting to the Faculty Council on a report on second-language training issued on 12 April 1973 by a study group in the Division of Education in DND's Directorate of Language Training, said that "the real problem is reduced to a choice between military training and language training." He argued that three years and eight months was too short a time for producing functional bilingualism and, at the same time, educating to an accepted degree standard. He noted also that many who stressed the importance of military leadership and professional skills, and who believed that language training could be acquired later, would not accept a preference for language training over military training. He agreed, however, that from an academic point of view, second-language training "would contribute to the future development of the individual."

Dacey suggested that phase 1 of the language training program should be moved to the summer before the recruits entered the college. If cadets needed further instruction, they should take second-language training courses after graduation. He confirmed the need for a more equitable balance between anglophones and francophones in RMC, and for more French-speaking faculty appointments, but said the report's proposal for training in French for nine hours a week in the first and second years was quite unrealistic. Dacey reminded the

RMC faculty that Garneau had recommended an overall reduction of the amount of formal classroom instruction. The Blackburn-Duplantie report had held that 1200–1600 hours were needed to achieve functional bilingualism, but this new report said only 500 hours would be required.[1]

In June, RMC complained to the Educational Council that DND's Directorate of Language Training had not provided it with clear objectives in functional bilingualism – a reference, no doubt, to the difference in hours between the two reports. The council agreed. At the same time, Dacey announced that the RMC French department had rejected the proposal that it should undertake audiolingual instruction and disputed the possibility of achieving functional bilingualism in the time available. The choice, Dacey said, was really between making a man bilingual and giving him a good engineering education.

The Directorate of Language Training informed the Educational Council that the reduction in the number of hours required for second-language training stemmed from the fact that an individual in DND was now expected to reach level two instead of level four. The Educational Council left the question of second-language training at RMC for future decision. Royal Roads had at one time stressed "conversational knowledge" in its language teaching,[2] and the council now agreed that audiolingual French teaching should be reintroduced there. Meanwhile, to prepare for an expected influx of francophones into RMC, the Public Service Commission, under contract with DND, provided second-language instruction for thirty anglophone members of the RMC faculty and administrative staff in summer courses in 1973, and arrangements were made for a series of more three-week courses to start in the fall. These courses were described loosely as "a correspondence course with students doing most of the work on their own time."[3]

Then, in July, General Dextraze circulated a letter throughout the forces instructing commanders at all levels to identify civilian positions that, by December 1978, were to be bilingual. The linguistic requirements in the military positions that supervised civilian personnel would also be examined. Since RMC came under this ruling, here was another impetus towards institutional bilingualism. This could no longer be delayed once the CDS gave the incoming commandant, Brigadier-General Turner, direct orders to proceed without further delay.[4]

In the following session, after RMC had placed bilingual signs throughout the grounds and in the museum in Fort Frederick, the official language commissioner received a complaint about one that had been overlooked. Meanwhile, although many cadets disliked the program, the Cadet Wing had already organized bilingual squadrons, and the cadets themselves had sponsored a French weekend. The French department had used the special second-language funding provided for the promotion of bilingualism and biculturalism to publish a new periodical, *Signum*, in both languages. It had also purchased apparatus for a new language laboratory at a cost of $42,000, which was intended to "give more flexibility" in French courses in the third and fourth years. Simultaneously, RMC explored ways to provide audiolingual French teaching for anglophones in the first and second years.[5]

What was more imperative, yet very difficult, was to duplicate instruction in various subjects in French. Dr Jones's plan to introduce bilingual teaching in RMC proposed a series of yearly stages for this development. The

Faculty Council wanted to avoid any large increase in faculty without a corresponding increase in cadets. On 3 October the council adopted phase 1 of the B&B program in principle, which was now to begin in September 1975. The council acted on an assumption that the extra space needed would be available by then, that there would be an intake of forty francophone recruits, and that RMC would be informed how many CMR cadets could be expected to enter the third year.

The council said that the introduction of phase 1 in 1975 would require second-language training for both English- and French-speaking cadets. The search for the academic staff to teach other courses in French should be approved by March 1974, and teaching should begin in September.[6] Nine professors were already giving third-year courses in French – in philosophy, history, English, economics, geography, mathematics, chemical engineering, electrical engineering, and physics – to ninety-nine anglophone and fifty-five francophone cadets.[7] The council noted that more language laboratory facilities were needed for recruits in phase 1, and that the calendar must be published in both English and French. Phase 2 of the program would begin in 1976–7, phase 3 in 1977–8, and phases 4 and 5 in 1978–80. Dacey suggested that, if more francophones were needed for the first year, it would be necessary to recruit from the Quebec Collèges d'enseignement général et professionnel (CEGEP), which had replaced the old classical colleges and the Ecoles supérieures.[8] The private and public systems were both sending students to RMC or to CMR's preparatory year as was appropriate, but those who came to RMC were usually already bilingual.

The RMC faculty was concerned that the addition of bilingual professors to duplicate courses in French, producing an RMC cadet-staff ratio of less than three-to-one, could open the college to more criticism on grounds of cost. Lieutenant-General Milroy, assistant deputy minister for personnel, explained at a special meeting of the Faculty Council on 17 October that the cost was justifiable because the Officer Development Board had requested more officers from the Canadian Military Colleges. To produce an adequate number of francophone officers, the colleges must become bilingual.

Faculty Council members then detailed difficulties they anticipated. Were future appointments and promotions, for instance, to hinge on scholarship or language? It would be difficult to attract enough French professors to Kingston, and facilities would be needed for francophone families. Milroy replied that the faculty needed to give courses in both languages would be funded.[9]

At an Advisory Board meeting later in October, Dacey again posed questions about the B&B program. He said that it had three related, but overlapping, objectives: to teach anglophones to speak French, to teach all subjects in French to francophones, and to create a bilingual and bicultural environment at RMC. In his opinion the first two could be achieved without the third. There was not enough time to achieve the first objective in the four-year course at RMC, and some other approach should be followed – for instance, leaving the second-language training program to be completed after graduation, or substituting second-language training for military training during the summer. Turner answered that his directive as commandant of RMC was clear and he must now find the way to implement it. He noted that two francophone instructors for mathematics and physics had been added to the staff and that two bilingual squadrons were now functioning effectively.[10]

When the Educational Council met in November, RMC feared it might not have enough francophone recruits to make the B&B program viable and suggested that, to secure them, CMR's intake should be increased.[11] Meanwhile, however, problems had emerged in the transfer of CMR cadets to RMC's third year: because CMR had concentrated on courses in administration, some cadets who wished to transfer to RMC lacked physics and chemistry, courses that were compulsory for RMC cadets in the first

two years. Others lacked principles of accounting. These courses would have to be made up later for certain honours programs in RMC.

Cadets who were short certain subjects could only be admitted to the RMC general degree course, and even then they would need an extra sixth subject. The CMR cadets who lacked subjects complained they had been misled by their advisers about requirements for RMC and talked of petitioning for redress of grievances and possible transfer to civilian universities. Dacey asserted that RMC had informed CMR of the requirements, but that CMR had assumed a temporary expedient arranged for 1972–3 would continue and had neglected to put the required prerequisites into its calendar. RMC held that it would be unfair to its own cadets to permit transfer from CMR without requiring the transferees to have the same prerequisites as RMC cadets. CMR argued for exemption.[12] This exchange between the two colleges shows how important a supervisory body was to coordinate the operations of the military colleges, especially since transfers were becoming even more important with the development of bilingualism.

By 1974, seventy-one RMC faculty and staff had taken three-week French courses supported by the Public Service Commission.[13] The next session saw the appointment of an RMC bilingual and bicultural coordinator, Ms J.P. Raynes, who worked to develop the program. The Directorate of Language Training financed two RMC staff members to attend an intensive immersion course at Université de Montréal. Two French-speaking professors in engineering and one in mathematics were added to the faculty to build the French science and engineering establishment up to seven. The French department acquired a colour TV camera and projector for use in language training.[14]

Turner reported to the Advisory Board that RMC had "given much attention to the implementation of B&B," but the Canadian Defence Educational Establishments (CDEE) were aware that some CMC faculty members still had doubts about the program. In justification, the commander of CDEE said that the cost of B&B at RMC was much less than that of duplicating all facilities and faculties (including engineering) at CMR. Furthermore, the separation of the CMCs into two unilingual streams would polarize the English- and French-speaking servicemen in the forces. The B&B program in the colleges was the "only course of action which will encourage and enhance the development of a professional officer corps truly representative of the nation through the sensitization of both founding groups." The program was not a "language-training programme but an institutional one." By making RMC a bilingual institution in the pattern of CMR, the teaching and acquisition of the second language would become a realistic and realizable goal.[15]

The Advisory Board was still not convinced and was uneasy about the B&B decisions. It saw two distinct and different objectives – to ensure a bilingual officer corps and to educate cadets in French or English according to their choice. Because of the high cost of the program, there should be further study. The board passed a temporizing resolution that recognized the substantial difficulties in implementing the program at RMC. At the same time, it urged the college to explore alternatives that would rapidly achieve the implementation of a program with minimal harmful impact on existing military, academic, and sports activities. The board also agreed that the achievement of a bicultural atmosphere at RMC was probably the key to success and should receive special attention.[16]

Dacey was concerned about finding time for extra French-language instruction to put the B&B program into effect in an already crowded timetable. He sought to secure a formal agreement by all the colleges that second-language training should not be included in their academic programs but recognized as a training activity.[17] This would mean that the time for it would be taken from time currently allocated to military training. CMR had, however, always used the direct conversational method, rather than the traditional grammar, reading, translation, and composition, in many of its classes in French for anglophones, and its courses were not classed as military training. Moreover, to make second-language training a part of military training would be unacceptable to those who thought that cadets needed more professional military training, not less.

In response, Lieutenant-Colonel A.C. Moffat of DPED's Division of Education asked whether the RMC French department was willing to teach the second language for up to four hours a week on Public Service Commission standards, and also whether it would take outside direction. Dacey replied that the department would not take outside direction, but that its teaching would be consistent with the PSC system. He said extra second-language instructors at RMC would not be members of the Faculty Council or the Faculty Board.[18] A few months later, Dr Bryan Rollason,* a new head of the RMC Department of French, proposed an intensive French course for recruits for four hours a day during a seven- or eight-week summer program prior to college entry in September.[19]

Major-General D.A. McAlpine, chief of personnel development, announced in April 1975 that the B&B program at RMC would commence in September and would be developed in three phases. In phase 1 in the 1975–6 session as many third-year courses as possible would be available in French for transferees from CMR. Phase 2, a year later, would see first- and fourth-year courses offered, and 30 per cent of RMC's recruits would be francophones. Phase 3 would introduce second-year courses in French and also more third- and fourth-year courses. Recruiting visits to Quebec would advertise that first-year courses at RMC would now be available in French. The

---

*Dr Bryan Rollason, formerly lecturer in modern languages, Massey University, New Zealand; assistant professor of French, RMC, 1972, head of department, 1975; resigned in 1982 to work with the Second Language Training Programme, Pedagogical Services, in Ottawa.

sum of $150,000 would be available to RMC for extra capital expenditures in support of the bicultural program. Cable TV would be installed to bring in French-language programs.[20] It had already been announced in the RMC Faculty Board that some engineering courses would be offered in French in September among the third-year courses.[21] Six weeks later, Dacey told the Advisory Board that RMC had undertaken to appoint seven French-speaking professors for 1975–6, that it would have no difficulty in finding those it needed in the humanities, but that it would not be easy to fill more positions in engineering.[22]

Up to this time the problem of providing extra space for language teaching had loomed in the background as a most formidable obstacle to the early introduction of a full B&B program at RMC. It had taken the college a dozen years to get the science and engineering buildings it urgently needed, and the first three modules of the Sawyer Building were not due to be completed until October 1976. In March 1975, however, Brigadier-General J.P.A. Cadieux,* director-general of recruiting, education, and training, announced that work would begin at once on a large extension of the Massey Library building (later a separate entity called "Girouard") to meet the needs of second-language teaching at RMC. This annex would be ready by July 1976.

The Kingston *Whig-Standard* called the proposed input of francophones and the building program a "Bilingual Bombshell." What had defence minister James Richardson, who was known to be cool to the government's bilingual program, been doing to let it develop? Construction already in progress at RMC, undertaken to enable the military colleges to catch up with civilian universities, would cost $25 million. Since every class was now to be offered in both languages, RMC's staff and facilities would have to be doubled. An RMC cadet-faculty ratio of 3:1 was in stark comparison with the Queen's University ratio of 13:1. What was being created was not really a bilingual institution, but two unilingual ones. As

The Girouard Building was provided in record time to house staff and classes in the bilingual program.

Flora Macdonald,** MP for Kingston and the Islands, said, anglophones who would normally have come to RMC would prefer to go to Royal Roads, now that it was to have a four-year program.[23]

---

*Brig.-Gen. John Paul Cadieux (no. 3814), RMC 1953–7, served in No. 423 All-Weather Fighter Squadron, RCAF, and took a degree at McGill University and an MBA at Harvard. He was lecturer in electrical engineering at RMC 1962–5, then commanded 433 Escadrille Tactique de Combat. From 1971 to 1973 he was commandant CMR, then DGRET, commandant Canadian Air Group NATO, and assistant comptroller-general and deputy comptroller from 1983. He is now a vice-president of Canadair.

**Flora Isobel Macdonald, MP for Kingston and the Islands after 1972, minister of external affairs 1979–80, minister of employment and immigration 1984–6, and minister of communications 1986.

The Sentinel, an organ of the Orange Order in Ontario, was also alarmed about the proposed increase in francophones in RMC to a "whopping 40%." A senior officer of the Frontenac County Orange Lodge, who was a veteran, saw the expansion as beneficial to francophones, who would be able to put after their names the small letters "rmcs" – meaning, "ready-made civil servants."[24] When the *Whig-Standard*'s adverse publicity about B&B was reported to the Advisory Board six weeks later, however, the matter had already been forgotten.[25]

When the Sawyer Building was completed for engineering and the sciences, space in the Currie Building, hitherto used for laboratories, became available for an excellent language teaching facility. With space for second-language teaching at RMC assured, it was possible to go ahead with arrangements to bring in the francophone recruits. In June the Educational Council arranged for RMC to sponsor eighty-two candidates from the CEGEPs whose academic backgrounds would be suitable for entry to RMC after they had successfully completed a prescribed course of study.[26] Acting on the advice of the "Dean of Engineering at Laval,"* who had told three RMC professors that some CEGEPs were "liberal in giving higher marks," the required minimum grade was set at 65 per cent. The RMC course was exacting, was only four years long as against four-and-a-half years in a Quebec university, and also included arts subjects, second-language training, and military studies.[27]

Seventy-one potential recruits enrolled in this CEGEP-ROTP program for RMC, fewer than were desired. They attended CEGEPs selected by DND and were subject to release if they did not follow prescribed courses or if they failed. They had no preparatory military training camp. It was expected that forty from this number would come to RMC in September 1976.[28] Twice during their session in their CEGEP, forty-seven were brought to Kingston for a first-hand view of RMC life.[29] This CEGEPs program did not work out well. Since there was no obligation to stay in the program, many potential recruits simply used their

subsidized year in a CEGEP as a step towards a civilian university, and by 1979 the system had collapsed. Instead, some preparatory course cadets from CMR were to be brought to RMC for the four-year course in engineering. This could have meant they would be subjected to a second year of "recruiting," so they were exempted from RMC's "recruiting," but had to follow certain first-year practices such as "running the square." In some later years, the number making the transfer declined. Attrition in the preparatory year at CMR was much the same as in the CEGEP scheme. CMR obtained extra appointments and new buildings for the program, but did not send as many cadets to RMC as was expected.[30]

Ms Raynes, B&B coordinator at RMC, had been admitted as a member of the Faculty Board in May 1975 to aid her work,[31] but she resigned during the summer and her replacement, Bernie Hamel,** a former RCAF officer, could not take up his duties until 1 January 1976. During this crucial time, implementation of the B&B program was directed by Mr C. Lumbers,*** assistant to the registrar.[32] Meanwhile, to give much greater impetus to the B&B effort, DND decided to have RMC appoint a French-speaking associate principal. Since the position failed to interest anyone of the required stature, Dr Pierre Bussières† of the engineering faculty was named

---

*Dr Paul Grenier, professor of chemical engineering, was dean of the Faculty of Applied Science at Laval at the time.

**B. Hamel was commissioned in the RCAF in 1951 as a fighter pilot and became a flying instructor. After service in Zweibrücken with No. 427 Squadron, he retired to take up the appointment at RMC as second-language coordinator.

***C.L. Lumbers (no. 9262), B.Eng. 1968–72, studied at Université de Pau-Pau, France, in 1974. In 1975 he was appointed as assistant registrar at RMC, in 1983 was a graduate student at Virginia Polytechnic, and is now on the Faculty of Administration of the University of Saskatchewan.

†Dr Pierre Bussières (no. 2864) was one of the New One Hundred who entered in 1948. He graduated in 1952 and was commissioned in the RCAF. He became professor of mechanical engineering at RMC and was appointed special assistant to the principal in 1975.

special assistant to the principal for the B&B program.[33]

Bussières's task was to help recruit staff to give courses in French. He found francophone academics from Quebec slow to apply. Nevertheless, rather than appoint French-speaking non-Canadians (who were available in considerable number), RMC persisted in its quest for Canadian citizens. Another plan, to use electronic transmission to bring classes in French from CMR, proved unsatisfactory for a live lecture program.[34] By the session of 1975–6, the appointment of French-Canadian professors was proceeding fairly satisfactorily, though there were a few vacancies.

Second-language training for anglophones, necessary if RMC was to be truly bilingual, was quite another matter. In October 1976 RMC informed CDEE and the Academic Council that the introduction of second-language training might be possible for third- and fourth-year arts students, but not for engineers. It contended that the official policy was second-language training, not bilingualism and biculturalism. Bilingualism and biculturalism would only be possible if proper plans were made and if the necessary resources were allocated. RMC's brief noted that in the summer of 1975, anglophone CMC graduates had been sent to language school immediately after graduation. Was this to be permanent policy? The Academic Council replied in the negative, stipulating that, during the academic session, five hours a week must be allocated to second-language training for all those cadets who had not yet reached level four. Instructors to bring them up to that level would be obtained from the Public Service Commission.[35] Captain (N) J.P. Côté of the Directorate of Professional Education and Development (DPED) told the RMC Faculty Council later that this standard was necessary because he was being required to defend the very existence of the CMCs, particularly in view of their current high cost of operation.[36] Bilingualism was thus a means of protecting the CMCs politically.

RMC must have been aware of this implied threat. Turner arranged for the RMC Senate, which hitherto had dealt only with the formal confirmation of degree results and with the selection of recipients of honorary degrees, to set up a committee to recommend a means of implementing second-language training. The minister of national defence, as president and chancellor of RMC, was a member of the Senate. This brought military and political influence to bear on the academics' handling of the problem. The Senate appointed a committee, chaired by Dean Clark Leonard, dean of engineering, to find the means.

Leonard's committee reported to a joint meeting of the Senate and the Faculty Council on 29 March 1976. He suggested two alternative plans. Plan A would provide five hours per week by reducing all periods from fifty minutes to forty-five, thus increasing the number of periods in a day from seven to eight. To make up for the loss of 10 per cent of teaching time, the academic year should be increased by three weeks by cutting the three-week period between final examinations and graduation down to two weeks, and taking two weeks from summer training. Plan B would keep the fifty-minute class periods and provide four hours a week for second-language training in all four years, with an additional four hours a day in the three weeks between final examinations and graduation. To make room for this training in the third and fourth years, military leadership and management, some arts electives, and drill would be dropped. The committee also recommended that in either plan, second-language training should be taught by locally appointed tutors under the direction of the Academic Wing.[37]

Since the introduction of plan A sequentially by years would produce timetable problems, Leonard favoured plan B. He said that because the mathematics and science elements in engineering had been reduced in the recent past, the time needed for second-language training should come from arts, military leadership and management, and drill. Turner preferred plan A. He said he would never agree to cutting military leadership and drill, and confirmed that Brigadier-General J.E. Vance, the director-general of recruiting, education, and training (DGRET), seemed well disposed to reducing the length of summer training. Turner also thought that second-language training should be introduced sequentially, so he too turned to plan B, with modifications to avoid cutting down military content.

The principal and Hutchison both protested that second-language training was being introduced too hastily and with insufficient study. Dacey argued that, because the twelve-week summer immersion course at the end of the first year was crucial, the program would have to be brought in sequentially. He said that the French department could handle the training in the first year by giving an introduction to French language and culture. Other members of the joint meeting of the Senate and Faculty Council stressed the need for sequential introduction because it would be impossible to recruit enough teachers for all four years at short notice.[38]

At this point, NDHQ set up an internal study group chaired by the chief of personnel development to examine the question of preserving the CMCs (see chapter 8). This group argued that officer cadets of each language group should have equal opportunity to acquire a functional level in the second language. Consequently, about 1000 hours of second-language training must be provided during the four-year pre-commissioning program. The group then took note of the plan to introduce the twelve-week concentrated second-language training program in the summer following the first academic year at a CMC, a program that would be supplemented by up to five hours of training per week during each academic year. It concluded that these arrangements would be more than adequate to satisfy Canadian Forces' bilingual requirements.[39] Cadets were to be required to take five hours of second-language training a week until they became functionally bilingual, and thereafter take at least one course in their second language. Cadets assessed as integrationally bilingual were to be freed of all second-language requirements.

Meanwhile, however, the introduction of second-language training for the first recruit year at RMC in 1976 had been, in the words of Dr Bryan Rollason, head of the Department of French, "disastrous," chiefly due to "the ineffectiveness of the Directorate of Language Training (DLT), NDHQ," which had been made responsible for the program. DLT had assumed that DND language teachers would be employed, but it had not acted quickly enough to hire them. It was also responsible for providing teaching materials and equipment, but had not done so until after the time when the course should have started. Even then the materials were in such poor condition that Rollason and his technician thought there might have been a deliberate attempt to make the program fail.

Rollason, appointed to head RMC's language training program, was largely responsible for its eventual success. Late in the summer, when DLT failed to act, he had personally initiated a direct approach to St Lawrence College, a local community college, to ask for tutors. After some classes had been cancelled, he managed to get the program started a month late. However, some members of his department, doubtful about their future at RMC because third- and fourth-year second-language training could not be considered an academic subject, had lost confidence and two of the best professors found employment elsewhere.

The Directorate of Language Training's contract for introducing second-language training at RMC specified that tutors would be given a three-week training session before commencing their duties, but the two instructors hired from Algonquin College had not arrived at RMC until 7 September, after the term had started. The training session could therefore be only five days long. Fortunately, the tutors "brought with them a devotion to the task, a sense of humour, and undeniable expertise." As a result, their abbreviated training session was very effective. Classes for cadets began during the week of 13–17 September, but after these experiences Rollason protested that "this year, 1976–1977, cadets have every reason to believe that NDHQ is not taking the programme seriously, so why should they take bilingualism seriously."[40] This was a serious charge. The probable explanation of DLT's failure is that the government's determination to push ahead with the bilingualization of the forces had not left enough time to develop an efficient system.

The RMC B&B coordinator, Bernie Hamel, confirmed Rollason's complaints about DLT in his quarterly report to the commandant. He added that registration for second-language training in 1976–7 was as follows: 189 officer cadets were taking French with seven tutors, and forty-one were taking English with three. In addition, thirty-five permanent staff, military and civilian, were enrolled in French classes, and nine in English. Hamel was able to assure Dr Rosario Cousineau, the academic adviser to ADM(Per), who visited the college on 7 December, that attrition of francophone recruits through voluntary withdrawal would not reach serious proportions. The mid-term academic results of cadets from the CEGEPs had been slightly higher on the average than those of the anglophones, and in every case they were satisfactory.[41]

The introduction of second-language training for the third year, due to begin in 1977, proved to be even more difficult, but the need for it was growing if a bilingual atmosphere was to be achieved at RMC. Although Dacey had been able to report a decline in overall academic attrition in June 1976, the number of francophones in engineering was also declining. The incoming fourth year had 25 in a class of 208, but the third year had only 12 in a class of 150. Furthermore, a smaller number of ROTP-CEGEP students had registered to come to RMC; only 58 out of a class of 247. It was decided that next year the allocation of ROTP-CEGEP students for RMC would be increased from 77 to 80.[42]

An ad hoc committee chaired by Dean Cairns to report on second-language training to the Faculty Council noted another difficulty in introducing it for the third-year class in 1977. That class had not had the summer immersion course at the end of the first year intended to be an integral part of the program. The committee therefore contended that compulsory second-language training for the class would be unproductive, or even counter-productive. The Cairns Committee also opposed the proposal to cut the length of class periods to forty-five minutes, and instead recommended the elimination or reduction of drill, physical education, military leadership and management, and Canadian military history.[43]

The Faculty Council used these problems as grounds for advising that compulsory second-language training for the third year should be postponed for a year. But Turner refused to agree to postponement, and also rejected reductions in drill, physical education, military history, and military leadership and management. He called for yet another effort to find a suitable and practicable solution.[44]

DND directed the college to introduce third-year second-language training in 1977–8 and also informed it that there could be no extension of the academic year. On 8 February, then, the Faculty Council reluctantly adopted plan A in principle, the plan that provided for eight forty-five minute periods in the day. Since two weeks were now not to be taken from summer training, it ruled "that a period of at least thirty-eight weeks should be provided for all College activities between opening the College in the Fall and graduation in the spring."[45] Thirty-eight weeks was to cover all holidays, reading periods, and examinations as well as classes. What must now be done was rearrange the calendar and timetable within that period.

The problem of carrying out plan A without the two extra weeks taken from summer training was referred to the Faculty Board's syllabus committee, chaired by

Professor D. Graham.* That committee saw its task to be "to minimize the harmful effects of this policy on staff, on academic and other programmes . . . and above all on the student body," and recommended that, because of the complexity of the problem, 1977–8 should be regarded only as a trial year. It suggested an extension in the number of teaching weeks from twenty-seven to twenty-eight and three days, mainly by cutting down the time between examinations and graduation. With eight forty-five minute periods in the day, this cut the loss of teaching time from 10 per cent to about 5 per cent. Some members were concerned that even this smaller loss might endanger engineering accreditation, but there was no other way out of the dilemma. The committee reiterated that, while it supported the principle of second-language training, it was concerned about its pedagogical value and about the harmful effect of so crowded a program for the cadets.[46] The Faculty Board approved these recommendations, but no one was completely happy with the compromise.[47]

By the time this compromise was reached, other controversies were threatening the academic system that had been established at RMC since 1948. These had been foreshadowed by Côté when he told the Faculty Council that DPED had been required to justify the expense of the CMC system and its value to the Canadian Forces. The controversy over UTPO/UTPM was examined in chapter 9, and another, over the enrolment of women, will be the subject of chapter 12. At this point, and by way of conclusion about the B&B program, something must be added about its operation over the next few years.

Outside the classroom, institutional bilingualism had already made considerable progress. In April 1977 the special funding to provide amenities to foster biculturalism was discontinued.[48] DPED apparently assumed it had now served its purpose. Steady progress had been maintained in the use of French in the Cadet Wing. From the original two squadrons that had operated bilingually as an experiment in 1968, six out of eight were now for-

mally bilingual. They worked in each language on alternate weeks.[49]

Dr Rosario Cousineau, academic adviser to the commander of the CMCs, Assistant Deputy Minister (Personnel) Lieutenant-General James C. Smith, reported on development in 1978. He said that second-language training at the military colleges was proceeding according to plan and that at RMC it had been established "honestly and intelligently." However, it was as yet too early to judge whether the efforts would reach the desired goal. Cousineau claimed he had "tried to be objective in a field where tradition and sentiment have an influence," and argued that since "the possession of a second language (and that means for Anglophones as well as Francophones) had been made an absolute requirement for promotion to higher rank [in the services as a whole], in a few years SLT would be considered [to be] as important as military training, and perhaps more important than an academic degree in a military college." He therefore recommended that functional proficiency should be regarded as "somewhat too little" for graduates of the CMCs: the higher "integral proficiency," known as "A level" or "level 5" – complete bilingualism – would be the goal in a few years time. He suggested that all cadet squadron commanders of bilingual squadrons should now be bilingual at that integral level. At present only one was, and he was French Canadian. Currently, 20 per cent of francophones and anglophones in the senior year had not yet reached level four, and 46 per cent of anglophones and 40 per cent of the smaller number of francophones had not yet been classed as "integral."[50] Since cadet squadron commanders are appointed at the end of the third year, Cousineau's recommendation would obviously have increased the chances of francophones to reach highly desired cadet appointments.

In the 1978–9 year following Cousineau's report, al-

---

*Dr David Graham was lecturer in French at RMC 1971–9.

ternate French and English weeks were standard procedure in the whole Cadet Wing, where "the French fact" was now accepted. On the academic side, five more bilingual professors had been hired, but there were still vacancies where suitable candidates had not yet been found.[51] RMC had set up a Second-Language Centre with thirty second-language instructors. The French department had no honours students and had to be restructured with an establishment of four professors.[52] The contract with St Lawrence College to supply language teachers had not been renewed, but DND was slow to authorize RMC to appoint others. Cousineau recommended that 60 per cent of the language teachers should be university professors. This was described by DPED as "impracticable."[53]

A year later, even though francophones lower down the list were being accepted by preferential treatment, francophones still made up only 26 per cent of the Cadet Wing at RMC, a decrease from the previous years when the percentage had been 28 per cent.[54] It now seemed unlikely that the 35 per cent, set by the deputy minister and the chief of the Defence Staff as a goal for 1982, would be reached. One reason for this shortfall was a decline in transfers from CMR, and also in the number of French applicants who had the equivalent of senior matriculation. Furthermore, many ROTP-CEGEP students who had originally committed themselves to go to RMC had later elected to go to CMR.[55] However, 95 per cent of RMC's compulsory courses were offered in French in 1980–1, the exceptions being metallurgy and structure of materials. Finally, in 1981, the Advisory Board endorsed the bilingual and bicultural program in the CMCs as good for national unity.[56]

RMC's second-language training programs, proposed in 1971 as "instant bilingualism" to be achieved by 1975, had taken five or six more years to implement because of the difficulties involved. The RMC administration and faculty had persistently and firmly pointed out those obstacles. Although many problems were subsequently overcome, Cousineau appeared to accept that, in the long run, bilingualism in the forces could only be at the expense of the CMCs' academic standards and, perhaps also, of their military training program. There is also no doubt that RMC's second-language training program greatly increased the financial cost of officer production.

Although the goal was functional bilingualism by the time of graduation and commissioning, it had from the first been necessary to modify this objective by accepting evidence of regular progress towards that end, especially for those recruits who started from scratch. In 1985 this modification was formalized. The Senate, at the behest of NDHQ, decided that the conditions for graduation, as expressed in Academic Regulation 29, would be amended "to demonstrate significant progress in Second-Language Training." The intention was not merely to monitor progress annually, but also to require by graduation a minimum proficiency level. This requirement, "if implemented," was to begin with the 1986 intake, which would graduate in 1990.[57] Regardless of the attainment of the functional level, students would continue second-language training until they reached "integral" level, or full bilingualism.[58] In 1977, 69 per cent of the graduates at RMC were functionally bilingual, and only one would have failed to graduate with his class if the new regulations were applied. Those who were not yet functionally bilingual went for further language training immediately after graduation and commissioning.[59]

To get to this stage had been a long struggle; the difficulties to be overcome had been numerous and complicated, and many more lay ahead. The reluctance of French-Canadian cadets to move into Kingston's anglophone environment was still a problem, especially since it was their presence that would help the implementation of effective institutional bilingualism. The government was, however, apparently satisfied. When Dr Dacey retired in 1978 he was made a Member of the Order of Canada for, among other things, his contribution to Canadian unity by developing RMC's institutional bilingualism.

Lady Cadets

When the Royal Commission on the Status of Women in Canada reported in 1970, it recommended that the government should give women the same support for a university education as it gave men through the Regular Officer Training Plan (ROTP). It also proposed that all trades in the Canadian Forces should be open to women and that women should be admitted to the Canadian Military Colleges (CMCs) on the same terms as men.[1] The minister of national defence, James Richardson, promptly ordered a study of the employment of women in the forces. The authors of this study recommended, and the Defence Council agreed, that women should be employed in all trades other than those with primary combat roles, in remote units, or on sea-going service. The study also advised, however, that women should not be admitted to the CMCs.[2]

By February 1973 the government had decided that women would be eligible for ROTP only in the universities. Dr Jack Hodgins, a former RMC professor, protested to the Advisory Board that women would cause less trouble in the CMCs than the authorities feared, and suggested a pilot project. The consensus on the board was, however, against admission.[3] In the House of Commons, George Hees* asked the minister of national defence whether the government would give women the same opportunity in the CMCs as men. When Richardson replied he was not ready to announce a plan, there were cries of "Shame." A month later Hees asked whether the govern-ment's recently announced policy to enable serving personnel to upgrade their educational qualifications in the CMCs would include women. Richardson apparently thought this was the same question Hees had asked before, so said he had nothing to add.[4] When the executive of the RMC Club met with him later that month, however, he asked them to sound out opinion on the admission of women to the colleges.

Approximately 200 women applied for ROTP in the universities, most of whom were well qualified.[5] In April Hees asked his question about female cadets again. Richardson replied that forty-two serving women were now in the universities, but that it was "not now our policy" that women be admitted to the CMCs. Again a number of members cried "Shame." Hees commented, "As a graduate of RMC I can tell the Minister that women would be welcome in those Colleges."[6]

The Advisory Board found that most of the respondents to its survey opposed the admission of women. One wrote, "While the pleasures of close contact between

*George Harris Hees (no. 1976) attended RMC 1927–31, served in the Second World War, and became brigade major of the 5th Infantry Brigade. Elected MP in 1950, he was appointed minister of transport in 1957 and minister of trade and commerce in 1960. He resigned from the Diefenbaker cabinet in 1963 over defence policy, but was re-elected MP in 1968 and became chairman of the Permanent Joint Board on Defence 1979–83 and minister of veterans' affairs 1984–8.

males and females are desirable – and probably necessary for most of us – I believe such associations should go on elsewhere than in military colleges or the officers' mess."[7] This is an indication of the male chauvinism in some of the opposition to admission. A more compelling reason for the exclusion of women from certain military classifications was the belief that women were unsuitable for combat, both psychologically and physiologically, and that men might be unduly protective of female comrades in emergencies, with consequent increased danger to all concerned.

Some opponents thought that the military colleges should be engaged primarily in preparing cadets for combat, but there was also a fear that the admission of women would destroy the Cadet Wing's homogeneity and its value as a training agent. General Lye reported RMC's view formally to NDHQ on 11 May 1973. He said that the Cadet-Wing rules had been developed to apply to young men: "RMC cannot envisage women students under any circumstances except as serving members in an extension program." Other women could be put in the university ROTPs.[8] RMC opposed the admission of women on the grounds that the CMCs were "not just universities attended by students in uniform." Women should not be admitted to the CMCs until they were acceptable as officers in combat units. They could be trained now through other agencies such as the University of Manitoba, using a nearby CF base. CMR, however, thought that serving women should be able to enrol in the CMCs as UTPO and UTPM; and Royal Roads believed that they could be non-residents like married UTPM, but not members of the Cadet Wing. Roads added that it could see no real argument against such an arrangement, but said that since the numbers at Roads would be small, perhaps women would best be educated outside the CMC system.[9] In July, the CMC Advisory Board formally recommended that "females not be admitted to the CMCs at this time."[10]

The Advisory Board repeated this recommendation in 1974, but added the phrase, "notwithstanding the real and moral justice of equal opportunity." It said that the "expense and projected disruption would be an unrealistic and unacceptable burden." It noted, however, that the Manitoba extension project proposed to include a female wing for women stationed in Winnipeg and vicinity.[11]

Refusal to consider women on equal terms with men was challenged by a few people in the forces. In an article in the *Canadian Defence Quarterly*, Colonel P. Charlton* declared that the concept of the limited employability of some personnel was "anathema in a military force," particularly in a small one with many tasks. The present policy of restricted employment of women must be closely and continuously reviewed and revised towards an ultimate goal of total equality of opportunity and employability.[12]

Women were already serving in most parts of the Canadian Forces, both in Canada and overseas, and some were with the peacekeeping forces in the Middle East. Indeed, in 1974 the Canadian Forces had a higher proportion of women than the forces of any other developed country, followed by the United States.[13] The policy of excluding women from getting commissions through the CMCs, then, was out of line with Canada's leadership in employing women in the forces as a whole. When the United States Congress overcame the service academies' opposition in 1975 and legislated that women would be admitted in all three,[14] Canadian authorities realized that change would come also at RMC. True to their military training to resist to the end, however, the opponents of the admission of women to the CMCs were not ready to give up meekly without further struggle.

At this point the commandant of Royal Roads wrote to the director-general of recruiting, education, and

---

*Brig.-Gen. (Commodore) P. Charlton was an engineer lieutenant with the Royal Navy in the 1950s who transferred to the RCN. He retired after integration.

training (DGRET) to show how his college would adjust to the admission of women. Adequate numbers would be necessary for mutual social support, he said, and suggested the minimum proportion of women should be 10 per cent. Women would be integrated into the present squadrons and flights, they could compete for places on representative teams or take part in female leagues, and they would be integrated on the parade square. They would wear male uniforms tailored to fit them, and would be permitted, like the men, to hold "stag" functions.[15]

During the winter of 1977–8 the minister of national defence, Barnett Danson, took an unusual step. He found two of the nominations for the CMC Advisory Board to be "unacceptable" and, in their place, he appointed two women: Kathleen Francoeur-Hendriks* for Quebec and M.D. Cameron** for Ontario.[16] This was obviously a deliberate move to swing opinion on the board, and it succeeded. At the next meeting the admission of women to the CMCs became the major item on the agenda, and the board discussed the question *in camera*. It recalled that its Executive Committee had recommended against admission, but noted that the minister had since stated on television that women would be in the military colleges within two years. Some members asked whether, if the decision had already been made, they were wasting their time in discussing it. General Smith assured them that no change in policy had yet been made and that NDHQ was still studying the problem. Women already received adequate opportunities in the forces, and the subject of their entry to the CMCs aroused strong emotions. The combat-function argument, however, was not conclusive: the CMCs were not restricted to the training of combat officers, and some male officers would have non-operational roles. Under the recent Human Rights legislation, moreover, there was a possibility that the colleges could be charged with illegal discrimination. This proved to be the decisive factor in favour of a decision to open the CMCs to women. The Advisory

Women's varsity volleyball, 1990

Board now reversed its previous stand and recommended that women should be admitted to the colleges.[17]

The chief of personnel development (CPD) then briefed the General Council of the CMCs about developments at NDHQ. The Armed Forces Council had been informed on 13 September 1978 that the CPD's opinion and the consensus in the CMCs was now that "women have

*Kathleen Francoeur-Hendriks, a graduate of Laval University, obtained a master's degree in education administration at the University of Alberta. After teaching school, she worked in the Quebec Department of Education, becoming director-general in 1970. In 1978 she became assistant deputy minister in the federal Department of Consumer and Corporate Affairs in Ottawa, retiring in 1986. She served on the CMC Advisory Board from 1977, becoming chairman of the CMR Regional Committee in 1984. She retired from the board in 1988.

**M.D. Cameron was appointed to the Advisory Board in 1977 and retired in 1981. She worked with the War Veterans Allowance Board in Charlottetown, PEI, in 1979.

Brigadier-General J.A. Stewart, CD, commandant, 1980–2

a right to attend the CMCs." A revised study report would recommend admission.[18] The council discussed the role of women in the forces and plans for their admission to the CMCs. It learned that one service woman would in fact attend RMC that fall, that more UTPO(W)s and UTPM(W)s would attend in 1979–80, and that in 1980 they would be joined by ROTP(W)s and RETP(W)s.[19] The first woman to take an RMC course, Lieutenant Valerie Spencer, was a reserve officer who was called out on duty. As a graduate student in war studies, she was accommodated at the Officers' Mess at Barriefield, so was with cadets only in class and in the library.

Recruits, fall term, 1980

In January 1979 Mr Danson publicly announced his timetable for the introduction of women into the CMCs, but the applications from serving women for the year 1979–80 were "not overwhelming." In April it was anticipated that one UTPO(W) and fifteen UTPM(W)s would register in September. The Advisory Board "noted with satisfaction" the steps being taken by DND and the colleges with respect to the admission of women. It recommended wide publicity to attract the largest possible number of applicants, distribution of information about the program to high schools, and careful monitoring of the program when it began.

The RMC Club fought a rearguard action. Allegations that the minister had attacked the club for opposing women cadets had appeared in the press and on 14 May 1979 its president, Walter B. Tilden,* counterattacked. The minister, he said, had acted too hastily, without preliminary experimentation. The restriction of women to non-combat roles would be difficult in practice, and the presence of women would be "prejudicial to the best interests of Canada, the armed forces, and the military colleges."[20] A club delegation that met with the minister to present this case received a polite hearing, but was told that the admission of women to the CMCs was now established government policy.[21] DND's long-range plan called for 10 per cent of the CMC enrolment to be women, and they were to be admitted on the same terms as men. Following American precedent, inter-year dating was to be restricted, especially between the senior class and juniors. The colleges would deal with pregnancies on an individual basis.[22]

Twelve serving members of the forces entered the three CMCs as UTPM(W) in the fall of 1979. Since they all lived off campus, they had only limited contact with cadets. In December the board learned that thirty-two

*Walter B. Tilden (no. RR00261) was a cadet at Royal Roads 1945–8. President of the RMC Club in 1978, he is president of Tilden Rent-a-Car and first vice-president of the Boy Scouts of Canada.

ROTP(W)-RETP(W) were to be admitted to RMC in 1980, four in each of the eight squadrons. The forces had designed a full set of uniforms for them that maintained RMC traditions but "recognized the female physiology." RMC had also made appropriate changes in its dormitory and washroom facilities, and had appointed females to its staff.[23]

Under the Official Languages Act, both francophone and anglophone women were eligible to receive instruction in their own language. Therefore CMR had taken part in the initial venture, but entry of women at Royal Roads was postponed for a time. What was most disappointing was that in 1979 no female officer had applied to follow Lieutenant Spencer in the UTPO program at RMC.[24]

A total of 215 Canadian women applied to be in the first group admitted to the CMCs in 1980. This was 17 per cent of the total of ROTP applicants. Seventy-two of these were good applicants for RMC's thirty-two vacancies. Seventy-one girls of similar high quality sought the twenty-one vacancies in the preparatory year at CMR. The aim was to graduate twenty female officers in the RMC class in 1984. Some of the details of the program were, however, still not clear. It was only now decided, for instance, that women cadets would take the same physical training program as men, and that they would be counselled by female officers of the Canadian Forces. As for career opportunities, in flying for example, women could not pilot fighter planes since they were excluded from combat roles, though they would be eligible to train as search-and-rescue pilots.[25] Unfortunately, this training was not available at the Canadian Military Colleges at that time.

Another matter of concern was that women applicants did less well than men in the CG2 (intelligence) test. In the Montreal centre, 40 per cent of the women failed, compared with 23 per cent of the men. Possibly this result reflected the general characteristics of the group that applied, since a lower percentage of females than men

Initiation: The recruit obstacle race

in the higher intellectual group saw the Canadian Forces as a career opportunity.[26]

Canada enrolled women as cadets in RMC in the same year that the American academies produced their first female graduates. This timing meant that Canada had the advantage of learning from the full American experience. One lesson was that traditional male attitudes are hard to change and that publicity about women in the colleges must be kept to a minimum lest it inflame male jealousy and chauvinism. The most important input from the United States was a confirmation of something Canada had already decided – that men and

Dr D.E. Tilley, BSc, PhD, director of studies and principal, 1978–84

women should be treated equally.[27] At RMC in 1980 all the women recruits completed the physically exacting recruit obstacle course, which was a great boost to their morale, though some of them afterwards complained about the time it took to get the grease out of their hair!

Brigadier-General John Stewart,* who was commandant when the first women entered RMC, attributed the success of that class to the quality of its women members who, in his words, were "a cross-section of Canadian society." They were determined to meet the challenge of pioneering and were not deterred by the hostility of some of their male comrades. Male harassment of female cadets did occur and had to be made a dischargeable offence. The RMC food services officer, Captain Lise

Berthiaume, together with RMC's first woman graduate student, Lieutenant Valerie Spencer, were detailed to listen to women cadets' questions and problems. Their services were invaluable in smoothing out difficulties, so much so that a "Minerva" Committee, established to take care of any problems that might arise as a result of the innovation, was dissolved when the second class of women was admitted.

By the end of the first session that enrolled women, 1980–1, thirty of the initial thirty-two recruits were still in the program at RMC and their first-year attrition rate was below the usual college average. At CMR, however, only fourteen of the twenty-one women who had entered at the same time were likely to go on from the preparatory year to the first-year proper. Attrition is always higher at this preliminary stage, but the CMR rules on fraternization were also more restrictive.[28] Dr Don Tilley,** RMC's principal, reported that all the females in the RMC program had aspirations to go on to engineering, but he believed half of them would probably end up in arts.[29]

In June 1982 the General Council agreed that, based on the experience of RMC and CMR, plans to enrol women in Royal Roads should go ahead earlier than had been expected.[30] The Advisory Board had learned, moreover, that cadets transferring from Roads in 1981 had a very negative attitude towards women in RMC and CMR. It was assumed that the introduction of women recruits in the western college would help to check this in the future.

Happy athletes: Brigitte Vachon and Marie-Pier Cloutier, 1984

*Maj.-Gen. John Arthur Stewart (no. 3173) was at Royal Roads 1949–51 and RMC 1951–3. He trained as a pilot in the RCAF and took a degree in civil engineering 1959–62. In 1968 he was technical officer in charge of constructional engineering, CFB Lahr, and in 1980 was appointed commandant, RMC. After leaving RMC he became chief of construction projects in NDHQ.

**Dr Donald Egerton Tilley served with the RCAF 1942–5. He was appointed assistant professor of physics at CMR in 1952, professor 1957, head of department 1961, and dean of science and engineering 1969. Appointed principal at RMC in 1978, he retired in 1984.

In 1982, however, the number of women applying to the CMCs was not as large as in 1981.[31] One problem was that females "gravitated" to arts in greater numbers than males, and few Canadian Forces' occupations were willing or able to accept women cadets enrolled in the arts program. With that academic qualification, they were mainly absorbed by Logistics/Air, which had only a limited quota of vacancies. As a result, very highly qualified arts women were being passed over in favour of less qualified engineering candidates. Stewart concluded, "This problem will continue until more classifications are opened to females."[32] In another development in keeping with Canadian Forces policy to avoid charges of discrimination, pregnant women cadets, instead of being automatically "separated," would be retained in the CMCs, "if at all possible."[33]

By 1983, the percentage of female applicants had steadied at around 17½ per cent. The total number of ROTP applicants had risen by 39 per cent (a reflection of the current economic depression), but the number of female applicants had increased by 72 per cent. The percentage of women in the Canadian Forces as a whole was 7 per cent, but if nurses were excluded from the reckoning, it was 5 per cent.[34]

The first female officers graduated from RMC in 1984. Twenty-four had completed the course or would eventually do so. They were described by General Stewart as being "as good as any previous RMC graduates."[35] In the CMCs as a whole, 59 per cent of the women were in the CELE occupation (electrical engineering) and 24 per cent in logistics; 14 per cent were to go to administration, 2 per cent were in air traffic control, and 1 per cent in air weapons control. The total number of women in the CMCs was now 145, including eleven UTPM(W)s. This was 9 per cent of the total CMC strength, or 11 per cent if Royal Roads, which had come late into the program, is excluded from the reckoning. At both RMC and CMR, women held command appointments at the squadron level as public relations, messing, and entertainment

After USMA became co-educational in 1976 and RMC in 1980, the traditional RMC–USMA weekend took on new dimensions. Saying goodbye, 7 February 1982

officers. Some were flight leaders and section commanders.[36]

The admission of women had been a success, but there were still some pockets of male chauvinism. For instance, the class that graduated in 1980 adopted as its slogan, "LCWB" (last class with balls), a more offensive designation than that of the equivalent class in West Point, which

Cadet Wing Commander Daryl L. Tremain, the first Lady Cadet to be appointed as CWC, winter term, 1986

Cadet Wing Commander Susan A. Whitley, the second Lady Cadet to hold the appointment and the first to be the Honour Slate CWC. She is shown receiving the Sword of Honour from the Reviewing Officer, Vice-Admiral H.M.D. MacNeil, DCDS, graduation, 1988.

Brigadier-General F.J. Norman, CD, commandant, 1982–5

Forces selected women for posting to a ship, to field ambulance sections in Europe, to search-and-rescue aircraft units, and to an all-male isolated unit.[38] By 1985, although the final conclusions had not yet been drawn from the experiment, the women who served with field units in Europe were being replaced by men, and women were no longer being recruited as aircrew. For the present, however, women were to continue to serve in HMCS *Cormorant* and at the remote Arctic base at Alert, and they could engage in communications research. Since the Human Rights Act, which came into force on 1 April, raised the possibility that selective employment based on sex might be illegal, it was necessary to proceed with caution. Final conclusions drawn from SWINTER, important for the forces generally, would obviously affect the future of women in the CMCs.[39]

The idea of women as cadets at RMC had been generally acceptable, including to most male cadets. By 1987 three women had been appointed as cadet wing-commander. At graduation, one of these cadets, Daryl Lynne Tremain,** won the college's most coveted trophy, the Van der Smissen-Ridout Memorial Award for the most distinguished all-round cadet who did not receive the Sword of Honour. (The Sword of Honour is normally given automatically to the CWC commanding the graduation parade; the Van der Smissen is partly based on peer evaluation.) Interestingly, the Van der

was *Omni Vir* (All Men). The media gave excessive attention to the female cadets during these experimental years, especially at the outset, and this caused some male resentment. When the first women graduated in 1984 the commandant, Brigadier-General Frank Norman,* suggested that the official attitude should be, "by the way there are also some female graduates."[37]

In 1984–5 the Canadian Forces drew preliminary conclusions from a study, "Serving Women in Non-traditional Environments and Roles" (SWINTER), set up by the Defence Council in 1979 to assess the validity of restrictions against the employment of women in potential combat situations and isolated units. The Canadian

*Maj.-Gen. Francis John Norman (no. 3572) entered Royal Roads in 1952 and graduated from RMC in 1956. Commissioned in the RCR, he was ADC to the Chief of the General Staff 1960–1, and attended the Royal Military College of Science at Shrivenham 1961–2. Commandant of RMC 1982–5, where he was said to know every cadet by his or her first name, he went to Ottawa as Associate ADM(Pol) and then became commandant of the National Defence College.

**Daryl Lynne Tremain (no. 15583), a reserve cadet, was at RMC 1982–6 and graduated with honours in engineering physics.

Smissen bears the name of Tremain's grandfather, K.H. Tremain (no. 1766).* In 1988 Susan A. Whitley,** like Tremain an RETP and an engineering physicist, was the first woman to command the graduation parade and to receive the Sword of Honour. Both of these graduates exemplify the modern bi-cultural and technically equipped women who have entered the Canadian Forces through the CMCs.

When a journalist asked the commandant, Brigadier-General de Chastelain, in 1980 what these newcomers should be called, he casually suggested "Lady Cadets."[40] The term "Gentlemen Cadets," revered by previous generations, is now rarely used except nostalgically. "Lady Cadets" was, however, adopted formally by de Chastelain's successor, Brigadier-General Stewart.

*Kenneth Hadley Tremain (no. 1766) was at RMC 1923–7. He played for RMC on two Dominion Championship Football teams. He graduated from McGill in mechanical engineering, worked as a professor engineer 1929–39, and then joined the Canadian army for service overseas and at HQ in Ottawa. In 1949 he commanded CFB Petawawa, resigning in 1951 to resume his business career.

**Susan A. Whitley (no. 16506) graduated in engineering physics in 1988 and was commissioned as an air reserve aerospace engineer.

University for the Canadian Forces

At a General Council meeting in November 1981, RMC broached the question whether the name Canadian Military Colleges should be changed to Canadian Military Universities, but the proposal was quickly dropped.[1] At the next meeting of the council, Brigadier-General J.A. Stewart, the RMC commandant, explained the rationale behind the suggestion and its abandonment. He said the word "university" enjoyed higher prestige in the minds of young people than "college" and a change of name might assist in recruiting. In the discussion at the previous meeting, however, it had become clear that legal and political implications made a new name not worth the complications that would ensue. The council decided that, instead of changing the name, the word "university" would be used frequently in advertising.[2] From then on, RMC was sometimes referred to as "Canada's major federal university." Since the CMCs were the only federally supported educational institution of this kind, RMC's claim could be justified because it had taken on many of the characteristics that distinguish a university from a college or academy. The RMC calendar and other promotional literature began to call it a "university with a difference."

Definitions of these terms need further elucidation. A secondary meaning of the word "college" is "an institution for special instruction, sometimes professional or military, often vocational or technical." In more general usage referring to higher education, the words "college" and "university" both have several overlapping connotations that differ from one country to another. In one such definition, a university is an institution that covers more areas of knowledge and levels of study than a college; in another, it is made up of one or more colleges or university colleges. A third usage conveys the concept that a university, as distinct from a college, is concerned with the expansion of knowledge as well as with its dissemination to students.[3]

A simple rule of thumb to distinguish most universities from most colleges in North America is that universities are more likely to expect their faculties to engage in research, and that they also undertake the education and training of postgraduate students who will, in turn, become researchers as well as teachers, or who will practise their profession at a more advanced level. University professors do research and also teach graduate as well as undergraduate students. While most recruits for the CMCs would not usually be interested in these attributes, they would still assume that a university with this potential has more prestige than a college, and therefore is preferable. Stewart's reasoning for proposing to use "universities" in place of "colleges" in the name of the CMCs was thus an attempt to attract the better students. The spread of community colleges in Canada had sharpened the distinction between "university" and "college" in the eyes of many high school students.

RMC advanced its proposal to change the name of the Canadian Military Colleges to Canadian Military Universities only a few years after the external campaign to "put the 'M' back into RMC" was at its height. That campaign had been based on an allegation that the college had become too much like a civilian university. The painful memory of that controversy may have been partly responsible for the way the proposal to change names was dropped in 1987. If a name change for the three CMCs were to take place, the singular form "Canadian Military University" might have been more logical for all three together than the plural. RMC, however, was already a degree-granting university in its own right. To change its name to one that incorporated all the military colleges might have seemed a lowering of its status.

By functioning as a university in the sense of expanding knowledge by research, RMC could offer a service to the Canadian Forces. Service of that kind had figured in the reopening of the college in 1948,[4] but that function had not at first been made explicit. The chief Canadian government institution to do research was, and still is, the National Research Council (NRC), a federal crown corporation that was set up in 1916 during the First World War. Between the wars, NRC was mainly devoted to pure scientific research of a general nature, but research on defence increased during the 1930s. During the Second World War it came to predominate. In 1947 NRC's laboratory activities in military research were transferred to a new agency, the Defence Research Board (DRB).[5] Like NRC, DRB was an autonomous organization that employed scientists and also made grants to university scientists. With a relatively small budget, there were strict limits to the support it could provide.

The creation of DRB was largely the work of General Charles Foulkes, chief of the General Staff.[6] Foulkes had at first opposed reopening RMC, but once the decision was made, he advised the deputy minister that the army required the RMC faculty to engage in high-level research.[7] Defence minister Brooke Claxton confirmed this seven years later in an answer to a question in the House of Commons.[8]

Colonel Sawyer, because he was determined to establish RMC on an academic par with Canadian universities, selected his faculty on the basis of their qualifications to do research as well as teach. Professors doing research, he said, were more likely to keep up with current thinking on their subjects and so would be more useful teachers. Sawyer required a record of publication, or a potential for it, as a qualification for a faculty appointment, and he always insisted on evidence of research as a requirement for promotion. Once appointed, RMC professors, like their colleagues in the universities, were left largely to their own devices to work on topics of their own choosing. The services rarely requested specific research at RMC, so there was thus no deliberate and close correlation of RMC's research with Canada's immediate defence needs.

Several of the new appointees to RMC in the early years had worked during the Second World War with NRC or other wartime research organizations.[9] After their appointment they usually continued to work on similar problems in laboratories improvised in the old college buildings. A chemical engineering laboratory, for example, was set up in an old coal cellar. The research done at RMC was often pure, rather than applied, science, with no specific orientation to defence. Because of government fiscal restrictions, RMC professors could not receive NRC grants, but some scientists were funded by DRB. Following university practice, RMC professors were not required to devote a stated proportion of their time to research, compared with teaching, but a 50-50 mix was considered desirable. Though their research was not necessarily defence oriented in the short term, RMC scientists did work on many projects that had long-term military value – for instance on low-temperature physics, on the use of aluminium for surface and subsurface vessels, and on sound projection.

The development of a scientific and engineering research program at RMC was hampered by the college's lack of graduate students to serve as research assistants. This was partly overcome in the early years by arrangements whereby service personnel, posted to RMC, could work at RMC on a thesis topic to be submitted for a degree at Queen's University; and DRB permitted RMC to use its grants to employ research assistants. Then, in 1959, when RMC received its degree-granting power, Dr Dacey immediately proposed regulations for the award of RMC graduate degrees to serving officers and RMC graduates. He suggested that, as in Canadian universities, candidates should enter the graduate program with an honours bachelor's degree with at least second-class standing. The Faculty Board added that, if the degree were long past, there should also be a qualifying examination. Candidates should be in residence for at least a year and should write a thesis. Degrees might also be awarded on the basis of courses taken, but in the sciences and engineering the preferred route should be a thesis supported by courses.[10] The Minister's Manpower Study in 1965 noted the existence of a small graduate program at RMC. In that same year the chief of personnel told the Advisory Board that DND was funding research in the CMCs, but that the total amount was only $14,400, of which RMC got $5000.[11] Nevertheless, RMC faculty publication in the sciences and the arts was already impressive, and RMC had established a Division of the Faculty for Research and Graduate Studies. In 1966 there were twelve graduate courses in science and engineering, and RMC awarded its first master's degree (in mechanical engineering) to Squadron-Leader Pierre Bussières. The following year Captain B.D. Hunt was awarded the first MA, and the first graduate degrees in science were earned by Captain Allen James Barrett* and Captain Paul Gordon Jefferson** in 1971. Hunt, Bussières, and Barrett later taught at RMC in history, mechanical engineering, and mathematics, respectively.

RMC faculty members in the humanities and social sciences were also able to spend a substantial portion of their time on research projects because they had teaching loads that made this possible. They had little financial aid for research, though the commandant had a small contingency fund that could be used for travel. Until the 1950s, when the Royal Society of Canada and the Canada Council began to make grants, there was no Canadian source of funds on which RMC research in the humanities and social sciences could be based. Nevertheless, RMC faculty publication was also significant in military history and other fields.

As a result of faculty inspiration and example, some graduates of the postwar RMC went on to further study on military and defence-related topics elsewhere. Canadian officers who wanted to pursue graduate work in military fields in the social sciences and humanities had to attend American or British universities, and depend on scholarships or grants to do so. Several went, for instance, to the University of London, to Duke University, and to the US navy's postgraduate school of Monterey. In 1966 twenty-eight graduating ROTP cadets in various fields, sixteen of them from RMC, applied for permission to go to graduate schools. A few RMC science and engineering graduates high on the graduation list were also being appointed to spend a year at DRB doing research.

*Allen James Barrett (no. 5992) was at CMR 1959–62 and RMC 1962–4, graduating in honours mathematics and physics. From 1964 to 1977 he was an RCAF radio navigator. Appointed lecturer in the RMC Department of Mathematics in 1967, he obtained a PhD in physics from King's College, University of London, in 1975 and was promoted to professor in the RMC Department of Mathematics in 1985. He became dean of arts and science in 1990.

**Paul Gordon Jefferson (no. 5491) was at RMC from 1958 to 1962 and then served with the RCAF as a radio navigator until 1969. He did graduate work at RMC from 1969 to 1971. In 1975 he was a long-range navigator with 415 Squadron. In 1976 he was a captain in the Logistics Branch and became responsible for procurement, communications, and electronics, and cost analysis at NDHQ. On retirement in the early 1980s he worked for Mel Defence Systems and then for Mitel in Kanata, Ontario.

Some senior officers saw these innovations as an interference with a cadet's military career. It was therefore decided that permission to postpone military service would be granted only for prestigious opportunities like the Rhodes and Athlone fellowships, and that cadets who wished to stay on at DRB after one year would first have to pay the cost of their tuition at RMC. In 1967 there were sixteen ROTP applicants for leave to do graduate work, eight of them from RMC.[12] This increase in the number of young officers wanting to interrupt their military careers to do more academic study sparked further discussion within the officer corps in the forces about a possible conflict of individual and service interest and achievement. The question was explored in a 1971 article in the *Canadian Defence Quarterly*, without, however, advancing a definite conclusion.[13]

There was some support in DND, however, for the premise that graduate work and research have value for defence development. During the winter of 1966–7, the need for further development of graduate studies in the arts departments at RMC became urgent when DND made a proposal to establish chairs of defence studies in several civilian universities. Here RMC had an obvious primary interest that could be exploited for the benefit of serving personnel. In 1967 the Educational Council discussed and approved a proposal to establish a program in war studies in the college that would offer an MA to qualified service officers. The recommendation noted the advantages that would ensue to the Department of National Defence and the forces: the program would provide officers with further professional training in a military atmosphere, and the resulting intellectual stimulation would encourage original thinking about defence problems throughout the services. It could also serve as preparation for the staff colleges and the National Defence College. Commodore D.S. Boyle, director-general of postings and careers, agreed that the department should employ officers with advanced degrees.[14]

RMC was, in fact, already planning graduate courses in history, political science, and economics that had military relevance. By 1967 eighteen officers were registered in RMC's graduate division, mostly in engineering, though teaching had also begun in war studies. The courses in the war studies program could be taught by existing RMC staff. Business administration was expected to follow in 1968, if a current freeze on hiring was lifted to allow the appointment of new staff.[15]

In 1968–9 General Rowley's Officer Development Board noted that Canada was unique among Western nations in not having a centre for strategic studies and research. He suggested the establishment of a government-supported centre associated with his proposed Canadian Forces University, and also the encouragement of independent centres elsewhere. These, he said, should all be interdisciplinary and should include economic, political, sociological, technical, and military facets. The government centre would function as the Canadian Forces' educational, doctrinal, and research centre, but it should be independent of government direction.

In an annex, Rowley gave details of a proposed graduate studies and research division of the Canadian Forces' university in management which "must be operated under government regulations governing the present operation of the Graduate Division of RMC." Rowley had already noted that in other countries, military research and teaching in government centres are usually not mixed, as they are in universities. He seemed to be suggesting in this annex that research should be directly associated with graduate teaching. Elsewhere in the report, however, there was no mention of scientific, engineering, or historical research, or of other graduate studies, except of strategy. It is clear that the Officer Development Board (ODB) did not contemplate the creation of a defence teaching university with which a strong graduate and research program would be associated.[16]

When General Sharp made his fruitless recommendation to the cabinet for establishing a university of the Canadian Forces he, too, appears not to have contemplated a university-type graduate and research function.[17] If RMC had been included with the relocated staff and defence colleges in Ottawa, as Rowley originally intended, however, the college might have automatically brought along with it from Kingston its developing research and graduate teaching capacity, not only in the sciences, but also in engineering, the humanities, and the social sciences.

While the ODB was planning its proposed revolutionary changes in the officer-development system, the basis of support for RMC's well-established research and graduate program had begun to dissipate. The chief reason was that the growth of the RMC faculty to cope with second-language education, and also with an increasing enrolment, had created a greater need for outside funding for research at a time when the rapidly expanding Canadian universities were also seeking more support from DRB. The amount of available support was therefore declining. There was a second reason. Although in the early 1950s NRC had again been drawn into defence research by the deteriorating international situation,[18] the Treasury Board still frowned on NRC grants to RMC on the grounds they would be a transfer of parliamentary-approved appropriations for one government department to another without legislative sanction, a contravention of sound government accounting. RMC was cut off from what could have been the most important source of grants for its research, and forced to turn more than ever to the Defence Research Board, the authorized administrative controller and financial support for all research in the Department of National Defence. From 1965 on, RMC regularly reported its growing plight in the Advisory Board annual reports.

When the college reopened in 1948, the administrative officers and faculty had had close and cordial relations with the Defence Research Board. DRB's founding president was Dr Omond Solandt,* who had organized the Canadian army's operational research during the war and worked with Colonel Sawyer and other RMC faculty

overseas as well as in Canada. Solandt became one of the four members of the Chiefs of Staff Committee. In that capacity, with his understanding of educational problems, he had been a tower of strength in the conceptual and formative years of the college. Thereafter he maintained an interest in RMC's development, and for a brief time a section of his Defence Research Board was provided with accommodation in unused space in the MacKenzie Building. By the mid-1960s, however, DRB's relatively small budget was unable, despite goodwill, to help solve RMC's problem of obtaining adequate support for research.

That problem was discussed at length in the Advisory Board in 1965. Dr A.H. Zimmerman,** who was now chairman of DRB, asked if the CMCs had tried to get support from industry. Colonel Sawyer replied they preferred not to do so, since they wanted to work on DRB projects related to defence needs and the future of the services. If they got industrial support, it would be at the expense of research for the services. He added that RMC had received $130,000 from DRB for the science and engineering faculties, which had been vital to its research program.

The 1965 meeting of the Advisory Board also received a detailed statement of the amount of DRB support for research at RMC. Since 1958 RMC had received 123 grants totalling $150,330. In 1965 it was receiving eighteen grants totalling $22,505. Support for research at the CMCs was approximately 10 per cent of all the grants that

*In 1943 Omond McKillop Solandt, then studying in England, was appointed to run operational research for the Canadian army overseas. He became its superintendent with the rank of lieutenant-colonel in 1944. He was appointed director of defence research in Ottawa in 1946 and was chairman of the Defence Research Board 1947–56.

**Adam Hartley Zimmerman Jr, educated at the Royal Canadian Naval College 1944–6, served six years in the RCN reserve and joined Noranda Forest Inc. in 1958; he became vice-chairman of Noranda in 1987.

DRB made to Canadian universities. In addition, RMC had granted $15,600 for travel to learned societies, an average of $65 per man, with a maximum of $800. The average individual travel grant in Canadian universities was $165. Zimmerman said he was gratified that DRB grants had preserved the RMC science and engineering faculties and had perhaps also permitted a modest program in postgraduate work.[19]

In 1967 the board learned that Dr R.J. Uffen, the new DRB head, had noted that the CMCs had no support for research other than DRB. It was difficult for his board to meet their current vastly increased need for $400,000 because its own budget had not been increased over the previous year's appropriation. DRB outside grants were based on two criteria: the scientific competence of the applicant and the extent of defence interest in the project. In that same year, DRB made its first two grants to RMC for economic and political studies.[20]

After it was established that RMC would not move to Ottawa and that its long-awaited science and engineering building (the Sawyer Complex) would be built, the funding of research became yet more urgent. Members of the Advisory Board met with the minister, James Richardson, in December 1972 and February 1973, and, although no solution was found, the board proposed that a joint committee be formed to recommend a policy for the further development and funding of research at RMC.[21]

This joint committee was chaired by Dr Howard Petch* of DRB and reported to Dr L.J. L'Heureux, chairman of the Advisory Board. It met in September 1973 and January 1974. The preliminary report stated that the CMCs had a "community of interest" with the universities that led them to engage in research and required funding. DRB was frequently asked why it funded research projects at the CMCs that had little or no relation to defence, while university applications were subject to ever more scrutiny for defence relevance. The report concluded that a higher proportion of the CMCs' research should be devoted to defence and military needs.[22]

RMC's case, as presented to the joint committee, noted that DRB was responsible for all research in the Department of National Defence, including the CMCs. Unlike the universities, RMC had no access to funds from the provinces, from the NRC, from other government departments, from industry, or from foundations that were restricted in their capacity to give to federal government organizations. It also had no PhD students and only a few MA students to assist in faculty research. The college was clearly inadequately funded. The brief suggested the creation of a standing committee – the CMCs' Research Review Committee – which should meet annually to review the research program and require a line-item in the DND budget.

RMC asked that the CMCs should receive research funds through individual faculty applications to DRB on the same lines as those on which university faculty applied, and that some means be found to transfer funds from DND to NRC, so that CMC professors could also apply to NRC for research grants. Furthermore, the RMC budget itself should include money specifically for research, "which was rightly part of our instructional budget." The sum of $10,000 per annum per department was suggested. The RMC brief concluded, "although research activity is required of a professor at the CMCs [and] in fact is assessed as equal in weight to teaching [for promotion], it has not been pursued with sufficient vigour . . . partly due to the fact that it has never been properly supported or funded. Many of our Masters [in Ottawa] look upon

---

*Howard E. Petch, physicist, served with the RCAF 1943–5, principal, Hamilton College of McMaster University 1963–7, vice-president academic, University of Waterloo 1967–8 and 1970–5, president and vice-chancellor, University of Victoria 1975.

research as a private matter to be grudgingly supported as a favour, rather than [as] an important part of a professor's duties for which he is paid his salary."[23]

DRB support for research at the CMCs was, in fact, substantially higher than in the early 1960s. By 1969–70 it had reached $223,260, but had since remained stable at that level, increasing to $276,000 for 1973–4,[24] barely enough to keep pace with inflation. What was necessary was a clear statement about RMC's research functions that would justify higher funding.

The Petch Committee produced a report that was to become the foundation of future policy. There was active research at the CMCs that was oriented to defence and military needs and that included the social sciences and humanities. Petch suggested that team and interdisciplinary research should be stepped up, that there should be joint programs with neighbouring universities and more CMC collaboration, including exchanges of faculty. CMC projects relevant to defence science should continue to be individually funded, but there should also be institutional grants called discretionary grants. Initial support for new major programs should be negotiated between DRB and the CMCs. The program would eventually require a budget in excess of $500,000.[25]

In December 1974 a team from DRB visited RMC to identify areas for research, particularly in chemistry and physics, and preferably interdisciplinary, where the college could meet defence requirements for research and expertise. This visit was intended to be the first of a series. The commandant, Brigadier-General Turner, told the visitors he would like to see RMC's research more directly aligned with defence needs than it was now. The visiting team found that the RMC faculty was anxious to follow that path. RMC faculty members complained that research projects were often set by a DND staff officer and that, before results could be obtained, many of them were declared no longer relevant. They asked for relevance to be judged against long-term defence requirements.

The committee reported that at RMC there were twenty graduate students, ten of whom were in electrical engineering and three in civil; the rest were doing an MA in arts. Twenty staff members were engaged in graduate training. It said that RMC needed, and could handle, more graduate students. The visitors suggested that RMC should build on its "established international reputation" in low-temperature physics by investigating superconductivity with sensitive detectors, and also acoustic emissions in extreme conditions. It also said that the electrical engineering program was strong and was meeting the needs of the services. But it could see no projects in chemistry or chemical engineering that could provide a nucleus for a unified research area. Research in mathematics, for instance in computer science for numerical analysis and operational research, could, however, be useful to an officer's career.

The visitors also suggested that the MA in war studies could include graduate specialization in the mathematical logistics of warfare and management, and that RMC needed to develop its programs in operational research and management science, which at present were inadequate. It concluded that the Petch Report provided a sound basis for future development.[26]

At the Advisory Board in April 1975, L'Heureux reported on the implementation of the Petch Report. He forecast that research funds available to the CMCs in 1976–7 would increase to $600,000. Dacey added that CMC faculty could now apply directly for grants, but L'Heureux warned that the problem of finding the money necessary for research was not yet solved.[27]

L'Heureux's warning was a portent of what seemed to be new difficulties. During 1975, by what Dr Solandt called a "secret" order-in-council, the government cut DRB down severely, leaving only a small staff to advise the minister of national defence on science policy. A budget decision by the deputy minister then removed from DRB the responsibility for external research funding. That function was transferred to the department itself, which would thus gain direct control over the type of research that was supported.[28] It now seemed that RMC's battle for research funding might have to be fought all over again.

The February meeting of the Advisory Board was briefed by Dr H. Sheffer,* chief of research and development at DND, who outlined the new procedures for research funding. The board stated that the present level of funding was adequate but should be escalated to take into account both inflation and staff growth, especially that caused by the bilingual and bicultural program. It said that the Petch Report should be brought up to date under the auspices of DND's Committee for Research and Development (CRAD). Funding was needed to sustain the academic standards of the staff and to support institutional research. The minister replied that he was in general agreement, but that research should be related to defence and not be "too general." The education of officers should remain the dominant RMC consideration.[29]

In October, Dacey wrote to Sheffer, outlining the history of research at RMC. He reminded him that in 1947 the college had been reopened partly to meet the army's research needs. He also claimed that statements made in the House of Commons, for instance by Brooke Claxton, had committed the government to support research in the CMCs on the same level as in the universities. He said that new francophone appointments, which were to be made in the five-year period beginning in 1976, would create a need for extra funding to support their research requirements on the same level as that of their colleagues.[30] The following March the Advisory Board again reported to the minister the need for a reasonable program of research, selective and, where possible, related to teaching responsibilities.[31]

CRAD's response was favourable to RMC's case. In his report for 1977–8, the commandant stated that, over the years, DRB and CRAD had stimulated research at the college and it now applied more directly to defence and military needs. He gave as illustration the fact that civil

engineering had eight DRB grants and one DND/CRAD grant.[32] A year later General Smith, ADM(Per), stated that continuing DRB grants to the CMCs for 1978–9 totalled $743,000, an increase of 16–18 per cent over the previous year.[33]

At the end of 1979, when the RMC dean of graduate studies and research outlined the sources of funds (small Canada Council grants for the arts, CRAD awards which were now the majority of RMC's research funding, and DND contracts in the form of financial encumbrances within the department), he stated that in real value the CRAD grants had halved in the past six years. They had not continued to increase with the growth of the faculty, inflation, and the high costs of research, and this had resulted in a severe cut in RMC's research activity. He expected only half of the applications now submitted would be funded, and indicated there was difficulty in administering the CRAD grants because they came under the department's capital vote and so could not be used to employ research assistants.[34]

By 1979 the Treasury Board had made substantial changes in the financial administration of what used to be called DRB grants. The Advisory Board asked Dr E.J. Bobyn,** CRAD, to explain them. Bobyn stated that a number of details still had to be worked out.[35] He explained the new research organization in the Department of National Defence (CRAD) and noted the increased participation of the Advisory Board, giving a list

---

*Dr Harry Sheffer served with the Canadian army 1942–7, director of research, DRB, 1952, superintendent, Kingston DRB Laboratory 1955–7, chief superintendent 1957–67, scientific assistant to the vice-chief of the Defence Staff, CFHQ, 1967–9, deputy chairman, DRB, 1969, vice-chairman 1969–78.

**Edward Joseph Bobyn served in the Royal Canadian Corps of Signals 1943–7, head of the Instrumentation Section, Canadian Armament, R&D Establishment, 1948, chief of systems research, SHAPE, in Holland, 1960–3, director-general, Defence Research Establishment, Valcartier, 1968–72, deputy chairman, DRB, 1972–4, chief of R&D, DND, 1974–83, and NATO Defence Research Group 1975–83.

of board recommendations that had been implemented.[36] The board again recommended the updating of the Petch Report to fit the new CRAD administration of research funding.[37]

Bobyn spoke to the board again in March 1980. He now explained that the government's reorganization of research funding had meant that the CMCs were excluded from university and industrial grant programs. Furthermore, DND had lost contact with the central granting councils, and was rapidly losing contact with the research and development community, because the central granting councils "do not get involved in military projects and most DND researchers are militarily oriented." However, his Research and Development Committee, which assessed research applications, was impressed by the improved quality of CMC submissions. He supported the board's recommendation that the Petch Report should be updated. Dr John Plant, dean of RMC's Graduate Studies and Research Division, suggested that RMC's problem in appointing research assistants might be solved by establishing an RMC Research Institute. The principal, Dr Tilley, added that funding should be indexed to inflation.[38]

Bobyn then went on to announce the allocation of departmental grants that had been made by an interdisciplinary committee on which RMC was represented. Of sixty grants to the CMCs totalling $856,000, RMC had received forty, amounting to $432,300. They had been indexed over 1979, but fell "far short of what the Review Committee thought were the real needs of the science, engineering, and mathematics faculties of the CMCs." Natural Science and Engineering Research Council (NSERC) grants, which had taken the place of NRC grants, had been increased, but CMC faculty could not compete for them. A major increase in staff at RMC since 1975, a result of the second-language program which had brought a 70 per cent increase in the RMC science and engineering civilian faculty, and also an increased interest in social science and humanities research, meant that

the CMC research budget now lagged far behind those of the universities. As branch chief of research and development at DND, Bobyn "endorsed strongly the rationale for an active research programme in the CMCs." He said it was highly desirable that research be carried out in a university-level institution, and that there was a potential spin-off from a defence-oriented technology base in the college. He added that it was also a means of attracting good staff.*[39] This encouraging statement by the officer who now managed funding for CMC research was made at a time when the fortunes of the whole science community in Canada had reached a low ebb. A Science Council of Canada report published by the government in 1980 argued that, because of declining enrolments, university research was in jeopardy.[40]

Bobyn's statement also came at a time when RMC was expected to increase its graduate program in at least one respect. In 1978 a committee to consider the possibility of a degree in computer science had reported that RMC did not have the resources to establish a full master's program in computer science, but could give an interdisciplinary MSc in digital systems by an arrangement between the Departments of Mathematics and Electrical Engineering. The computer science content could be increased by courses at Queen's University, but not by more than 40 per cent. It should also be coordinated with an AERE computer course.[41]

In June 1980 the RMC principal, Dr Tilley, proposed to the Academic Council an RMC graduate program for an MSc in computer systems that would be tailored to the requirements of the forces. This proposal had already been enthusiastically endorsed by engineering classifications in the forces, and also by CFHQ, because it

---

*NSERC grants now permit doctoral fellows to work at RMC if they choose. Their stipend is paid to the fellow and RMC is not involved financially.

RMC Redmen football, 1979. This was the last year in which RMC won a championship before football was dropped as a varsity sport in 1983.

addressed areas not covered by programs in other institutions. It would require two years for the master's degree and would enrol twelve to sixteen students, half of whom would be satisfied with one year's work and a diploma. It would be implemented in the summer of 1981, but this depended on new automated data processing facilities being made available for RMC. The council approved the program, subject to the proviso that it require no extra personnel.[42]

RMC's graduate work in war studies was also under scrutiny. The chief of personnel development at NDHQ had called for a larger review of the Canadian officer development program, to follow up such major studies

as the *ODB Report* (1968) and the chief of the Defence Staff's Seminar on Professionalism in the Armed Forces (1971–2). In 1981 Professor Nils Örvik* of Queen's University published an article in the *Canadian Defence Quarterly* which argued there was a need for Canadian programs in strategic studies and national and international security for service officers and personnel in External Affairs. At present there was demand for such a program only in the three military colleges. He proposed that the colleges should therefore have an interdisciplinary four-year undergraduate honours course, a two-year masters' program, and a science/engineering option that had a small strategic studies input. There should be a director of strategic studies at each of the "military universities."[43] His article provoked replies from two officers, both of whom said that strategic studies should be taken only at the graduate level, and that, as prerequisite, there should be a requirement of an undergraduate program in international studies. Both of these critics said that RMC already offered a satisfactory undergraduate course of that kind and also had a master of arts in war studies that covered most of the areas Örvik had recommended.[44]

By December 1979 the graduate program in war studies had, in fact, been reviewed and tailored to make it serve the forces' requirements more closely by the addition of two new courses, quantitative analysis and Canadian government and administration. Progress had been halted then, partly because of the illness of Dean James Cairns, but even more because the Academic Council had directed that the option be presented without a large bill for personnel. The RMC commandant, Brigadier-General John Stewart, then asked the Academic Council about the likelihood of an expansion of

*Nils Örvik, born in Norway, served in the Norwegian army in the Second World War. He was directing staff, Norwegian National Defence College, 1951–62, director of the Centre for International Relations, Queen's University, 1975–85.

postgraduate opportunities for officers from operational classifications. Colonel Savard,* commandant of CMR, suggested that the new program might be offered in two stages separated by a number of years of active duty. The chairman of the council said that expansion might be possible by September 1981, with twenty-two officers ranging from captain to lieutenant-colonel, but that not all of these would be at RMC.[45]

The development of RMC's facilities for research and graduate work led in 1981 to a memorandum of understanding signed by Queen's University's principal and RMC's commandant: the two institutions would henceforth provide reciprocal courses and services to graduate students, without imposing a charge. Queen's also agreed to act as an agent for hiring research assistants for RMC. As further recognition of the increased degree of co-operation of the two Kingston schools, on 21 March 1983 the Iron Ring Ceremony to induct graduates as engineers, traditionally held at Queen's, took place at RMC. In the future it was to alternate between them annually.[46]

Dr W. Furter, formerly the professor in charge of the chemical engineering program and now dean of the Canadian Forces Military College (RMC's Division of Continuing Education), suggested in February 1983 that one source of difficulty in dealing with NDHQ on research-related problems was that faculty research was not mentioned in the official statement of the college's role in *Queen's Regulations and Orders for the Canadian Military Colleges (QR Canmilcols).*[47] Major-General J.A. Fox,** CPD, agreed it was lacking. Commenting on the fact that the average grant per applicant in civilian universities was now $30,000, while in the CMCs it was only $20,000, Fox said there was indeed need for a clear statement about principles to legitimize CMC research. The Association of Universities and Colleges of Canada suggested that such a statement should be short. However, the director of the Directorate of Professional Education and Development disliked making research an objective of the CMCs, and declared that the conduct of research would be tied to

Brigadier-General W. Niemy, CD, commandant, 1985–7

Prince Philip, 21 May 1980

---

*Col. Charles-Eugène Savard (no. 3759) entered CMR in 1952 and withdrew in 1955 to join an armoured regiment. He was commandant of CMR, 1978–81.

**Lt-Gen. James A. Fox (no. 3818) entered Royal Roads in 1953 and graduated from RMC in 1957. He was commissioned in the Canadian Hussars. In 1982 he was chief of personnel development and in 1986 commanded the Mobile Force HQ at St-Hubert, Quebec. He then became assistant deputy minister, personnel, and, later, vice-chief of Defence Staff.

Commodore E.R.A. Murray,
OMM, CD, commandant, 1987–

available budget resources.[48] As a result, only a brief reference to research was added to the *Queen's Regulations and Orders.*[49]

There was, however, more awareness now in headquarters that research was not merely a valuable, but also an essential, element in the CMC system. Funding by DND was facilitated by an Engineering and Research Standing Committee chaired jointly by the chief of engineering and maintenance and RMC's dean of engineering. Also the chief of construction and properties in Ottawa provided research support to RMC's Department of Civil Engineering. Mr P. Hall-Humpherson, director of administration training and education support (DATES), announced a new submission of Treasury Board for an allocation of discretionary funds that were to be primarily used as arts research grants, which had not been updated since 1985. The proposed increases were: RMC by $5000 to $25,500; CMR by $6250 to $21,800; and RRMC by $3150 to $16,000. Principal Eric Graham of Roads commented that the arts faculty felt they had not been well treated in the past; and Principal Benoit* of CMR said the increase merely reflected the increase of staff since 1965. Principal Tilley of RMC, cynical as a result of previous experience with Treasury Board's cheeseparing, interjected that by putting the proposal up, "we might lose all."[50]

Nevertheless, Dr D. Schofield,** acting chairman of DND's Committee on Research and Development, reporting to the Advisory Board on research a few weeks later, stated that the total research funding for the CMCs for 1982–3 had been $1.375 million, and that the total funding for 1983–4 was expected to rise to $1.458 million, making an increase of 70 per cent over 1979–80, when it had been only $856,000. There had been sixty-seven applications for grants, involving one hundred faculty members. They had requested just under $2 million. He commented, "The acquisition in recent years of young, keen, and well-trained faculty members has greatly improved the potential research environment at

the Military Colleges, and means must be found to encourage and stimulate these valuable young people."[51] To a large extent the increase in the faculty at RMC was due to the bilingual and bicultural program, and many of the "bright young people" were francophones.

The Advisory Board was told in March 1984 that "total funding for the academic research programs [in the CMCs] in the fiscal year then ending was just under $1.5 million of which about fifty percent was allotted to RMC."[52] CRAD's total research grant funding for the CMCs in 1983–4 was $2.25 million, of which RMC's share was approximately $1.5 million. Social sciences and humanities received approximately 3 per cent, science 48 per cent, and engineering 49 per cent. RMC's share of the total CMC funding was proportionately bigger than those of the other two colleges, because it was the only one with an engineering faculty.[53]

In DND's Academic Research Program the primary requirement was academic excellence. But the chief of research and development added that another requirement was defence relevance, though relevance to Canadian and North American society as a whole might be acceptable. RMC's research funding for 1985–6 for science and engineering was $1,836,000. It came from DND/ CRAD grants, from DND agency contracts, and from other contracts. Another $36,000 was received for arts research.[54] RMC was now in receipt of almost $2 million a year. The transfer of RMC's research funding from DRB to DND/CRAD had thus proved to be advantageous.[55]

Publicity from one major achievement helped to bring about this improved understanding of the value of RMC

---

*Marcel Benoit, a physics professor, was principal and director of studies at CMR 1965–87.

**Derek Schofield, a PhD in Physics from Sheffield University, was engaged in defence research in Canada from 1955, principally in anti-submarine warfare. From 1968 to 1976 he was chief scientific adviser to the vice-chief of Defence Staff, and then became deputy chief of research and development (laboratories), DND.

research. In 1980 Dr Tilley had explained to the Advisory Board that Dr T.G. Barton,* a former major who was now an associate professor of civil engineering at RMC, was doing research on a "Plasma Torch" that might prove to be an effective way to destroy polychlorinated biphenals (PCBs).[56] His research had originally been funded under the CRAD Research Allocation program and later was supported by DND development funds. It had involved students as assistants. In 1981, when the Ontario Minister of the Environment provided $400,000 for the work, that grant brought RMC wide publicity because of popular concern about pollution.[57]

In addition to producing officers with advanced degrees for the Canadian Forces, and also undertaking many other research projects that had defence relevance, RMC had begun to make another contribution facilitated by the specialized qualifications of its faculty. It offered short courses for serving officers and, in some cases, also for federal civil servants. In 1977–8 the commandant reported on "out-service" courses of this kind in AERE computer programming, advanced radiation safety, and non-destructive testing. He also said that similar offerings in defence resources management were in the planning stage.[58]

Next year 107 officers and public service employees attended five different short out-service courses given by various RMC departments under the direction of the Division of Graduate Studies and Research.[59] Then in 1979, at the insistence of Major-General Loomis,** chief of program at CFHQ, RMC established a Centre for Defence Resources Management in the Department of Political and Economic Science. It was directed by Dr Jack Tredennick*** and was designed to do research and act as a consultant for NDHQ.[60] That centre also offered short courses.

In April 1985 the Faculty Council confirmed RMC's policy on such short courses. They would be less than one term in length and the minimum level for acceptance would be that for university undergraduates. Admission

Recruit arrival, 1990: The basic officer training course at CFB Chilliwack is completed before beginning at RMC and, on arrival, the "recruits" have become members of the Canadian Armed Forces.

would require at least a high school graduation certificate. The courses would be conducted by RMC faculty, and there would be no outside hiring, except for guest lecturers. Grades would be given. Credits for RMC faculty would be the same as the Canadian Association of University Teachers' "Equivalents," or something similar.

*Dr Thomas Gordon Barton (no. 6590), Royal Roads 1961, RMC BEng. 1965, was commissioned in the RCAF. A lecturer in civil engineering in 1970, he resigned from the forces to become professor of civil engineering in 1983.

**Maj.-Gen. Dan Gordon Loomis (no. 2861), Royal Roads 1948–50, RMC BSc 1952, Queen's University BSc 1954, MA RMC 1969, served in the Korean War, winning the MC. CO 1RCR, Cyprus, 1970, ADM(Policy) staff 1972 as director of strategic planning, CFHQ, deputy commander and chief of staff, Canadian Contingent, Vietnam, 1973, special policy analyst, ADM(Policy), chief of staff operations, Mobile Command, 1974.

***Dr John Macauley Tredennick (no. 4824) graduated from RMC with honours in economics in 1965. Commissioned in the navy, he served in HMCS *Saskatchewan*. In 1976 he was appointed associate professor of economics at RMC. In 1988 he was selected to serve for three years on the staff of the NATO Defence College, Rome.

The Dr Walter S. Avis UTPM *Honour Shield: Lieutenant-Commander P. Avis presents the award (named after his father, the late Dr W.S. Avis, dean of the* CFMC, *1974–80) to Officer Cadet Barbara Anne Messier, graduation, 1990.*

The vastly increased research program that RMC had achieved by the mid-1980s was the coping stone on the university edifice that had been founded in 1948 and had built up so substantially in the following years. During the turbulent 1970s, RMC had reshaped its undergraduate courses to ensure they meet the requirements of Canadian military professionalism. By the mid-1980s it had become institutionally bilingual, a mirror image of what had first been introduced at CMR. The CMCs could now meet the essential requirements of both founding linguistic groups in Canada to enable their young men and women to participate as officers in the defence forces of their country. In this way, the colleges served the cause of national unity. They had also extended their courses to serving personnel to provide an opportunity for them to improve their academic qualifications, and had thereby increased the colleges' output of educated officers. In all these things, except in bilingualism and biculturalism where CMR had been the pioneer, RMC had given leadership.

The development of graduate and research programs, and of "out-service" short courses at RMC, was indicative of what the colleges, and especially RMC, had now become. RMC is in a very real sense a university for the Canadian Forces. As such it is almost unique, a military university that, with its two sister affiliates, not only educates and trains potential officers, but also renders advanced scientific and educational services for the forces.* RMC had made all these changes without weakening its earlier longstanding commitment to produce well-educated professional officers, the primary reason for its existence "to serve Canada."

These short courses were seen as an inexpensive way of upgrading departmental and forces' expertise; they would give the RMC faculty more exposure, and they would increase the military presence in the college and so create more military awareness and a greater DND appreciation of RMC's contribution to DND goals. They would also serve to accrue resources to RMC's future advantage.[61] In 1985 an Atomic Energy of Canada SLOW-POKE II nuclear reactor was installed in the chemical engineering laboratories at RMC. This facility was for teaching and research at RMC, but was also used by various DND agencies and by Queen's University, which contributes a share of the cost of its operation.[62]

*The Australian Defence Force Academy is the closest equivalent. It has no sister military colleges but is complete in itself. As a college affiliated with the University of New South Wales, it gives UNSW undergraduate and graduate degrees.

Historic Hockey: RMC *vs Queen's, 1986*

Notes

Introduction: Tradition and Change in Military Education

1 Michael Lewis, *England's Sea Officers: The Story of the Naval Profession* (London: Allen and Unwin 1948); Samuel P. Huntington, *The Soldier and the State: The Theory and Politics of Civil-Military Relations* (Cambridge: The Belknap Press 1957); Morris Janowitz, *The Professional Soldier: A Social and Political Portrait* (Glencoe, Ill.: Free Press 1960); Allan R. Millett, *The General: Robert Bullard and Officership in the United States Army 1881–1925* (Westport, Conn: Greenwood Press 1975), 3–10

2 Huntington, *The Soldier and the State*, 9–10

3 Richard A. Preston, *Perspectives in the History of Military Education and Professionalism* (United States Air Force Academy, Harmon Lecture no. 22, 1980), esp. 7

4 Millett, *The General*, 4–5

5 For example, Earl McGrath, Introduction to William E. Simons, *Liberal Education in the Service Academies* (New York: Columbia Teachers' College 1965), xi

6 Alfred T. Mahan, "The Military Rule of Obedience," *Retrospect and Prospect: Studies in International Relations, Naval and Political* (London: Sampson Low, Marston and Co 1902), 282–3

7 John W. Masland and Laurence I. Radway, *Soldiers and Scholars: Military Education and National Policy* (Princeton: Princeton University Press 1957), 170–1, 198–9; John P. Lovell, *Neither Athens nor Sparta: The American Service Academies in Transition* (Bloomington and London: Indiana University Press 1979), 58; Lovell, "Modernization and Growth of Service Academies," in Franklin D. Margiotta, ed., *The Changing World of the American Military* (Boulder, Colorado: Westview Press 1978), 309

8 Lovell, *Neither Athens nor Sparta*, 58; Lovell, "Modernization and Growth of Service Academies," Margiotta, ed., *The Changing World*, 309

9 Preston, *Perspectives*, 4–27; W.H. McNeill, *The Pursuit of Power: Technology, Armed Forces, and Society since AD 1400* (Chicago: University of Chicago Press 1982), 254

10 Maj.-Gen. H.R.S. Pain, director of training, army, in Michael Howard and Sir Cyril English, *Education in the Armed Forces: Report of a Seminar Held at the Royal United Services Institute for Defence Studies*, Wednesday, 15 Nov. 1972 (London: Royal Institute for Defence Studies 1973), 7

11 Oscar Browning, *Wars of the Century and the Development of Military Science* (Philadelphia: Lippincott 1903)

12 Gen. James M. Gavin, *War and Peace in the Space Age* (New York: Harper and Row 1958), 34–5, 38–9

13 John Wilkinson, MP, in Howard and English, *Report*, 3

14 John C. Toomay, Richard H. Hartke, and Howard L. Elman, "Military Leadership: The Implications of Advanced Technology," in Margiotta, *Changing World*, 243–80

15 Raymond Garthoff, *The Soviet Image of Future War* (Washington, DC: Public Affairs Press 1959), app. B

16 Maj.-Gen. G.I. Pokrovskii, *Science and Technology in Contemporary War* (New York: Praeger 1959), 33–53, 101–2, 116

17 Thomas Schelling, *Arms and Influence* (New Haven and London: Yale University Press 1966), passim; Klaus Knorr and Oscar Morgenstern, *Science and Defense: Some Critical Thoughts on Military Research and Development* (Princeton: Center for International Studies, Policy Memoranda no. 37, 1965), 37; Klaus Knorr, *On the Uses of Military Power in the Nuclear Age* (Princeton: Princeton University Press 1966), 172; Stanley Hoffman, *The State of War: Essays on the*

Theory and Practice of International Politics (New York: Praeger 1965), 236; Raymond Aron, *The Great Debate: Theories of Nuclear Strategy* (New York: Doubleday 1965), 27

18 William E. Simons, *Liberal Education in the Service Academies* (New York: Columbia Teachers' College 1965), xi, 2–3

19 Ibid., 9

20 Howard and English, *Report*, 3 (my italics)

21 Andrew J. Goodpaster, "Educational Aspects of Civil-Military Relations," in Goodpaster and Samuel P. Huntington, *Civil-Military Relations* (Washington, DC: American Enterprise Institute for Public Policy Research 1977), 34–5

22 Simons, *Liberal Education*, 20–30

23 Charles C. Moskos, Jr, "Armed Forces and American Society: Convergence or Divergence," in C.C. Moskos, ed., *Public Opinion and the Military Establishment* (Beverly Hills: Sage 1971); Moskos, "The Emergent Military: Civilianized, Traditional, or Pluralistic?" (Paper in the Inter-university Seminar on Armed Forces and Society, Chicago, May 1972); Peter Kasurak, "Civilianization and the Military Ethos: Civil-Military Relations in Canada," *Canadian Public Administration* 25, 1 (spring 1982): 108–29

24 Lt-Gen. Sidney B. Berry, "The United States Military Academy in the 1970s and 1980s: Change within Tradition," *Signum* 3, 2 (1976): 23, 33

25 Lovell, *Neither Athens nor Sparta*, 274

26 Richard Gabriel and Paul Savage, *Crisis in Command: Mismanagement in the Army* (New York: Hill and Wang 1958), 30

27 Lovell, *Neither Athens nor Sparta*, 47–9; Gene M. Lyons, "Defense Policy Making," in David B. Bobrow, ed., *Components of Defense Policy* (Chicago: Rand McNally 1956), 124–5

28 *Military Review* 60, 7 (July 1980); Lieutenant-Colonel Sam Sarkesian, "Military Leadership: Time for a Change," ibid. (Sept. 1980): 16–24

29 Richard A. Preston, "Military Academies in a Changing World: Possible Consequences of the Student Protest Movement," in M.R. Van Gils, ed., *The Perceived Role of the Military* (Rotterdam: University Press 1971), 3–18

30 Oscar Grunsky, "Education and Military Commitment," *Armed Forces and Society* 6 (fall 1979): 144–5

31 Simon Raven, "Perish by the Sword," *Encounter* (May 1959): 37–59

32 Lovell, *Neither Athens nor Sparta*, 244–8

33 John P. Lovell, "The Professional Socialization of the West Point Cadets," in Morris Janowitz, *The New Military* (New York: Wiley 1967), 116, 119

34 Edward Glick, *Soldiers, Scholars, and Society: The Social Impact of the American Military* (Pacific Palisades, Cal: Goodyear 1971), 104

35 [Maj.-Gen. J.J. Paradis], *CPD Study: Rationalization of the Canadian Military College System* (Ottawa: Department of National Defence, October 1976), 43–51

36 Howard and English, *Report*

37 Hugh Smith, ed., *Officer Education: Problems and Prospects* (Duntroon: Royal Military College, September 1980)

38 Eg, Frederick C. Thayer, "AFROTC and USAFA: Time for a Change," *Air Force Magazine* (July 1972): 58; William J. Taylor and Donald F. Bletz, "A Case for Officer Education," *Journal of Political and Military Sociology* (fall 1974): 251–67; Roger Beaumont and William P. Snyder, "A Fusion Strategy for Pre-commissioning Training," *Journal of Political and Military Sociology* 5, 2 (fall 1977): 259–77; Glick, *Soldiers, Scholars, and Society*

1 The Old RMC and the New Canadian Services Colleges

1 Richard A. Preston, *Canada's RMC: A History of the Royal Military College* (Toronto: Published for the Royal Military College Club of Canada by University of Toronto Press 1969)

2 John Boyd, *Sir George Etienne Cartier: His Life and Times. A Political History of Canada from 1814 to 1873* (Toronto: Macmillan 1914), 291

3 Mackenzie to Dufferin, 5 Aug. 1969, Dufferin Papers, microfilm reel A 111, National Archives of Canada (NA); Preston, *Canada's RMC*, 20–47. Documentation and further details on the history of the prewar RMC can be found in *Canada's RMC* when they are not given in this chapter.

4 G. Walker, "The Royal Military College," *RMC Review* 15 (June 1927): 26

5 Victoria, Parliamentary Papers, 57/1889, 19; Deputy-Ministers' Papers, A 10346, 19 Sept. 1890, RG 9, NA; Sir Arthur P. Douglas, *The Dominion of New Zealand* (London: Pitman 1909), 278; *Army Review* 4 (Oct. 1902): 331; Melbourne *Argus*, 30 April 1910; Joseph E. Lee, *Duntroon: The Royal Military College of Australia, 1911–1946* (Canberra: Australian War Memorial 1952); Chris Coulthard-Clark,

Duntroon: The Royal Military College of Australia, 1911–1986 (Sydney: Allen and Unwin 1986), 2–3

6   Preston, *Canada's RMC*, 96

7   Carl Berger, *The Sense of Power: Studies in the Ideas of Canadian Imperialism, 1867–1914* (Toronto: University of Toronto Press 1970)

8   Canada, House of Commons, *Debates*, 3 July 1895, II, 3782

9   Ibid., 18 April 1896, II, 6273. The number in West Point in four annual classes without repeaters would have been about 218. General G.W. Cullom, *Biographical Register of the Officers and Graduates of the United States Military Academy* (Cambridge: Riverside 1901), IV, 459–549

10   *Debates*, 1895, II, 3792; 1896, II, 6762

11   Preston, *Canada's RMC*, 67

12   *Debates*, 1899, III, 7022–4

13   Preston, *Canada's RMC*, 93, 96–7, 111, 117–19, 121, 170

14   Correlli Barnett, *Britain and Her Army, 1505–1970* (New York: Morrow 1970), 331, 333–4; compare, Allan R. Millett, *The General: Robert Bullard and Officership in the United States Army, 1881–1925* (Westport, Conn. and London, Eng.: Greenwood Press 1975), 8–9

15   Preston, *Canada's RMC*, 74, 100n, 118–19, 170n

16   Canada, House of Commons, *Debates*, 1895, II, 3781–813; 1896, II, 6683–763

17   Preston, *Canada's RMC*, 101–16, 130–49

18   Ibid., 177

19   Richard A. Preston, *Canadian Defence Policy and the Development of the Canadian Nation, 1862–1917* (Ottawa: Canadian Historical Association Historical Booklet no. 25, 1970), 16

20   Preston, *Canada's RMC*, 200

21   *Canada in the Great War*, V: *The Triumph of the Allies* (Toronto: United Publishers 1920), 357

22   John S. Moir, ed., *History of the Royal Canadian Corps of Signals* (Ottawa: Privately printed 1962), 1–7; Preston, *Canada's RMC*, 208

23   Preston, *Canada's RMC*, 220–25

24   Ibid., 252

25   "Attachment of RMC cadets to Permanent Force units during their summer vacation," DND, HQ 74/37/22, vols. 1 and 2, NA

26   War Office to Adjutant-General, DND, Aug. 1928, Film C 4852, NA

27   "Recommendations re. introduction of special and optional courses in the Final Year at RMC," 24 June 1937, A.G.L. McNaughton to Brigadier H.H. Matthews, 13 May 1937, 113.1009 (D5) DHist; J. Fergus Grant, "RMC Revisited," *RMC Review* 19 (Dec. 1938): 29

28   Canada, House of Commons, *Debates*, 1937, II, 1135

29   Nigel Hamilton, *Monty: The Making of a General, 1887–1945* (New York: McGraw Hill 1981), 501, 507, 513; David Fraser, *Alanbrooke* (New York: Atheneum 1982), 188

30   James Eayrs, *In Defence of Canada: Peacemaking and Deterrence* (Toronto: University of Toronto Press 1972), 84

31   Ibid., 86

32   Post-War Plan H, 20 Sept. 1945 [secret], DHist.

33   Preston, *Canada's RMC*, 305–7

34   Col. S.H. Dobell, "The Re-opening of the College," in R. Guy Smith, *As You Were* (Kingston: RMC Club 1984), II, 214, 223; and information from Lt-Gen. W.A.B. Anderson

35   Brooke Claxton, "Unpublished Autobiography," NA, MG 32, B5, vol. 201, brown volume 5, 986–98

36   Preston, *Canada's RMC*, 321–2; NAC 1700–121/2, vol. 1, NA

37   Requirement of officers, Air Vice Marshal H.L. Campbell, Aide Memoire, 20 March 1947, RCAF, C 320-103, DND; Preston, *Canada's RMC*, 322–5

38   Preston, *Canada's RMC*, 310

39   Eayrs, *Peacemaking and Deterrence*, 92, quoting the King diary

40   Ibid., 96

41   Ibid., 110–19; Preston, *Canada's RMC*, 325; RCN, 1700, 121/1, vol. 5, DND

42   Brooke Claxton to Air Vice-Marshal E.W. Stedman, 30 May 1947, Stedman Committee RR128-4, RG 24.11886, NA; Claxton, "Autobiography," 988

43   Stedman Committee RR1282-4

44   RCAF 895-70/1, DND

45   Admission Requirements, Joint Services Colleges, 112.352.009 (D 27), DHist.

46   Cadetships, RMC and Royal Roads, 171.066 (D 1), DHist.

47   Preston, *Canada's RMC*, 333

48   NS 1280-43, RG 24.8102, NA

49 Claxton, "Autobiography," 980; W.B. Sawyer, "The Present Four Year Course at the Royal Military College of Canada," *RMC Review* 32 (1951): 1–4

50 RCN 2140-82/2, vol. 3, RG 24.19228, NA

51 Canadian Services Colleges Co-ordinating Committee, NS, 1282-14, RG 24.8122, NA

52 Liaison correspondence between history departments of RMC and RR, 30 Jan./Sept. 1951, 171.009, D 259, DHist.

53 Chief of Staff Committee, 27 April 1948, DHist.

54 Preston, *Canada's RMC*, 330

55 Ibid., 331

56 RMC Faculty Board Minutes, 15 Nov. 1948

57 Address by the prime minister to the Royal Military College Club of Canada, Kingston, 30 Sept. 1950, 111.009, D 1, DHist.

58 Herbert Fairlie Wood, *Strange Battleground: The Operations in Korea and Their Effects on the Defence Policy of Canada* (Ottawa: Queen's Printer 1966), 206n, 217n, 219

2 Winds of Change: The Regular Officer Training Plan and Collège Militaire Royal

1 Herbert Fairlie Wood, *Strange Battleground: The Operations in Korea and Their Effects on the Defence Policy of Canada* (Ottawa: Queen's Printer 1966), 16–19

2 *Canada's Defence Programme, 1949–50* (Ottawa: Queen's Printer 1950), 8, 13

3 Ibid., 24; *Report of the Department of National Defence for the Year Ending 31 March 1952* (Ottawa: Queen's Printer 1952), 86, app. 6

4 *Canada's Defence Programme, 1949–50*, 36

5 Macklin to Cabinet Defence Committee, nd, RCAF, S320-103, vol. 3, DND

6 This is an estimate based on the proportions given in the 1952 defence program. *Canada's Defence Programme, 1952–1953* [Ottawa 1953], 11

7 G/C Ripley to CAS, 30 Oct. 1947, RCAF, S320-103, vol. 2, DND

8 Claxton, "Unpublished Autobiography," MG 32, B5, vol. 201, 990, NA

9 G/C Millward to AMP, 11 Jan. 1949, RCAF, S320-103, vol. 2, DND; *Canada's Defence Programme, 1952–1953*, 10–11

10 Preliminary Advance Signal, AFHQ to NWACTA, 7 Oct. 1950, RCAF, S320-103, vol. 3, DND; *Canada's Defence Programme, 1952–1953*, 10–11

11 Macklin to Cabinet Defence Committee, nd, RCAF, S320-103, vol. 3, DND

12 Canadian Services Colleges Co-ordinating Committee, 11 May 1948, RR1282-4, RG 24.11886, NA

13 Canadian Services Colleges Committee, 2nd Meeting, 20 Oct. 1949, NS1280-53, RG 24.8102, NA

14 Snow to secretary, PMC, 9 Nov. 1949, NS1290-53, RG 24.8102, NA

15 Canadian Services Colleges, Memo to CNP, RMC brief on Obligatory Active Force Service, nd, RR1280-53, RG 24.8102, NA

16 Chiefs of Staff Committee, 31 July 1950, DHist.

17 Claxton, "Autobiography," 989

18 Chiefs of Staff Committee, 28 Sept. 1950, DHist.

19 Canadian Services Colleges – Commandants and Directors of Study Conferences, 17–18 May 1951, RR1280-53, RG 24.8120, NA

20 Wood, *Strange Battleground*, 258–60

21 Enrolment of Officers, 4 Oct. 1951, RCAF, S320-103, vol. 4, DND

22 COS Committee, 6 Nov. 1951, DHist.

23 CGS to CAS, 20 April 1950, RCAF 895-70/1, DND; Richard A. Preston, *Canada's RMC: A History of the Royal Military College* (Toronto: Published for the Royal Military College Club of Canada by University of Toronto Press 1969), 342

24 *Report on Certain "Incidents" Which Occurred on Board HMC Ships ATHABASCAN, CRESCENT, and MAGNIFICENT, and on Other Matters Concerning the Royal Canadian Navy* (Ottawa, Oct. 1949), 51

25 Maj. C. Falardeau for Maj.-Gen. W.H.S. Macklin, chairman PMC, to secretary, Chiefs of Staff Committee, 6 March 1952, RCAF, 801-100, vol. 272, DND

26 Canada, House of Commons, *Debates*, 20 March and 4 April 1952, I, 625–6, 1137. See Jean-Yves Gravel, "La fondation du Collège Militaire Royal de Saint-Jean," *Revue d'Histoire de l'Amérique Française* 27 (Sept. 1973): 257–80

27 Brig. M.P. Bogert, DGMT, to VCGS, 27 March 1952, Army 2001-82/3, vol. 1, DND

28 Lt-Col. W.S. Hunt, D Org, 16 April 1952, Army 2001-82/3, vol. 1, DND

29 Minister of national defence, Memo for Cabinet Defence Committee, 21 April 1952, Committees, Joint Canadian Services Colleges Committee, RCAF, 801-100, vol. 272, DND

30 CGS to minister, 26 May 1952, Army 2001-82/3, vol. 1, DND; Chiefs of Staff Committee, 13 March 1952, DHist.

31 Canada, House of Commons, *Debates*, 12 June 1952, III, 3149–50

32 Brooke Claxton to deputy minister, 13 June 1952, Army 2001-82/3, vol. 1, DND

33 ROTP, Special Meeting of the PMC on "Implementation of the Plan for the Production of Junior Officers," 13 June 1952, Items from Meetings, 113.3M3.009 (D 5), DHist.

34 [PMC], 376th Meeting, Item 12 [1952], ibid.

35 Regulations for the Services Colleges, 1950, 112.3H1.009 (D 258), DHist.

36 Instructor-Commander C.H. Little and Instructor-Lieutenant-Commander J.C. Mark to secretary, PMC, 14 Jan. 1952, RCAF, 801-100, vol. 272, Committees – Joint Services Canadian Services Colleges Committee, DND; RCN, 1282-77, RG 24.8124, vol. 2, 1 Dec. 1951, NA

37 RCAF, s895-100-ROTP, Joint Universities and Military Advisory Committee on ROTP, 19 May 1955, DND; Items from Meetings and Conferences, ROTP, 9, 14, 113.3M3.009 (D 5), DHist.

38 PMC, Terms of Reference of JSUCC, 15 Oct. 1954, RCN 1282-77, vol, 2, RG 24.8124, NA; Decisions and Conferences, CSC Advisory Board, 13 Feb. 1956, RCN 1282-5, vol. 1, RG 24.11886, NA

39 ROTP, 1952, PMC Committee, 14 Aug. 1952, Items from Meetings, 113.3M3.009 (D 5), DHist.

40 Secretary PMC to chairman, 18 Jan. 1953, ROTP, 405th Meeting, PMC, RCAF, 895-70/1, vol. 5, DND

41 Chiefs of Staff Committee, 13 March 1953, DHist.

42 Lt-Gen. G.G. Simonds to chairman, Chiefs of Staff, 22 Dec. 1953, Joint Services University Colleges Committee, Committees RCAF, 801-100, vol. 272, DND

43 Foulkes to chiefs of staff, Jan. 1954, RCAF 801-100, vol. 272, DND

44 Faculty Board Minutes, 11 March 1954, RMC

45 Faculty Council Minutes, 18 Jan. 1952, RMC

46 Blatchford, secretary PMC to commandant RMC, 15 April 1954, DM, 2-490-82/0, DM, Secretary 241-1300-1, DND

47 Percy Lowe, "Is ROTP a Failure?," *RMC Review* 39 (1958): 187–96; Preston, *Canada's RMC*, 343–4

3 A Two-Year Military Training Course or a Degree?

1 Information from Air Vice-Marshal Bradshaw

2 Wilson Smith, 10 Oct. 1957, ORG and EST. CSCs, HQ 2001-82/3, DND

3 JSUCC-RMC Special Meeting, 23–24 Nov. 1954, JSUCC, 28 Oct. 1954, RR1282-4, RG 24.11886, NA; Minutes JSUCC, 21 Nov., 5 Dec. 1955, 9 Jan. 1956, 113.3M3.003 (D 2) DHist; Faculty Council, 10 Feb. 1956, RMC

4 Fisher and Tedman Report, Oct. 1954, RMC Administrative Officer's files, Development Plans, 1954

5 Information from Air Vice-Marshal Bradshaw

6 Faculty Council, following 11 Nov. 1956

7 Advisory Board, 8–9 March 1957, RCAF 895/70/0, vol. 2, DND

8 Faculty Board, 18 April 1952

9 RMC Correspondence and Comments on Timetables, 25 Oct. 1948, 171.009 (D 528), DHist.

10 Ibid., Sept. 1948–Aug. 1950, 171.009 (D 528), DHist.

11 Faculty Board, 5 Jan. 1954

12 Memo from director of studies to all staff, 13 Oct. 1953, Faculty Council, 6 Oct. 1953, RMC

13 Faculty Council, 9 Jan. 1955

14 MG 32, B5, vol. 201, 994, NA

15 Faculty Council, 23 Oct. 1953, and appendices

16 *Commandant's Report*, 1948–53, 9.01/1; 1953–4, 8.08

17 Faculty Board, 4 Nov. 1954

18 Ibid., 28 Oct. 1955

19 RCAF, 895-100-ROTP (513, PMC 19 May 1955), DND

20 Faculty Council, Appendix to 3 Jan. 1956

21 JSUCC, 17 Oct. 1955, RG 24.11886, NA

22 Faculty Council, 10 Feb. 1956

23 Minutes of JSUCC, 14 Dec. 1955, 113.3M3.003 (D 2) DHist; Faculty Board, 28 Oct. 1955

24 Advisory Board, 1955, RR1282-5, RG 24.11886, NA

25   Faculty Board, 3 Jan. 1957

26   Faculty Council, 4 Jan. 1957; *Commandant's Report*, 1953–4, 1955–6

27   Advisory Board, 9 March 1957, DPED

28   Report of the Ad Hoc Committee on the Regular Officer Training Plan [Landymore Committee], Dec. 1957, RMC Massey Library

29   Advisory Board, 1958, DPED

30   Chiefs of Staff Committee, Special Meeting, 20 Jan. 1958, Massey Library, RMC

31   Faculty Board, 3 Jan. 1958

32   JSUCC 24/58, 19 June 1958, 113.3M3.009 (D 4), DHist.

33   Meeting of commandants with PMC, 9 Dec. 1958, Minutes, JSUCC, RR1282-7, RG 24.8124, NA

34   Faculty Board, 10 Dec. 1958

35   Advisory Board, 4 Oct. 1958, DPED

36   Faculty Council, 31 Jan. 1953

37   Richard A. Preston, *Canada's RMC: A History of the Royal Military College* (Toronto: Published for the Royal Military College Club of Canada by University of Toronto Press 1969), 347

38   Copy of AGI, 4 Nov. 1953 dated 17 December 1953, Faculty Council, 31 Jan., 10 March 1953

39   PMC 418th Meeting [1952], 113.3M3.009 (D 5), DHist.

40   Faculty Council, following #282 of 13 April 1953

41   Ibid., 23 Oct. 1953 and appendices

42   Legislature of Ontario, *Debates*, 9 March 1954

43   Establishment of the Services Colleges, Professional Staff, 10 June 1957, app. I, 2140-82/0, RG 24.19227, NA

44   Faculty Council, 21 Nov. 1955; Bradshaw to Chief Secretary DND, 23 Nov. 1955, DM 2-490-82/0-1, DND

45   Faculty Council, 26 April, 26 May 1954 and item #352; *Commandant's Report*, 1956–7, 8.02/2

46   *Queen's Regulations*, 1957, app. X, 2.10; 2.11

47   CSC Advisory Board, 1st Meeting, RMC, 31 Jan., 1 Feb. 1955, DPED

48   Faculty Council, 15 Feb. 1957; Advisory Board, 23 Nov. 1955, RR1282-4, RG 24.11886, NA

49   Faculty Council, 15 Feb. 1957, app.; JSUCC, 15 Nov. 1956, 113.3M3 (D 2) DHist.; "The Report of the Ad Hoc Committee on RCN Personnel Structure (1957)" (Tisdall Report), DHist.

50   Faculty Council, 15 Feb. 1957; Advisory Board, 23 Nov. 1955, RR1282-4, RG 24.11886, NA

51   Minutes of the Canadian Services Colleges Advisory Board, 9 March 1957, DM 2-490-82/0-1

52   Report of Ad Hoc Committee on ROTP, JSUCC 7/58, 13 Feb. 1958, 113/3M3.009 (D 4), DHist.

53   JSUCC Minutes, 9 Dec. 1958, RR1282-2, RG 24.8124, NA

54   Report of an Ad Hoc Committee on Honours Courses, 17 Feb. 1959, Faculty Council, 1 April 1959

55   Preston, *Canada's RMC*, 399–400

56   DMT, Brief on the Canadian Services Colleges, 27 April 1960, Army 3245-82/0, DND; A.S. Duncan, secretary to the Defence Council, 28 April 1960, chairman, Chiefs of Staff Yellow File, "Canadian Services Colleges and Advisory Board," vol. 1, DHist.

57   Information from Commodore W.P. Hayes

58   Advisory Board, 1973, Item VII A, DPED

59   Faculty Council, 10 Nov. 1952

60   *Commandant's Report*, 1951–5, 8.04/6

61   Ibid., 1956–7, 8.02/1

62   Richard A. Preston, Sydney F. Wise, and Herman O. Werner, *Men in Arms: A History of Warfare and its Interrelationships with Western Society* (New York: Praeger 1959; 5th ed., New York: Holt, Rinehart and Winston 1991)

63   Faculty Council, 16 Sept. 1954

64   JSUCC, Special Meeting, 28 Oct. 1954, RR1282-4, RG 24.11886, NA

65   Bradshaw to chief secretary, 5 Oct. 1956, 895/70/2, vol. 2, DND

66   Faculty Council, 15 Sept. 1953

67   Minutes of a Special Meeting of PMC and Commandants of the Services Colleges, 9 Dec. 1958, RR1281-2, RG 24.8124, NA

68   Minutes, JSUCC Meetings, 28 May 1956, 113.3M3.003 (D 2), DHist.

69   Faculty Council, 5 Feb. 1957

70   Minutes, JSUCC, 4 July 1956, 113.3M3.009 (D 4), DHist.

71   *Commandant's Report*, 1958–9, app. B

72 Neil McArthur, "The Canadian Military Colleges," *Canadian Geographic Journal* 77, 3 (Sept. 1968): 106

73 Bradshaw, "Problems of Being a Military College," part II, appendix, 20 May 1955, 2140-82/2, vol. 6, Establishments, RMC, RG 24.19228, NA

74 ISEC, 18 March 1955, RCAF 895/70/2, vol. 1, DND

75 Kitching to secretary, Vice-Chiefs of Staff Committee, 3 Jan. 1957, HQ, 2001/82/2, vol. 1, DND

76 *Queen's Regulations*, 1957, appendix, 2.03 (3)

77 Information from Air Vice-Marshal Bradshaw

4 Under the Director of the Regular Officer Training Plan

1 Chairman of Chiefs of Staff Yellow File, "Canadian Services Colleges and Advisory Board," vol. 1, HQ 2-170-2131, 6 Jan. 1959, DHist.

2 *Commandant's Report*, 1957–8

3 Richard A. Preston, "Broad Pennants on Point Frederick," *Ontario History* 50 (1958): 81–90, and *RMC Review* 39, (1958): 165–77

4 Commodore D.W. Piers to vice-commandant, 2 Dec. 1958, RMC Faculty Board Minutes, 10 Feb. 1959, Appendix

5 *Commandant's Report*, 1957–8

6 Piers to vice-commandant, 2 Dec. 1958; Faculty Board, 10 Feb. 1959, RMC

7 Faculty Board, First [sic] Meeting, 10 Feb. 1959, app. C

8 Faculty Board, 10 Feb. 1959, app. A; RMC Faculty Council Minutes, 29 Feb. 1959

9 Faculty Board, 5 May 1959

10 *Commandant's Report*, 1959–60; Faculty Council, 7 March 1962

11 Faculty Board, 1 June, 28 Sept. 1960

12 Faculty Board, 23 Nov. 1959, 30 Nov. 1960

13 Army 2001/82/0, vol. 2, 29 July 1959, DND

14 DROTP Directive, no. 5, 29 July 1959, army 2001/82/0, vol. 2, DND

15 Commodore Piers to Advisory Board, 30 June 1960, Rear-Admiral Piers Papers; Richard A. Preston, *Canada's RMC: A History of the Royal Military College* (Toronto: Published for the Royal Military College Club of Canada by University of Toronto Press 1969), 330; Chiefs of Staff Committee, 27 April 1948, RCAF 895/70/2, vol. 1, DND

16 JSUCC, 10 April 1958, 113.3M3.009 (D 14), DHist.

17 ISEC, 16 June 1964, RCAF 895/70/2, vol. 3, DND

18 PMC, Responsibility for Canservcol Policy, 27 Jan. 1955, RCAF 895/70/2, vol. 1, DND

19 Brig.-Gen. W.A.B. Anderson, End-of-Tour Report, *Commandant's Report*, 1961–2, Addendum

20 Terms of Reference, Joint Universities Military Advisory Committee (JUMAC), nd, RCAF 895/100-ROTP, JSUCC, 13 Feb. 1958, 113.3M3.003 DHist; JSUCC, 26 Sept. 1958, 113.3M3.009 DHist; JSUCC Minutes, 5 Jan. 1959, RR1282-2, RG 24.8124, NA

21 W/C J.A. Stephens to AMP, 3 June 1958, RCAF 895-100-ROTP, DND

22 JSUCC, 13 Feb. 1958, 113.3M3.009 (D 4) DHist.

23 Piers to Duncan, 22 July 1958, RMC 1625-1, Piers Papers; RMC Faculty Salaries, *Commandant's Report*, 1957–8, app. A

24 JSUCC, Minutes of special meeting of PMC and commandants of the services colleges, 9 Dec. 1958, RR1282-2, RG 24.8124, NA

25 *Statutes of Ontario*, 1959, c. 131

26 *Commandant's Report*, 1958–9; President's message, *RMC Club Newsletter*, 30 Jan. 1959

27 Faculty Council, 26 Jan. 1959

28 Faculty Board, 13 Oct. 1969

29 Faculty Council, 21 Oct. 1959

30 *Commandant's Report*, 1963–4

31 President's message, *RMC Club Newsletter*, 30 Jan. 1959

32 Canada, House of Commons, *Debates*, 1958, 1510, 1636

33 Douglas Fisher, MP, to Arthur Smith, chairman, Committee on Estimates, 7 July 1957, Ministerial Enquiries, General, PMC 1540-1, DND; Smith to Pearkes, 8 July 1958, ibid.; F.R. Miller, deputy minister, to Brigadier Rothschild, 10 July 1958, ibid.; Duncan to Commodore Piers, 11 July 1958, ibid.; Piers to Duncan, 22 July 1958, Piers Papers

34 House of Commons, *Debates*, 1959, 3610–11

35 Advisory Board, 1960, 1961, DPED

36 Canada, *Royal Commission on Government Organization*, 3 Dec. 1962, III, 165–6

37 Commodore H.V.W. Groos, DROTP, "Glassco Commission Report," 15 Jan. 1963, chairman, Chiefs of Staff Yellow File, Canadian Services Colleges and Advisory Board, vol. 2, May 1960–November 1963, DHist.

38 [CDEE] to chairman, Chiefs of Staff Committee, 21 May 1963, "Analysis of Glassco Commission Report," ROTP 241-17, in CSC Advisory Board, 8–9 Nov. 1963, DPED

39 Advisory Board, 1960, DPED

40 Ibid., 1963; Commodore H.V.W. Groos to chiefs of staff, 21 May 1963, ibid.

41 House of Commons, *Debates*, 1963, 932

42 Advisory Board, 1963, DPED

43 House of Commons, *Debates*, 1963, 5 Dec., 1963, 5454

44 A.J.G.D. de Chastelain, "Canada's Germany Brigade," in R. Guy Smith, ed., *As You Were* ([Kingston]: RMC Club 1984), 320–3

45 *Commandant's Report*, 1963–4, 9.01; ibid., 1963–4, 9.01

46 Ibid., 1958–9, 2.01; ibid., 1961–2, 2.01

47 Commodore D.W. Piers, "Commandant's Annual Report: General Conclusions and Recommendations, 30 June 1960," Piers Papers

48 Information from Lt-Gen. W.A.B. Anderson

49 *Commandant's Report*, 1952–3, para 5; ibid., 1958–9, 4.02/1-7, 9, 9159-60, 4.01

50 President's message, *RMC Club Newsletter* (summer 1966)

51 *Commandant's Report*, 1963–4, 8.01, 18

52 President's message, *RMC Club Newsletter* (Dec. 1965)

53 Lt-Gen. W.A.B. Anderson, "Address to the Senior Staff and Cadet Wing," 20 March 1961, *RMC Review* 43 (1962): 138–41

54 Information from Lt-Gen. Anderson

55 *Commandant's Report*, 1961–2, addendum A, Conclusions and Recommendations (not in bound edition)

56 Information from Maj.-Gen. G.H. Spencer; *Commandant's Report*, 1961–2, addendum A (not in bound edition)

57 *Commandant's Report*, 1962–3, 6.01

58 House of Commons, Special Committee on Defence, *Interim Report*, 20 Dec. 1963, 9–11

59 *Commandant's Report*, 1961–2, Annex D to ch. 11

60 House of Commons, *Reports of the Special Committee . . . on Defence Expenditures Session, 1964–1965*, 15, table 2

61 *Snowy Owl*, 1962–3, 3

5 Manpower and Integration Problems

1 David P. Burke, "The Unification of the Armed Forces," *Revue internationale d'histoire militaire* 51 (1982): 307, and "Armed Forces Unification in Canada," unpublished paper, Inter-university Seminar on Armed Forces and Society, Chicago, 21–23 Oct. 1983

2 Desmond Morton, *Canada and War* (Toronto: Butterworth 1981), 169, 192–7; J. Eayrs, *In Defence of Canada*, vol. 3: *Peacemaking and Deterrence* (Toronto: University of Toronto Press 1972), 3–16

3 Richard G. Ross, *A Paradigm in Defence Organization: Unification of the Canadian Armed Forces* (Fort Lee, Virginia: United States Army Logistics Management Center 1968); Vernon J. Kronenberg, *All Together Now: The Organization of the Department of National Defence in Canada, 1964–1972* (Toronto: Canadian Institute of International Affairs, Wellesley Papers No. 2, 1973)

4 RMC Club *Newsletter*, Nov. 1982, Feb., March, Aug., Nov. 1983, Feb. 1984. The story about Winston Churchill's reference to Air Commodore Birchall is from a copy of a letter from Prime Minister L.B. Pearson to Birchall, 7 July 1967, and printed in ibid., Aug. 1967.

5 Advisory Board, 1965, DPED; *Commandant's Report*, 1963–4, Addendum A and Annex A. This addendum is not bound with the report as distributed. A copy is to be found with an unbound set of the report in the Massey Library.

6 House of Commons, *Interim Report of the Special Committee . . . on Defence* (20 Dec. 1963); *Reports of the Special Committee . . . Defence* (Session 1964–5), table 1, 14

7 *Reports of the Special Committee on Defence* (1964–5), table 2, 15

8 Ibid., 13–17

9 Commodore W.M. Landymore, "Report of the Personnel Structure Review Team," June 1964, 76–7, DHist.

10 Spencer to DROTP, 19 March, 24 April 1963, Org and Est, CSCs, RCAF 895-70/2, vol. 2, DND

11 Information from Air Commodore Birchall

12 Report of the Minister's Manpower Study, May 1963, 75/ 519, DHist.

13 See note 5 above

14 Morton, *Canada and War*, 178–99

15 Advisory Board, 1963, Item 40, DPED

16 Faculty Board Minutes, App. B to Minute 900, RMC; RMC Paper No 2 by Director of Studies, RMC, Educational Council, 5 July 1966, DPED

17 Minister's Address, Advisory Board, 1961, DPED; Faculty Council, 6 Nov. 1968, RMC; Brigadier-General Spencer, director-general of recruiting and training, Advisory Board, 12/13 Nov. 1965, DPED

18 Morton, *Canada and War*, 174–7

19 RMC Club *Newsletter*, March 1965

20 Ibid., summer 1966; Advisory Board, 1967, DPED and DM 1150-110/C 53-67, DND

21 ROTP commitment to serve, Advisory Board, 1963, DPED; Minister's Address, Advisory Board, 1965, DPED

22 Canadian Services Colleges Advisory Board, 1965, Address of the Director of Cadets, DPED; Advisory Board, 1967, 1968, Addresses of DGRTP, Commodore Leir, DPED

23 Groos to Chiefs of Staff, Analysis of the Glassco Commission Report, 21 May 1963, Advisory Board, 1963, DPED; Canada, House of Commons, *Debates*, 5 Dec. 1963, 5454

24 Faculty Board, 13 Jan. 1965; Faculty Council, 13 Jan., 7 April 1965; Faculty Council, 1966, Annex B to Minute 1753, 21 Dec. 1966

25 Faculty Board, 13 Jan., 7 April 1965; ibid., RMC Director of Studies, "Pass Arts and Pass Science Courses in the First Year," 14 Jan. 1965; Faculty Council, 12 Jan. 1965, 21 Jan. 1966, 9 June 1967, 12 April 1969, Educational Council, 5 July 1966, RMC Paper No 2, DPED; ibid., June 1968

26 Faculty Council, 17 June 1968

27 Advisory Board, 1965, DM 1150-110/C53-65, DND

28 Advisory Board, 1967, DPED

29 Faculty Council, 5 June 1964

30 Ibid., 20 July 1966

31 See Charles C. Moskos, ed., *Public Opinion and the Military Establishment* (Beverly Hills: Sage 1971), especially Laurence I. Radway, "Military Education and Civilian Values,"

9–30, where RMC's primacy in academic specialization is twice noted with approval (ii, 18)

32 Faculty Council, 20 July 1966

33 A.D. Chant, "The State of the Union: The RMC Moral Crisis Revisited," *The Marker*, March 1967, 7

34 A/M E.M. Reyno to CDS, 6 March 1967, CDS Yellow Files, Canadian Services Colleges, DHist.

35 Allard to Reyno, 13 March 1967, ibid.

36 Faculty Council, 3 May 1967

37 Maj.-Gen. George Spencer to Maj.-Gen. Bruce F. MacDonald, 24 May 1967, Spencer MS

38 Faculty Board, 4 March 1963

39 Advisory Board, 1967, DM 1150-110/C53-67, DND; Neil McArthur, "The Canadian Military Colleges," *Canadian Geographic Journal* 77, 3 (Sept. 1968): 94–101

40 Educational Council, Jan. 1967, DPED

41 Advisory Board, 1967, DM 1150-110/C53-67, DND

42 Ibid.

43 Ibid.

44 Ibid.

45 Ibid.

46 Faculty Board, 4 March 1963

47 General Council, 9 June 1965, DPED

48 Advisory Board, 1968, DPED

49 RMC Club *Newsletter*, May 1967

50 Lt-Col. P.T. Nation, director of administration, to Gordon S. Adamson and Associates, 22 Aug. 1967, RMC 7605-1

51 RMC 7605-1, 7, 26 April and 29 Sept. 1965; nd [1965?]; and 4, 5, 8 May 1966

6 Unification, the Officer Development Board, and Professionalism

1 Vernon J. Kronenberg, *All Together Now: The Organization of the Department of National Defence in Canada, 1964–1972* (Toronto: Canadian Institute of International Affairs, Wellesley Papers No. 2, 1973), 68–85; David P. Burke, "Hellyer and Landymore: The Unification of the Canadian Armed Forces and an Admiral's Revolt," *American Review of*

Canadian Studies 8 (autumn 1978), 3–27, and "The Unification of the Armed Forces," *Revue internationale d'histoire militaire* 51 (1982): 302–27

2   RMC Club *Newsletter*, Feb. 1968

3   RCAF 895/80/2, vol. 3, 18 June 1964, DND

4   Canadian Military Colleges Advisory Board, 27 Nov. 1968, DPED

5   *Commandant's Report*, 1967–8, 10; Group Capt. G.R. Truemner, director of recruiting, DM 1150-110/C53-67, CMC Advisory Board, 1968, DPED

6   RMC Commandant to Advisory Board, 1967, DM 1150-110/C53-67, DND

7   Capt. C.E. Durant to E.C. Russell, 20 Nov. 1968, Advisory Board, DM 1150-110/C53-68, DND

8   Group Capt. G.R. Truemner, 4 Oct. 1967, DM 1150-110/C53-67, DND

9   Suggested Outline Report by the Director of Training to the Advisory Board, 17 Nov. 1967, ibid.

10   RMC Commandant to Advisory Board, 1967, ibid.

11   Advisory Board, 1968, DPED

12   RMC Club *Newsletter*, Nov. 1968

13   Faculty Council, 4 Dec. 1963, no. 1423

14   *Commandant's Report*, 1967–8, 7; Advisory Board, 1968, DPED

15   Faculty Board, 7 Oct. 1968

16   Advisory Board, 1968, DPED

17   General Council, June 1968, DPED; CMC Advisory Board, 1968, DPED

18   Advisory Board, 1968, DPED

19   *Commandant's Report*, 1967–8, 22–30; Advisory Board, 1968, DPED; General Council, June 1968, DPED

20   Advisory Board, 1968, DPED

21   William E. Simons, *Liberal Education in the Service Academies* (New York: Institute of Higher Education 1965), xi, 16; Amos A. Jordan, Jr, "Officer Education," in Roger W. Little, *Handbook of Military Institutions* (Beverly Hills: Sage 1971), 218; Laurence I. Radway, "Recent Trends at American Service Academies," in Charles C. Moskos, Jr, *Public Opinion and the Military Establishment* (Beverly Hills: Sage 1971), 3–35

22   *Commandant's Report*, 1967–8, para 17

23   Adrian Preston, "Canadian Newsletter: Military Education and Canadian Policy," *Army Quarterly* 96 (July 1968): 172

24   RMC Club *Newsletter*, 1968

25   Advisory Board, 1968, DM 1150-110/C53-68

26   RMC Faculty Board, 1 April 1969

27   DGRTP Report, Advisory Board, 1968

28   Educational Council, June 1969, DPED

29   Report of the Commandant's Time-Utilization Study Group, RMC 1967–8, 31 Oct. 1968, 4, 12, 14

30   Director of Training to the Advisory Board, 17 Nov. 1967, HQ 1150-110/C53-67, DND

31   James Jackson, "Mr. Hellyer and the Officers: Why Unification Is Not the Issue," *Saturday Night* (April 1967): 22–5

32   Preston, "Canadian Newsletter," 166–74

33   Gen. Roger Rowley to the RMC Faculty Board, 28 Nov. 1967; Général Jean Allard, en collaboration avec Serge Bernier, *Mémoires* (Ottawa: Les Editions de Mortagne 1985)

34   Maj. G.A. Zypchen, "The Role of the Professional Soldier in Canada," *Snowy Owl: Yearbook of the Canadian Land Forces* (1968): 11–16

35   Col. A.S.A. Galloway, "Some Thoughts on the Subject," ibid., 17–22

36   Suggested address for the Minister of National Defence, Advisory Board, 12 Jan. 1967, HQ 1150-110/C53-67, DND

37   Advisory Board, ibid.

38   Faculty Board, 28 Nov. 1967

39   Ibid.

40   Suggested Address for the Minister of National Defence, Advisory Board, 12 Jan. 1967, DM 1150-110/C53-67, DND

41   Department of National Defence, *Report of the Officer Development Board* (Ottawa, March 1969), I, 5. For a discussion of Rowley's original contribution to military thought see Richard A. Preston, "Military Education, Professionalism, and Doctrine," in *Revue internationale d'histoire militaire* 51 (1982): 273–301

42   Morris Janowitz, *The Professional Soldier: A Social and Political Portrait* (Glencoe, Ill.: Free Press of Glencoe 1960), 264–77

43   ODB, I, 15

44   Ibid., 45

45   Ibid., 32–7

46   RMC Club *Newsletter,* May 1969

47   Advisory Board, Special Meeting, 1 Nov. 1969, DPED

48   Report of the Meeting held on 1 Nov. 1969 DM C1130/110/C53-70, DND

49   H.D. Smith to Leo Cadieux, 29 Nov. 1969, Advisory Board, 1969, DPED

50   Allard, *Mémoires,* 453. Guimond was ill and retired from the board before it completed its work.

7 The Canadian Defence Educational Establishments and Canadian Military Professionalism in RMC

1   Desmond Morton, *Canada and War* (Toronto: Butterworth 1981), 182–3, 188

2   Rear-Adm. R.W. Murdoch, *Report of a Study of Professionalism in the Canadian Forces* (Ottawa: Canadian Forces Headquarters 1971), 4–5

3   CMC Advisory Board, 27 Nov. 1968, DPED

4   Address by CDEE to CMC's Advisory Board, Jan. 1971, DM 1150-110/C53-70, DND

5   Faculty Board, 10 Nov. 1960; Hayes to CDEC [sic], 7 Jan. 1970, annex A to Faculty Board Minutes, 1013B, 2 Dec. 1969

6   Faculty Board, 30 Nov. 1970, annex A

7   *Commandant's Report,* 1969–70, 6

8   Milroy to Commandants, 25 Feb. 1970, Lye, Miscellaneous RMC Letters (henceforth Lye Papers)

9   Commodore W.P. Hayes to CDEE, 11 June 1970, DM 1150-110/C53-70, DND

10   Faculty Board, 30 Nov. 1970, annex A

11   [Baird Report] Draft (no title or date) supplied by Dr David Baird

12   Lye to Milroy, 4 Jan. 1971, Lye Papers

13   Lye to CDEE, 26 Jan. 1971, DM 1150-110/C53-70, DND

14   Information from Dr David Baird

15   Milroy, Academic Policies at the Canadian Military Colleges, 7 Dec. 1970, Faculty Council, 6 Jan. 1971, annex E

16   Faculty Council, 6 Jan. 1971

17   Lye to CDEE, 24 Aug. 1971, Lye Papers

18   Address by the commander CDEE to the twelfth meeting of the Canadian Military Colleges Advisory Board [1971], DM 1150-110/C53-70, DND

19   *Commandant's Report,* 1970–1, 3

20   Ibid., 4

21   Address by Commander, CDEE to Advisory Board [1971], DM 1150-110/C53-70, DND; Faculty Council, 6 and 10 May 1971

22   Lt-Commander J.G.M. Smith, secretary, Canadian Military Colleges Advisory Board [to members of the board], Dec. 1970, DM 1150-110/C53-70, DND; Brig.-Gen. W.K. Lye to Col. L.G. MacNevin, Prince Edward Island, 6 Jan. 1971, Lye Papers

23   Faculty Council, 3 Feb. 1971

24   Joint announcement by E.J. Benson, minister of finance, and Donald S. Macdonald, minister of national defence, 23 Nov. 1971, Lye Papers

25   Educational Council, 30 June 1971, DPED

26   Faculty Council, 6 Jan. 1971; Milroy to CMCs, 20 July 1971, DM 1150-110/C53-72, vol, 2, DND; Advisory Board, Increased Tasks, Final Recommendations to the Minister, 12–14 April 1972, ibid.

27   DND, "Report on the study of professionalism in the Canadian Forces, Ottawa, April 1972, Papers" (Ottawa: CFHQ Sept. 1972), 2–3; Richard A. Preston, "Military Education, Professionalism and Doctrine," in *Revue internationale d'histoire militaire* 51 (1982): 288–91

28   *ODB Report,* vol. 1, sect. 3; see James Jackson, "Mr. Hellyer and the Officers: Why Unification Is Not the Issue," *Saturday Night* (April 1967): 23–5

29   *Draft Report on Military Professionalism,* 13, 66–7

30   Morton, *Canada and War,* 190

31   *Draft Report on Military Professionalism,* 73

32   *Military Review* 9 (1980)

33   Col. A.P. Wills, "A Personal Creed of Officership," [Air Force] College, Canadian Forces Extension School, Toronto, *Extension Bulletin* 20 (Oct. 1969): 45–7

34   Lt-Commander J.G.M. Smith, "What Is Wrong with Professional Education?" *Canadian Defence Quarterly* 2 (summer 1972): 8–12

35   Robin McNeil, "The Post-graduate as an Officer: Is There, Must There Be, a Conflict?" *Canadian Defence Quarterly* 1

(1971): 51–2; Capt. J.C. Dendy, "Needed: Direction, Support, Recognition of Professional Self-improvement," ibid., 1 (winter 1972): 42–5

36   Other presentations in this extensive debate in Canada are Pierre E. Coulombe, "The Changing Military Career in Canada," in Morris Janowitz and Jacques Van Doorn, eds., *On Military Ideology*, Vol. III (Rotterdam 1971), 176–96; R.B. Byers and Colin S. Gray, eds., *Canadian Military Professionalism: The Search for Identity* (Toronto: Canadian Institute of International Affairs, Wellesley Papers No. 2, 1973); Col. G.K. Murray, "The Military Profession in Canada: *Quo vadis?*" *Canadian Defence Quarterly* 5, 1 (1975): 7–11; Lt-Col. E. Patrick, "The Need for Better Selection and Training for Future Military Leaders," ibid., 2 (1975–6), 34–41; Adrian Preston and Peter Dennis, eds., *Swords and Covenants* (London: Croom Helm 1976); and R.B. Byers, "Understanding SALT Part I: The Context of the Debate," *Canadian Defence Quarterly* 9 (1979): 11

37   Faculty Council, 10 March 1971, annex A

38   Morton, *Canada and War*, 190

39   Donald S. Macdonald, *White Paper on Defence; Defence in the Seventies* (Ottawa: Department of National Defence, August 1971), 12–16

40   Gen. F.R. Sharp, "The Challenge of Change," *Canadian Forces Staff School Extension Bulletin* 29 (June 1972)

41   Coulombe, "The Changing Military Career," 171, 193

42   C.S. Gray, "Defence in the Seventies: A White Paper for All Seasons," *Canadian Defence Quarterly* 4 (spring 1972): 134

43   C.P. Stacey, "The Staff Officer: A Footnote to Canadian Military History," *Canadian Defence Quarterly* 3 (winter 1973–4), 43, 49

44   RRMC, "CMC Academic Programmes: Brief for Educational Council," 18/19 Oct. 1971, DPED

45   Commandant, CMR, to CDEE, 15 Sept. 1971, Briefs for Educational Council, Oct. 1971, DPED

46   RRMC, CMC Academic Programme, ibid.

47   Philip A. Lapp, John W. Hodgins, and Colin B. Mackay, *Ring of Iron: A Study of Engineering Education in Ontario* (Report to the Committee of Presidents of Universities of Ontario, Dec. 1970), 82–3; Faculty Council, 3 Feb. 1971

48   Murdoch to Commandant, RMC, 26 April 1972, annex A, Faculty Council, 15 May 1972

49   Garneau papers on Academic Review, 14 April 1972, annex B, Faculty Council, 15 May 1972; also in Lye Papers

50   Faculty Council, 15 May 1972

51   Ibid.

52   Lye to Garneau, 7 June 1972, annex to Educational Council Minutes, June 1972, DPED; Lye to Garneau, 29 July 1972, Lye Papers

53   Lye to Garneau, 25 July 1972, Lye Papers

54   Educational Council, Nov. 1972, DPED; Faculty Council, 13 Sept. 1972

55   Educational Council, 1972, DPED; Lye to Garneau, 25 July 1972, Lye Papers

56   *Commandant's Report*, 1972–3, 11

57   Advisory Board, 1971, DPED

58   Faculty Council, Special Meeting, 16 April 1970

59   Ibid., 3 Feb. 1971

60   *Commandant's Report*, 1970–1

61   Faculty Council, 3 Oct. 1973; information from General Lye

62   Lye to CDEE, 9 Aug. 1972, including semi-annual report, Unit Drug Educational Program, Lye Papers; Senate Minutes, 16 May 1972

63   Lye to CDEE, 30 May 1972, Lye Papers

64   R.A. White, "Change and Fealty to Tradition in Today's Royal Military College of Canada," *Canadian Defence Quarterly* 1 (winter 1971): 35–41; Report of the Director of Cadets, *Commandant's Report*, 1970–1, 19–26

65   RMC Faculty Council, 22 April 1970

66   Ibid., 7 Oct. 1970

67   Ibid., 10 March 1971

68   White, "Change and Fealty to Tradition"; *Commandant's Report*, 1970–1

69   W.A.B. Anderson to Donald S. Macdonald, 31 Aug. 1971, Lye Papers

70   White, "Change and Fealty to Tradition"; *Commandant's Report*, 1970–1

71   Advisory Board, 22–23 Jan. 1971, annex B, DM 1150-110/C53-71, vol. 2, DND

8 The Directorate of Professional Education and Development and the Rationalization of the Canadian Military Colleges

1 Notes on the Educational Council meeting, 7 June 1973, DPED

2 *Queen's Regulations and Orders* (revised 1978), A 2.75

3 Information from RMC registrar

4 Advisory Board, 1963

5 Educational Council, 7 June 1963

6 *Who's Who Biographical Service* (Toronto: Who's Who Publications 1973) 8, 1, June 1973

7 Milroy to Lye, 28 June 1973, Lye Papers

8 Advisory Board, 1973

9 Ibid., 1971

10 Ibid., 1973; General Council, 1973, DPED

11 Faculty Council, 5 April 1974

12 Information from Brig.-Gen. Turner

13 General Council, 8 June 1973

14 Faculty Board, 21 June 1973

15 Ibid., 24 Sept. 1974

16 Ibid., 21 June 1974

17 Ibid., Syllabus Committee, 23 April 1974

18 Ibid., Syllabus Committee, 24 Sept. 1974

19 Ibid., 3 March, 6 June 1975; information from Brig.-Gen. Turner

20 Faculty Board, 3 March 1975

21 Ibid., 23 April 1975

22 Ibid., 10 May 1973

23 DGRET Briefing to the Advisory Board, Annex B to Minutes of the Summer Meeting, 1974, DND 1150-110/c53-78, vol. 1

24 Col. W.G. Svab, CO 202 Workshop Depot, to DGLEM, NDHQ, 16 May 1975, 4840-1, vol. 6, Production of Officers, General, DND

25 Brig.-Gen. J.E. Vance to DGLEM, 20 June 1975, ibid.

26 Vice-Adm. D.S. Boyle, commander Maritime Command, to ADM(Pcr), 20 Aug. 1975, 4840-1, vol. 6, DND

27 Advisory Board, 1973

28 *CPD Study, Rationalization of the Canadian Military Colleges System* (Oct. 1976), 1–3

29 *Commandant's Report*, 1975–6; Advisory Board, 18–20 March 1977, DPED; information from Brig.-Gen. Turner

30 *CPD Study*, 90

31 Ibid., 13–17, 63; Academic Council, Oct. 1976; "Brief for the Advisory Board: Substantiation of the Requirements of the CMCs," Advisory Board, 21 Jan. 1977, DPED

32 Advisory Board, Summer Meeting, Camp Borden, 1977, Annex

33 "Report of Lieutenant-General James C. Smith, Assistant Deputy Minister (Personnel) to the CMC Advisory Board, 18 March 1977," Annex A, Advisory Board Minutes, 18–20 March 1977, DPED

34 Advisory Board, 22 March 1977; *Commandant's Report*, 1977–8

35 ADM(Per), Military Development Task Force, *Report on Military Task Force* (March 1977), 3–17

36 ADM(Per), Academic Development Task Force, *Report on Academic Development at the Canadian Military Colleges* (March 1977)

37 Ibid., 8–9

38 Faculty Council, 17 Jan. 1977, Annex B

39 Academic Task Force, 30

40 Ibid., 11

41 Ibid., 30

42 Ibid., 18–21

43 Ibid., 21

44 Ibid., 2; Military Task Force, ii

45 Faculty Council, 9 May 1977, and Annexes

46 Ibid., 16 May 1977

47 G.F.G. Stanley, "Comments on the Report of the Task Force on Academic Development at the Canadian Military Colleges," Advisory Board, Summer 1977, Annex C, 1150-100/c53/78, CMC Advisory Board, 1978, vol. 1, DND

48 W.D. Shuttleworth, DGPR, to Assoc. ADM(Per), 12 Dec. 1977, DND 1150/110c, Committees and Boards, CMC Advisory Board, 1977

49 Information from Lt-Gen. de Chastelain

50 Advisory Board, 1978

51 Ibid. (summer), 28–9 July 1978

52 Philippe Garigue, à Général St Aubin, 20 oct. 1978, enclosing "Educating Officers," 13 sept. 1978, DPED

53 Advisory Board, 5–6 April 1979

54 NDHQ Report to Advisory Board [Gen. Smith], March 1980, Part 4, p. 1, 1150-110/c53-80, vol. 1, DND

55 Dr Rosario Cousineau to Lt-Gen. James C. Smith, ADM(Per), 8 Nov. 1977; Advisory Board, 1981–2

56 David N. Solomon, "The Royal Military College of Canada: Programmes in the Department of English and Philosophy and the Department of French" (28 June 1979); Advisory Board, 1979, Regional Committee

57 Advisory Board, 1978

58 Ibid., 1979

59 "Report of RMC Commandant to Advisory Board, 1979"

60 Advisory Board, 1978; ibid., 1979; ibid., 1981–2; Faculty Board, 11 Dec. 1979, Annex A; Faculty Council, 21 Nov. 1979; information from Lt-Gen. de Chastelain

61 Commandant's Report, 1978–9; Advisory Board, 1979, Annex A

62 Advisory Board, 1980

63 Ibid., 1981–2

64 Academic Task Force, chap. 6, 66–7

65 Advisory Board, April 1978, Item XII, 51, 62, Annex A, 16

66 Ibid., 19–20 March 1980

67 Ibid., Annex F

68 Ibid., Nov. 1978, March 1979

69 Faculty Council, 2 Sept. 1981

70 Ibid., 3 Nov. 1980

71 Advisory Board, Nov. 1978

72 Faculty Board, Annex C to Minute 1633, 2 June 1981

73 MARE Training to Classification Qualification (MOC 44A), MS Subclassification Training and Development, and Captain (N) W.J. Broughton, "Get Well Project: Maritime Engineering MARE Classification," Maritime Engineering Journal, Aug. 1984, 1–28

74 Commandant's Report, 1980–1

75 Ibid., 1981–2

76 Faculty Council, 4 March 1981; Advisory Board, 1982

77 Commandant's Report, 1982–3

9 Serving Personnel at RMC

1   Lye to NDHQ, 11 May 1973, Enrolment of Women at CMCs, quoting Advisory Board, 23–25 Feb. 1973, Item VII B, Lye Papers

2   R. Guy Smith, ed., As You Were: Ex-cadets Remember, 2 vols. (Kingston: RMC Club 1984)

3   Thomas L. Brock, The RMC Vintage Class of 1934, 3 vols. (Victoria, BC: Published by the class, 1985)

4   Advisory Board, 6–8 April 1978, DPED

5   RMC Club Newsletter, Nov. 1976

6   Information from Air Commodore Birchall

7   DEdn to CDS, 5 Jan. 1973, Academic Training, General, vol. 1, NDHQ 4508-1, DND

8   See CDEE Directive, 5/70, 19 Dec. 1970, Faculty Council, 6 Jan. 1971

9   Ibid.

10 Faculty Council, 10 Jan. 1967

11 Radley-Walters to commandants, CMCs, etc., 10 May 1967, Academic Training, General, NDHQ 4508-1, 5 Jan. 1973, DND

12 Rowley, Presentations to CDS Staff Meeting, 6 Sept. 1967, and to Defence Council, 30 Oct. 1967, Annexes B and C to DPED, c5570/cODS, 30 Oct. 1967, DND

13 s1715-410/03, ODS, 31 Oct. 1967, Annex B, and NDHQ 5570-1, ODS, 3 Oct. 1971, filed after Faculty Council, 18 Oct. 1967

14 Sharp, Memo to chairman, DCECPG, 15 Sept. 1969, Advisory Board, NDHQ, 1150-110/c53-69, DND

15 ODB, II, 170, para. 57f

16 Faculty Council, 20 Oct. 1968

17 Advisory Board, 1 Nov. 1968, NDHQ c1150-110/c53-70, DND

18 Faculty Council, 10 April 1970, Annex A, Cairns to Hayes, 17 Feb. 1970

19 DPED, Briefs for Educational Council, 18–19 Jan. 1972

20 Information from Dr Steve Harris, 14 Jan. 1987

21 DPED, Briefs for Educational Council, 18–19 Jan. 1972

22 Advisory Board, 22–23 Jan. 1971, Report by General Lye

23 Faculty Council, 6 Jan. 1971

24 Lye to principal, etc., 5 Feb. 1973, Lye Papers

25 General Council, June 1971, DPED

26 Educational Council, 18 Oct. 1971, DPED

27 RMC, Brief for the Educational Council, 18–19 Nov. 1972, Advisory Board, 1972, Annex A, Item VI, DPED; Faculty Council, 12 Jan. 1972, Annex B to Minute 2342/2

28 Advisory Board, 1972, DPED

29 Brief for the Educational Council, 18–19 Nov. 1972, DPED

30 Telex, NDHQ to CMCs, Nov. 1972, Lye Papers

31 Boyle to CMCs, 31 Jan. 1973, ibid.

32 Lye to Macdonald, 20 Sept. 1972, ibid.

33 Macdonald to Lye, 6 Oct. 1972, ibid.

34 Telex, NDHQ to RMC, etc., 2 Nov. 1972, ibid.

35 Faculty Council, 31 Jan. 1973; Lye to commandants, RR and CMR, 31 Jan. 1973, Lye Papers

36 Advisory Board, 19 Feb. 1973, Item VI, DPED

37 Lye to principal, etc., 5 Feb. 1973, Lye Papers; Faculty Council, 5 Feb. 1973; Educational Council, Feb. 1973, DPED

38 Faculty Council, 14 Feb. 1974

39 Anderson, "Mature Student Residency at CMCs," Item VII B, Advisory Board, 1973, DPED

40 "Notes on the Educational Council Meeting," 7 June 1973, Advisory Board, 1973; NDHQ 1150-110/C53-73, vol. 4, DND

41 Educational Council, 7 June 1973, DPED; Faculty Council, 3 Dec. 1973, 9 Jan. 1974; Dacey to NDHQ, 10 Sept. 1973; Educational Council, June–November 1973, Annex B, DPED; Advisory Board, 1974, NDHQ 1150-110/C53-74, vol. 4, DND

42 Dextraze to Lye, Dacey, and Col. W.W. Turner, commandant-designate, RMC, 13 March 1973, Lye Papers

43 Lye to principal, etc., copies to commandants, RR and CMR, 2 April 1973, ibid.

44 Boyle to CDS, 13 April 1973, Advisory Board, 1973, NDHQ 1150-110/C53-73, vol. 4, DND

45 Report on UTPO/UTPM and CMCs, Advisory Board, 15–17 March 1974, DPED

46 Turner, Program Evaluation Report, UTPM/UTPO at the CMCs, Academic Year 1973–4, Advisory Board, 1974, NDHQ 1150-110/C53-74, vol. 1, DND

47 Anderson, Statement at Advisory Board, 15–17 March 1974, Item V; Report on UTPO/UTPM at CMCs, Report on UTPM, Advisory Board, 1975, Annex B, NDHQ 1150-110/C53-74, vol. 1, DND

48 Report on UTPM, Jan. 1975, Advisory Board, 1975, NDHQ 1150-100/C53-74, vol. 1, DND; information from Mrs W.S. Avis

49 Academic Council, 27–28 Oct. 1975, DPED

50 Faculty Council, 5 May 1976

51 Review of UTPM Program, General Council, 18 Oct. 1977, Advisory Board, 1977, NDHQ 1150-110/C53-78, DND

52 *Commandant's Report*, 1978–9

53 Ibid., 1980–1

54 Faculty Council, 14 April 1982

55 *Commandant's Report*, 1981–2

56 Information from Major-General Norman

57 *Commandant's Report*, 1982–3

58 RMC Club *Newsletter*, Aug. 1984

10 Francophone Representation and Bilingualism

1 See above, page 127

2 CDEE Directive No. 4/71, 24 Feb. 1971, P5570-187 D 0244 (CDEE) and annexes, B&B Admin. file, RMC

3 Lt-Commander J.V. Andrew, *Bilingualism Today, French Tomorrow: Trudeau's Master Plan and How It Can Be Stopped* (Richmond Hill: BMG Publishing 1977), 2, 11

4 Desmond Morton, "French Canada and the Canadian Militia," *Histoire sociale* 3 (April 1969): 44–6

5 J.L. Granatstein and J.M. Hitsman, *Broken Promises: A History of Conscription in Canada* (Toronto: Oxford University Press 1977), 22–8

6 J. Pariseau, "Le bilinguisme et le biculturalisme au Ministère de la Défense nationale 1946–1963" (Ottawa: Service historique QGDN, 22 déc. 1980), 15

7 Jean-Yves Gravel, *L'armée au Québec: un portrait social 1868–1900* (Montréal: Boréal Express 1974), 18–51

8 Pariseau, "Le bilinguisme," 5–11

9 *Debates*, 1950, I, 880; 1951, I, 490–3, 569; Jean Allard avec Serge Bernier, *Mémoires du Général Jean Allard* (Ottawa: Les Éditions de Mortagne 1985), 248, 250–1; Pariseau, "Le bilinguisme," 5–11

10 Allard, *Mémoires*, 251

11 Paul Mathieu, associate deputy minister, 15 April 1965, in NDHQ 1211-1, DND, vol. 1, Official Languages, General, including a press cutting from the *Globe and Mail*, no date, and a statement dictated by the chief of staff, Gen. J.V. Allard, 16 Aug. 1966; Paul Mathieu, Minister's Information Book, Sect. 6.8, Use of English and French Languages within DND, I, para 4, 15 Nov. 1968, ibid.

12 Richard A. Preston, *Canada's RMC: A History of the Royal Military College* (Toronto: Published for the Royal Military College Club of Canada by University of Toronto Press 1969), 59

13 *Debates*, 1878, IV, 4 March 1878, 737

14 Research by T.L. Brock, RMC Club historian

15 Morton, "French Canada and the Canadian Militia," 40

16 Information from T.L. Brock

17 Preston, *Canada's RMC*, 58–9, 69–70, 164, 189–90, 216, 263

18 Lt-Col. T. Gelley, registrar, to Lt-Commander J.S. Mark, chairman, Canadian Services Colleges Committee, 22 Feb. 1950, RG 24.8102, National Archives of Canada (NA)

19 C.M. Mooney to Lt-Commander J.S. Mark, 22 Dec. 1949, NS1280-53, RG 24.8102, NA

20 Faculty Board, 15 Nov. 1948

21 G. Tougas, "Report on the Teaching of French and on the Position of French-speaking Cadets in the Services Colleges," Faculty Council, April 1962

22 Advisory Board, 1963, 1150-110, C53-68, DND

23 JSCC, 10 Sept. 1956, 113.3M3.003 (D 2) DHist; JSCC, 19 June 1958, ibid.

24 Advisory Board, 1961, 1150-110/C53-68, DND

25 Ibid., 1963

26 Allard, *Mémoires*, 349–51, 373

27 Hayes to Allard, 30 Dec. 1968, and enclosure, Official Languages, General, 1211-1, vol. 2, DND

28 *Commandant's Report*, 1964–5, 11

29 Faculty Council, 6 April 1966

30 Canada, *Royal Commission on Bilingualism and Biculturalism* [*Report*] (Ottawa, 19 Sept. 1969), III, 340

31 Faculty Board, 16 Sept. 1969

32 Faculty Council, 14 Jan., 4 Feb. 1970

33 Educational Council, June 1970, DPED; Faculty Council, Sept. 1970

34 CDEE Directive No. 4/71, Bilingualism at CMCs, 24 Feb. 1971, and B&B Admin. file, RMC

35 Advisory Board, 22–23 Jan. 1971, DPED

36 Milroy, Program for B&B in the CMCs, 24 Feb. 1971, B&B Admin. file, RMC; Maj. M.C. Stewart, HQCDEE, 22–23 Jan. 1971, Advisory Board, DPED

37 Faculty Council, 14 April 1971

38 Educational Council, Oct. 1971, DPED

39 Murdoch to Garneau, 21 Dec. 1971, B&B Admin. file, RMC

40 Lye to principal et al., 28 Dec. 1971, Faculty Council

41 Faculty Council, 5 Jan. 1972

42 T.S. Hutchison, Francophone Increase to CMCs, 14 Feb. 1972, B&B Admin. file, RMC; Lye to CDEE, 24 Feb. 1972, Lye Papers

43 Laubman to Lye, 4 April 1972, Lye Papers

44 Commander, CDEE, to Advisory Board, 13 May 1972, Annex E, DPED

45 RMC Club *Newsletter*, Aug. 1972

46 Maj.-Gen. W.W. Carr to CDS, 13 March 1970, 4840-1, vol. 2, DND

47 RMC Club *Newsletter*, Aug. 1972

48 General Council, 12–13 June 1972, DPED

49 Maj. R.K. Wilson for CDS to commandants RMC and CMR, 18 Oct. 1972, Lye Papers; Chief of Personnel Report, Advisory Board, 1973, Annex A, DPED

50 Associate Deputy Minister's Report, Advisory Board 1974, Annex A, DPED

51 Faculty Council, 8 Nov. 1972

52 Educational Council, Nov. 1972, DPED

53 Faculty Council, 6 Dec. 1972; Educational Council, Feb. 1973, DPED

54 Chief of Personnel Report, Advisory Board, [April] 1973, Annex A, DPED

55 Advisory Board, 1973, DPED; Advisory Board, Report on Bilingualism and Biculturalism, April 1973, 1150-110/C53-73, vols. 2 and 3, DND

56 Dextraze to Lye, Dacey, and Turner, 13 March 1973, Lye Miscellaneous RMC Papers

57 Lye to principal, 20 March 1973, Lye Papers

58 R.E. Jones to principal and commandant, 1 June 1973, ibid.

59 Lye to Anderson, 20 June 1972, and Dextraze to Lye, 28 June 1973, ibid.

11 Institutional Bilingualism

1 Faculty Council, 9 May 1973

2 Educational Council, June 1973, DPED; Advisory Board, 7 June 1973, 1150-110/C53-73, DND; Gilbert Tucker, *The Naval Service of Canada: Its Official History* (Ottawa: Queen's Printer 1952), II, 260

3 *Commandant's Report*, 1972–3; Maj.-Gen. D.S. Boyle, chief of personnel, to CDS, 15 April 1973, Advisory Board, 1973, NDHQ 1150-110/C53-73, DND

4 Gen. J.A. Dextraze, CDS, to all Commanders of Commands, and all Base Commanders, 12 July 1973, 1211-10 (CDS), B&B Admin. file, RMC; Dextraze to Lye, Dacey, and Turner, 13 March 1973, Lye Papers

5 *Commandant's Report*, 1973–4

6 Faculty Council, 3 Oct. 1973

7 Jean Pariseau, *Le Bilinguisme et le biculturalisme au sein des institutions d'enseignement supérieur du MDN, Etude No. 9* (Service historique QGDN [Ottawa, 20 mai 1983]), 15; Faculty Council, 3 Oct. 1973

8 Faculty Council, 3 Oct. 1973

9 Ibid., Special Meeting, 17 Oct. 1973; Armand Letellier, *Reforme linguistique à la Défense Nationale: La mise en marche des programmes de bilinguisme 1967–1977* (Ottawa: Service historique de la Défense nationale 1987), 178

10 Advisory Board, 26 Oct. 1973, NDHQ 1150-110/C53-74, vol. 1, DND

11 Educational Council, Nov. 1973, DPED

12 Faculty Council, 3 Oct. 1973

13 *Commandant's Report*, 1973–4

14 Ibid., 1974–5; Pariseau, *Le Bilinguisme*, 18

15 Rationale for B&B at RMC, Advisory Board, 1974, Annex E, DPED

16 Advisory Board, 15–17 March 1974, Annex E and Motion, DPED

17 Educational Council, 4 June 1974, DPED

18 Ibid.

19 Ibid., 2 Oct. 1974

20 Advisory Board, 18–20 April 1974, DPED

21 Faculty Board, 3 March 1975

22 Advisory Board, 18–20 April 1974, DPED

23 *Whig-Standard*, 7, 8, 13 March 1975

24 *The Sentinel*, May 1975, 12–13

25 Advisory Board, 18–20 April 1975, DPED

26 Educational Council, June 1975, Item XI, DPED

27 Academic Council, 27 Oct. 1975. The Educational Council was henceforward called the Academic Council.

28 Ibid., Sept. 1975, Item V, DPED

29 *Commandant's Report*, 1975–6

30 Information from Lt-Gen. de Chastelain and Dr Pierre Bussières

31 Faculty Council, 7 May 1975

32 *Commandant's Report*, 1975–6, 3

33 Ibid.

34 Academic Council, 27–28 Oct. 1975, DPED

35 Ibid., 27 Feb. 1976

36 Faculty Council, 7 April 1976

37 Report to the Senate on second-language training at RMC, Joint Senate and Faculty Council meeting, 29 March 1976, Faculty Council Minutes, Annex A

38 Joint Senate and Faculty Council meeting, Faculty Council Minutes, 29 March 1976

39 Academic Council, Oct. 1976, DPED

40 Bryan Rollason, "A Report on the Implementation of Second-Language Training at RMC: The First Stage, 9 February 1976–15 September 1976, submitted to Dr. Rosario Cousineau, Université de Sherbrooke, Academic Advisor to ADM(Per)," B&B Admin. file, RMC

41 B.R. Hamel, B&B coordinator, 10 Dec. 1976, ibid.; Faculty Board, 5 Oct. 1976

42 Academic Council, June 1976, DPED

43 "Faculty Council Committee on Third-Year Second-Language Training, Interim Report for 1977–78 Academic Year Only," Annex C, Faculty Council, 17 Jan. 1977

44 Faculty Council, 17 Jan. 1977

45 Implementation of second-language training in the third and fourth years, Annex A, Faculty Board, 1 March 1977

46 "Report on the Syllabus Committee to Faculty Board Concerning Implementation of Second-Language Training (SLT) in Third Year, 1977–8," 28 March 1977, Annex A, Faculty Board, 5 April 1977

47 Faculty Board, 5 April 1977

48 *Commandant's Report*, 1976–7

49 RMC Club *Newsletter*, Nov. 1977

50 Rosario Cousineau to ADM(Per), "Report on the Implementation of Second Language Training at the Canadian Military Colleges," 25 May 1978, Minutes of the Meeting of the Canadian Military Colleges Academic Council, 13–14 June 1978, DPED

51 *Commandant's Report*, 1978–9, 13

52 Ibid.; Faculty Board, Dec. 1979

53 Cousineau, "Report"; *Commandant's Report*, 1978–9, 3

54 Information from Maj.-Gen. John Stewart and Dr Pierre Bussières

55 *Commandant's Report*, 1980–1

56 Advisory Board, 1981, NDHQ 1150-110/C53-78, vol. 1, DND

57 Faculty Board, 1 Oct. 1985

58 Principal John Plant to members of the Faculty Board, 2 Nov. 1985, 47051 (Princ) RMC

59 Information from Dr Leopold Gauthier, assistant to the principal

12 Lady Cadets

1 *Report of the Royal Commission on the Status of Women in Canada* (28 Sept. 1970; Ottawa: Information Canada 1971), 180, 404–5

2 Col. P. Charlton, "Women in the Canadian Forces," *Canadian Defence Quarterly* 4, 1 (summer 1974): 34–41; "Report on the SWINTER Investigation," General Council, 14–16 March 1984, DPED

3 Advisory Board, 1973, DPED

4 House of Commons, *Debates*, 1973, 16 Feb., 1362; 28 March, 2699

5 Advisory Board, 1973

6 *Debates*, 1973, 13 April, 3279

7 General Council, 8 June 1973, DPED

8 Lye to NDHQ, 11 May 1973, Lye Papers

9 Advisory Board, 1973, Notes on the Educational Council Meeting, 7 June 1973, NDHQ, 1150-110/C53-73, vol. 4, DND

10 Advisory Board, 17 July 1973

11 Advisory Board, Recommendation, 17 July 1974, NDHQ 1150-110/C53-74, vol. 1, DND

12 Charlton, "Women," 36–41

13 Advisory Board, 6–8 April 1978, Annex C. The statistics given are as follows (per cent): Australia 5.0, Canada 5.7, France 1.7, FRG less than 1, Israel 5.0, Russia less than 1, Britain 4.3, USA 5.1

14 Maj.-Gen. Jeanne Holm, *Women in the Military: An Unfinished Revolution* (Novato, CA: Presidio 1982), 305–12

15 General Council, 17 March 1978, Annex E

16 Lt-Col. J.S. Cantlie, SO/CDS, 11 Jan. 1978, "Women in Military Colleges," Advisory Board, 1977, NDHQ 1150-110/C53-77, DND

17 Advisory Board, 6–8 April 1978. See also Yvon Gagnon, "A Short Essay on Equality," *Signum* 4 (1976): 1–6

18 CPD Brief, General Council, 11 Oct. 1978

19 General Council, 11 Oct. 1978, Annex B

20 RMC Club *Newsletter*, Aug. 1979

21 Information from Lt-Gen. Anderson

22 Advisory Board, 5–6 April 1979

23 Ibid., 13 Dec. 1979

24 General Smith, Advisory Board, 1980

25 Advisory Board, 22 March 1980

26 Ibid., 1980, App. 8 to Annex A

27 Advisory Board, 1980

28 Information from Lt-Gen. Stewart

29 Advisory Board, 1981

30 General Council, 17 June 1982

31 Advisory Board, 3–6 March 1982

32 *Commandant's Report*, 1981–2, DPED

33 General Council, 23 March 1983

34 Advisory Board, 27–30 April 1983

35 Information from Lt-Gen. Stewart

36 Advisory Board, 21–23 March 1984

37 Ibid.

38 General Council, 14–16 March 1984

39 Advisory Board, 14–16 March 1985

40 Information from Gen. de Chastelain and Lt-Gen. Norman

13 University for the Canadian Forces

1 General Council, November 1981, DPED

2 Ibid., 17 June 1982

3 *Webster's New International Dictionary* (2nd ed., unabridged. Springfield, Mass. 1955), 525, 2782; (3rd ed., 1961), 415, 2302

4 Dacey to Dr H. Sheffer, 13 Oct. 1976, 7345-Princ, RMC

5 W. Eggleston, *National Research in Canada: The NRC, 1916–1966* (Toronto: Clarke Irwin 1975), 271–5; Captain D.J. Goodspeed, *A History of the Defence Research Board of Canada* (Ottawa: Queen's Printer 1958), 1–44

6 Eggleston, *National Research*, 473–4; Robert Bothwell, "Defence Research," *Canadian Encyclopedia* (Edmonton: Hurtig 1985), I, 479

7 CDS to DM, 12 Sept. 1947, HQS 24-8, vol. 3, quoted in J.R. Dacey to Dr H. Sheffer, DRB-CRAD, 7345-Princ, RMC

8 House of Commons, *Debates*, 21 June 1954

9 For example, Doctors J.R. Dacey, S. Naldrett, J. Hodgins, T. Hutchison, and Andrew Elliott

10 Faculty Board, 13 Oct. 1959

11 Advisory Board, 1965, DM 1150-110/C53-65, DND

12 Ibid., 1967

13 Robin McNeill, "The Post-Graduate as an Officer: Is There, Must There Be, Conflict?" *Canadian Defence Quarterly* 1, 2 (autumn 1971): 50–2

14 Educational Council, 8–10 Jan. 1967, filed in Faculty Council Minutes after 10 Jan. 1967

15 Faculty Council, 8 June 1967 and 12, 17 June 1968; Proposed Addendum for RMC's submission to the Advisory Board, 1967, 1150-110/C53-67, DND

16 Report of the Officer Development Board, I, 88, 95; III, 265, 267, 324, Annex F

17 Sharp, Memo to the cabinet, 15 Sept. 1969, 1150-110/C53-69, DND

18 Eggleston, *National Research*, 283

19 Extract, Minutes of the Advisory Board, Nov. 1973, Annex A to 1150-1210/C53-73 (DEdn), 4 Sept. 1973, DND

20 Maj. J.D. Snowball, Aide-Memoire for the Advisory Board, 4 Sept. 1973, in Advisory Board 1967 in Annex to Advisory Board, 1973, 1150-110/C53-73, vols. 2 and 3, DND

21 Hodgins to the minister of national defence, 6 March 1973, Advisory Board, 1973, 1150-110/C53-73-80, vol. 3, DND

22 "Report on Research," Annex C, Advisory Board, 1974, DPED

23 "Recommendations Regarding Research at CMCs," Annex K to 1150-110/C53-73 (DEdn), 4 Sept. 1973, DND. From internal evidence it is apparent this report was prepared by Dr J. Dacey

24 DRB Support of Research at Military Colleges, Annex to 1150-110/C53-73 (DEdn), 4 Sept. 1973, DND

25 "Research Policy for the CMCs," Advisory Board, 15–17 March 1974, Annex C, in DPED and also 1150-110/C53-74 (DEdn), DND; Dr E.J. Bobyn, "Recommendations of the Petch Report," General Council, 1980

26 G.D. Watson, C Plans, to vice-chairman, DRB, 19 Dec. 1974, 1150-110/C53-74, vol. 2, DND

27 Advisory Board, 18–20 April 1975, DPED

28 O.M. Solandt, "The Defence Research Board's Untimely End: What It Means for Military Science," *Science Forum* 8, 5 (Oct. 1975): 19–21; Gordon D. Watson, "Why the Bureaucrats Secretly Carved Up the DRB: It Worked Too *Well*," ibid., 22–5

29 Advisory Board, 27–29 Feb. 1976, 1150-110/C53-76, DND

30 Dacey to Dr H. Sheffer, CRAD, DRB, 13 Oct. 1976, 7345-4 (Princ), RMC

31 Advisory Board, 18–20 March 1977, 1150-110/C53-81, vol. 2, DND

32 *Commandant's Report*, 1977–8, 16–20

33 Advisory Board, April 1978, DPED

34 Ibid., Regional Committee, 13 Dec. 1979

35 General Smith, Advisory Board, 24 April 1979, Annex B

36 CPD Report to Advisory Board, 1979, Annex A

37 Advisory Board, 1979, 1150-110/c53-79, DND

38 Ibid., 22 March 1980, 1150-110/c53-81

39 Ibid., 19–22 March 1980, Annex E, 24 April 1980

40 Science Council of Canada, *Report No 31* (Ottawa: Government of Canada 1980)

41 Faculty Council, 3 May 1978, Annex

42 Academic Council, 19 June 1980, DND 1151-110/c100, DPED

43 Nils Örvik, "A Plan for a Programme of Strategic Studies at the Canadian Military Colleges," *Canadian Defence Quarterly* 8 (summer 1981): 38–42

44 Captain Rychard Brûlé, "On a Plan for a Programme of Strategic Studies at the Canadian Military Colleges," *Canadian Defence Quarterly* 8, 3 (winter 1978–9): 46; Captain I.F. Malcolm, "On a Programme of Strategic Studies at the Canadian Military Colleges," ibid., 46–7

45 Academic Council, 19 June 1980, 1151-110/c100, DPED

46 Information from Capt. (N) P. Fortier

47 Faculty Council, 8 Feb. 1983, Annex B

48 Advisory Board, 27–30 April 1983, DPED

49 Academic Council, 17 April 1984

50 Ibid., 23 March 1983

51 Advisory Board, 27–30 April 1983, DPED

52 Ibid., 21–23 March 1984, Annex B, 30 April 1984

53 Ibid., 14–16 March 1985, 80

54 Ibid.

55 Budget information from Dr W. Furter

56 Advisory Board, 1980, DPED

57 Ibid., 25 May 1981, Annex D, CRAD Report, 27 March 1981

58 *Commandant's Report*, 1977–8, 16

59 Ibid., 1979–80, 2

60 Ibid., 1981

61 Faculty Council, 3 April 1985, Annex A

62 Information from Capt. (N) P. Fortier

Appendices

APPENDIX A

An Act Respecting National Defence, assented to 30 June 1950 (excerpt). Statutes of Canada, chapter 43, National Defence Act, 14 Geo. VI, 486

EDUCATIONAL INSTITUTIONS

Establishment.

45. (1) The Governor in Council, and such other authorities as are prescribed or appointed by the Governor in Council for that purpose, may in the interests of national defence establish institutions for the training and education of officers and men, officers and employees of the Department and of the Defence Research Board, candidates for enrolment in the Canadian Forces or for employment in the Department or by the Defence Research Board and other persons whose attendance has been authorized by or on behalf of the Minister.

Administration.

(2) The institutions mentioned in subsection one shall be governed and administered in the manner prescribed by the Minister.

APPENDIX B

An Act Respecting the Royal Military College of Canada (excerpt). Statutes of Ontario, 7–8 Eliz. II, 1959, 503–4

WHEREAS the Royal Military College of Canada by its petition has represented that it was created by *An Act to establish a Military College in one of the Garrison Towns of Canada*, being chapter 36 of the Statutes of Canada, 1874, to be an institution for the purpose of imparting a complete education in all branches of military tactics, fortification, engineering and general scientific knowledge in subjects connected with and necessary to a thorough knowledge of the military profession and for qualifying officers for command and for staff appointments; and whereas, under authority of section 46 of *The National Defence Act* (Canada), being chapter 43 of the Statutes of Canada, 1950, the Governor in Council by P.C. 2512, dated the 19th day of May, 1950, entitled "Regulations for the Canadian Services Colleges," designated the College as one of the Canadian Services Colleges for the purpose of the education and training of officer cadets for the Royal Canadian Navy, the Canadian Army and the Royal Canadian Air Force, with the government, conduct, management and control of the College and of its work, affairs and business being vested in the Minister of National Defence; and whereas the College by virtue of the *National Defence Act* (Canada) is now governed and administered according to the Queen's Regulations for the Canadian Services Colleges, P.C. 20/848 and P.C. 2/971, 1957, and is thereby empowered to grant diplomas, certificates and awards; and whereas the College has prayed for power to grant university degrees; and whereas it is expedient to grant the prayer of the petition;

Therefore, Her Majesty, by and with the advice and consent of the Legislative Assembly of the Province of Ontario, enacts as follows . . .

2. (1) The Senate shall have the power to grant degrees and honorary degrees in arts, science and engineering.

(2) The Senate may also confer degrees in arts or science upon any person who successfully completed the curriculum in arts or science at the College during the period from the 1st day of September, 1948, to the 1st day of January, 1959.

3. The Chancellor or, in his absence, the President or the Commandant shall have the power to confer degrees and honorary degrees upon candidates to whom such degrees have been granted by the Senate.

4. This Act shall be deemed to have come into force on the 1st day of January, 1959.

5. This Act may be cited as *The Royal Military College of Canada Degrees Act, 1959.*

APPENDIX C

Presidents and Chancellors of the Royal Military College of Canada

PRESIDENTS

1946–54	Hon. B. Claxton
1954–7	Hon. R.O. Campney
1957–9	Hon. G.R. Pearkes VC

PRESIDENTS AND CHANCELLORS

1959–60	Hon. G.R. Pearkes VC
1960–3	Hon. D.S. Harkness
1963	Hon. G. Churchill
1963–7	Hon. P.T. Hellyer
1967–70	Hon. Leo A. Cadieux
1970–2	Hon. D.S. MacDonald
1972	Hon. E.J. Benson
1972–6	Hon. J.A. Richardson
1976–9	Hon. B.J. Danson
1979–80	Hon. A. McKinnon
1980–3	Hon. J.G. Lamontagne
1983–4	Hon. J.J. Blais
1984–5	Hon. R.C. Coates
1985–6	Hon. E.H. Nielsen
1986–9	Hon. Perrin Beatty
1989–	Hon. W.H. McKnight

APPENDIX D

Commandants

1947–54	Brigadier D.R. Agnew, CBE, CD, LLD (no. 1137)
1954–7	Air Commodore D.A.R. Bradshaw, DFC, CD (no. 2140)
1957–60	Commodore D.W. Piers, DSC, CD (no. 2184)
1960–2	Brigadier W.A.B. Anderson, OBE, CD (no. 2265)
1962–3	Brigadier G.H. Spencer, OBE, CD (no. 2424)
1963–7	Air Commodore L.J. Birchall, OBE, DFC, CD (no. 2364)
1967–70	Commodore W.P. Hayes, CD (no. 2576)
1970–3	Brigadier-General (L) W.K. Lye, MBE, CD, (no. 2530)
1973–7	Brigadier-General W.W. Turner, CD (no. 2816)
1977–80	Brigadier-General A.J.G.D. de Chastelain, CD (no. 4860)
1980–2	Brigadier-General J.A. Stewart, CD (no. 3173)
1982–5	Brigadier-General F.J. Norman, CD (no. 3572)
1985–7	Brigadier-General W. Niemy, CD (no. 3543)
1987–	Commodore E.R.A. Murray, OMM, CD (no. 4459)

APPENDIX E

Principals (Directors of Studies)

1948–67	Colonel W.R. Sawyer, OBE, CD, psc, rmc, BSC, MSC, PhD, LLD, DScMil, FCIC (no. 1557)
1967–78	Dr J.R. Dacey, MBE, BSC, MSC, PhD, FCIC
1978–84	Dr D.E. Tilley, BSC, PhD (McGill)
1984–	Dr J.B. Plant, OMM, CD, ADC, ndc, PhD (MIT), PEng (no. 3948)

APPENDIX F

Registrars

1948–63	Lieutenant-Colonel T.F. Gelley, MA, LLD
1963–81	R.E. Jones, MA, PhD
1981–	Captain (N) (Ret'd) P.C. Fortier, CD, BA, DipBusAdm (UWO) (no. 3210)

APPENDIX G

Staff Adjutants

1948–50	Major E.G. Brooks, DSO (no. 2517)
1950–1	Wing-Commander H.C. Vinnicombe, CD
1951–4	Major P.T. Nation, CD (no. 2472)
1954–8	Squadron-Leader A.C. Golab, CD
1958–61	Lieutenant-Commander J.G. Mills, CD
1961–4	Lieutenant-Colonel J.C. Gardner, CD (no. 2632)
1964–5	Lieutenant-Colonel J.M. Brownlee, CD (no. 2441)

APPENDIX H

Directors of Cadets

1965–7	Lieutenant-Colonel J.M. Brownlee, CD (no. 2441)
1967–9	Wing-Commander A. Pickering, CD (no. 2908)
1969–72	Lieutenant-Colonel (A) R.A. White, CD (no. 2893)
1972–4	Lieutenant-Colonel (L) D.M. Youngson, CD (no. 3439)
1974–5	Lieutenant-Colonel J.A.R. Gardam, CD
1975–7	Lieutenant-Colonel C.E.S. Ryley, CD (no. 3927)
1977–81	Lieutenant-Colonel J.A. Annand, CD (no. 4154)
1981–3	Lieutenant-Colonel L.R. Larsen, CD (no. 5573)
1983–6	Commander R.H. Thomas, CD (no. 5300)
1986–90	Commander B.R. Brown, CD (no. 8389)
1990–	Lieutenant-Colonel P.D. Mansbridge, CD (no. 9391)

APPENDIX I

Administrative Officers

1947–9	Major Peter Mumford
1949–56	Lieutenant-Colonel G.W. Swartzen, CD
1956–63	Lieutenant-Colonel L.J. Perry, MC, CD
1963–4	Lieutenant-Colonel P.T. Nation, CD (no. 2472)

APPENDIX J

Directors of Administration

1964–8	Lieutenant-Colonel P.T. Nation, CD (no. 2472)
1968–70	Lieutenant-Colonel (L) I.D. Macdonald, CD (no. 2777)
1970–2	Lieutenant-Colonel (L) D.M. Youngson, CD (no. 3439)
1972–7	Commander P.C. Fortier, CD (no. 3210)
1977–80	Lieutenant-Colonel F.A. Hlohovsky, CD (no. 3608)
1980–2	Commander D.G. Oke, CD (no. 6285)
1982–3	Commander R.H. Thomas, CD (no. 5300)
1983–6	Lieutenant-Colonel A.H.C. Smith, OMM, CD
1986–90	Lieutenant-Colonel J.W. Stow, CD
1990–	Lieutenant-Colonel J.R. Dicker, CD

APPENDIX K

Chairmen/Deans of RMC Academic Divisions

a) GRADUATE STUDIES AND RESEARCH

1959–64	J.R. Dacey, MBE, MSc (Dalhousie), PhD (McGill), FCIC
1964–72	T.S. Hutchison, BSc, PhD (St Andrews), FInstP, FAPS, FRSE
1972–84	Captain (N) (Ret'd) J.B. Plant, OMM, CD, ndc, PhD (MIT), PEng (no. 3948)
1984–	W.F. Furter, ndc, BASc (Toronto), SM (MIT), PhD (Toronto), FCIC, PEng (no. 3045)

b) SCIENCE

1959–60	Percy Lowe, BA (Toronto), PhD (Queen's)
1960–3	T.S. Hutchison, BSc, PhD (St Andrews), FInstP, FAPS, FRSE
1963–72	J.R. Dacey, MBE, BSc, MSc (Dalhousie), PhD (McGill), FCIC
1972–80	T.S. Hutchison, BSc, PhD (St Andrews), FInstP, FAPS, FRSE
1980–90	D.C. Baird, BSc (Edinburgh), PhD (St Andrews)
1990–	A.J. Barrett, CD, BSc (RMC), MSc (RMC), PhD (London) (no. 5992)

c) ARTS

1959–70	G.F.G. Stanley, BA (Alberta), MA, BLitt, DPhil (Oxon), D-ès-L (Laval), DLitt (Mount Allison), LLD (RMC), FRHistS, FRSC
1970–2	J.P. Cairns, BA (Toronto), MA (Columbia), PhD (Johns Hopkins)
1972–3	R.E. Waters, BA, MA (Toronto), PhD (Wisconsin)
1973–80	J.P. Cairns, BA (Toronto), MA (Columbia), PhD (Johns Hopkins)
1980–90	H.H. Binhammer, BA (Western), MA (Queen's), PhD (McGill)
1990–	B.D. Hunt, ndc, BA (RMC), MA (Queen's), PhD (Queen's) (no. 4919)

d) ENGINEERING

1959–62	G.W. Holbrook, BSc(Eng) (London), MSC (Queen's), PhD(Eng) (London)
1962–74	J.W. Dolphin, BASc (British Columbia), MSC (Minnesota) (no. 1963)
1974–84	A.C. Leonard, MBE, CD, BSC (Saskatchewan), MSE, PhD (Michigan), PEng
1984–	W.C. Moffatt, ndc, BSC, MSC (Queen's), ScD (MIT), PEng (no. 3342)

e) CANADIAN FORCES MILITARY COLLEGE AND THE EXTENSION (CFMC)

1973–4	Captain (N) (Ret'd) J.B. Plant, OMM, CD, ADC, ndc, PhD (MIT), PEng (no. 3948)
1974–80	W.S. Avis, BA, MA (Queen's), PhD (Toronto)
1980–4	W.F. Furter, ndc, BASc (Toronto), SM (MIT), PhD (Toronto), FCIC, PEng (no. 3045)
1984–	D.W. Kirk, CD, ndc, BSC, MSC, PhD (Queen's), PEng

APPENDIX L

Cadet Wing-Commanders (CWCs) and Sword of Honour

1951–2	J.I.B. Williamson (no. 2890)
1952–3	A. Hampson (no. 3055)
1953–4	J.A. Marshall (no. 3300)
1954–5	G.M. Kirby (no. 3403)
1955–6	P.D. Manson (no. 3528)
1956–7	B.L. Rochester (no. 3833)
1957–8	P.P.M. Meincke (no. 4106)
1958–9	G.F. Williamson (no. 4406)
1959–60	R.B. Morris (no. 4800)
1960–1	R.D. Byford (no. 5051)
1961–2	J.G. Allen (no. 5533)
1962–3	T.B. Winfield (no. 5851)
1963–4	R.B. Harrison (no. 6182)
1964–5	J.D.S. Harries (no. 6464)
1965–6	R.S.J. Cohen (no. 6898)
1966–7	J.M.P. Goineau (no. 7246)
1967–8	D.O.C. Brown (no. 7637)
1968–9	K.R. Moulden (no. 8162)
1969–70	R.B. Mitchell (no. 8059)
1970–1	M.R. Grinius (no. 8816)
1971–2	R.C. Smith (no. 9294)
1972–3	D.S. Heath (no. 9673)
1973–4	G.B. Mitchell (no. 10155)
1974–5	S.P. Tymchuk (no. 10329)
1975–	D.V. Jacobson (no. 10419)
1975–6	J.O.M. Maisonneuve (no. 1966)
1976–7	R.S. Richards (no. 11551)
1977–8	L.N. Stevenson (no. 11721)
1978–9	T.J. Lawson (no. 12192)
1979–80	T.W. Sweeney (no. 12464)
1980–1	J.J. Lund (no. 13136)
1981–2	J.P.A.P. Bedard (no. 12915)
1982–3	J.F. Lafortune (no. 13893)
1983–4	A.J. Howard (no. 14274)
1984–5	J.D.G. Falardeau (no. 14725)
1985–6	P.F. Wynnyk (no. 15706)
1986–7	J.M.S. Dubois (no. 15765)
1987–8	Susan A. Whitley (no. 16506)
1988–9	A.J.R. Nicolle (no. 16889)
1989–90	D.R.L. Ludlow (no. 17362)

APPENDIX M

RMC Ex-Cadets Who Served in Korea with Canadian or British Forces during Hostilities or Immediately Afterwards, with Decorations They Won in Korea

College Number	Name	Unit	Decorations for Korea
a) Pre-World War II Cadets			
1890	M.P. Bogert	Brigade Commander	CBE
2105	W. Moogk	Commander at Base	
2175	H.M. Millar	RE	
2234	P.M.H.D. McLaughlin	RCR, HQ 25015	m.i.d.
2249	H.W. Sterne	RCA	DSO
2280	T.R. Gemmell	RA	
2399	W.M. Landymore	RCN	m.i.d.
2400	J.A.D. Lantier	RCN	DSO
2444	J.G. Charles	RCN	
2464	J.E. Leach	PPCLI	MBE
2465	F.P.O. Leask	RCA	
2510	E.A.C. Amy	RCAC	OBE
2517	E.G. Brooks	RCHA	OBE
2541	J.S. Orton	RCHA	
2576	W.P. Hayes	RCN	
2577	G.E. Henderson	PPCLI	
2600	L.E.C. Schmidlin	RCE	MC
2640	C.H. Lithgow	RCR	
2648	J.G. Price	RCD	
2657	J.C. Stewart	RCA	
2662	R.M. Black	RCE	
2686	W.D.C. Holmes	RCE and RE	
2796	W.H. Pope	R 22nd R	
b) Graduating Class of 1952			
2832	J.G. Forth	RCE	
2839	G.C. Coops	RCE	
2840	D.W. Strong	RCCS	
2853	R.P. Bourne	RCHA	
2861	D.G. Loomis	RCE	MC
2872	C.J. Crowe	RCHA	
2875	J.K. Devlin	RCHA	
2896	J.D. McDougall	RCEME	

College Number	Name	Unit	Decorations for Korea
2897	H.C. Pitts	RCAC	MC
2890	J.I.B. Williamson	RCE	
2905	M.A. Fereday	RCE	
2915	J.O. Ward	RCHA	
2929	K.R. Black	RCD	
2931	J.A. Keane	RCOC	
2932	G.F. Hammond	RCHA	
2939	B.F. Simons	RCCS	
2942	R.J.M. Bell	RCAC	
2948	A.M. King	RCE	MC
2949	D.C. Patterson	RCAC	
2951	R.M. Withers	RCCS	
2967	C.D. Carter	RCE	MC
2973	R.W. Bull	RCAC	
2981	A.C. Moffat	RCHA	
3003	R.S. Peacock	PPCLI	

College Number	Name	Unit	
c) Graduating Class of 1953 (posted to Korea during hostilities who arrived after the armistice)*			
2896	J.M. Scott	RCA	
2899	W.A. Ferguson	RCAC	
3015	R.J.G. Adams	RCCS	
3026	W.L. Conrad	RCAC	
3042	W.B. Fisher	RCE	
3044	J.F. Fulton	RCE	
3059	F.J. Joyce	RCASC	
3087	R.V.A. Roe	RCCS	
3108	A.J. Bremer	RCCS	
3111	H.R. Bohne	RCE	
3129	R.D. Gross	RCAC	
3135	J.R. Jefferies	RCIC	
3137	R.D. Keen	RCE	
3144	C.A. Lowry	RCE	
3162	P.F. Pinsonneault	RCE	

*Naval cadets who had not attended RMC are not listed here.

APPENDIX N

High School Averages of Students Entering Ontario Universities* (Fall 1986)

Comparative statistics, taken for the most part from COU Circuletter 4048, dated 10 June 1987, show the academic quality of RMC entrants in the previous year by comparison with entries to other universities in Ontario. The table shows the percentage of Ontario students entering the first year of various universities with final high school marks on six grade 13 subjects or OACs with an average of 80 per cent or better.

University	Entering Average, Arts	Entering Average, Engineering
RMC (Ontario registrants only)	79.6 (17)	83.9 (76)
Queen's	68.0	87.4
Toronto	41.8	87.7
Waterloo	40.8	90.6
Western Ontario	29.0	42.1
Ottawa	22.6	50.0
Wilfrid Laurier	19.8	–
McMaster	16.3	74.0
York	15.4	–
Laurentian	14.1	14.3
Trent	12.8	–
Lakehead	11.0	–
Windsor	9.3	42.0
Brock	8.7	–
Guelph	8.7	38.3
Carleton	8.5	30.3

*Information received from Captain (N) P. Fortier, RMC registrar.

APPENDIX O

RMC's Rhodes Scholars* (to 1987)

Year of Competition	College Number	Name	Province of Competition
1946	2565	A.W. Duguid	Khaki University
1959	4393	D.P.D. Morton	Ontario
1962	5417	W.K. Megill	Ontario
1964	6219	R.W. Broadway	Saskatchewan
1964	6182	R.B. Harrison	Ontario
1965	6508	J.L. Adams	New Brunswick
1967	7291	T.A.J. Keefer	Ontario
1975	10419	D.V. Jacobson	Ontario
1976	10941	G.M. Gibbs	Ontario
1985	15040	P.E. Stanborough	Ontario
1987	15595	W.D.E. Allan	British Columbia

*Currently, eleven Rhodes scholarships are awarded annually in Canada for study at the University of Oxford – three in the Western Region (Manitoba, Saskatchewan, and Alberta), two each to Ontario, Quebec, and the Maritime Region (New Brunswick, Nova Scotia, and Prince Edward Island), and one each to British Columbia and Newfoundland.

APPENDIX P

Masters Degrees Granted by RMC (to 1987)

Master of Arts

English	1
History	3
Economics	1
War Studies	42
Total	47

Master of Science

Physics	3
Mathematics	13
Computer Science	4
Materials Science	6
Total	26

Master of Engineering

Civil Engineering	25
Computer Engineering	16
Electrical Engineering	100
Engineering Management	4
Mechanical Engineering	37
Nuclear Engineering	1
Total	183
Combined Total	256

APPENDIX Q

Presidents of the RMC Club of Canada since 1945

College Number	Name	Year Elected
982	Major D.W. MacKeen	1945
1841	Brigadier-General D.G. Cunningham, CBE, DSO, ED, QC	1946
1230	Colonel S.H. Dobell, DSO	1947
1855	Brigadier-General I.S. Johnston, CBE, DSO and Bar, ED, QC	1948
1625	Lieutenant-Colonel D. Watt, OBE, ED, KC	1949
1542	Captain E.W. Crowe, LLD, FAS, FAIA	1950
1860	Lieutenant-Colonel N. Kingsmill, ED, QC	1951
1828	Brigadier E. Beament, OBE, ED, QC	1952
1620	Lieutenant-Colonel R.R. Labatt, DSO, ED	1953
1766	Colonel K. Tremain, OBE, ED	1954
1474	Lieutenant-Colonel De L. H.M. Panet	1955
2034	Group-Capt. P.Y. Davoud, DSO, OBE, DFC	1956
1954	Lieutenant-Colonel W.P. Carr	1957
1945	Brigadier-General G.D. DeS. Wotherspoon, DSO, ED, CD, QC	1958–9
1379	H.A. Mackenzie, OBE, FCA	1960
2157	Brig.-Gen. J.H.R. Gagnon, OBE, CD	1961
2183	J.E. Pepall	1962
2336	Lieutenant-Colonel J.H. Moore, FCA	1963
2351	Major G. Savard, MBE	1964
2749	J.B. Cronyn	1965
2691	J.F. Maclaren, ED	1966
2791	J.P.W. Ostiguy	1967
RCNC90	Commodore J.F. Frank	1968
2859	Professor J.G. Pike	1969
2494	Major G.E. Ward	1970
2761	Colonel C.S. Frost, CD, QC	1971
2544	Major T.A. Somerville	1972
3641	W.I. McLachlan	1973
2954	J.H. Farrell	1974
3661	T.E. Yates	1975
5533	J.G. Allen	1976
3172	M.M. Soule	1977
RR261	W.B. Tilden	1978
3672	C.C.M. Powis	1979
3251	J.W. Tremain	1980
2897	Major-General H.C. Pitts, MC, CD	1981
2576	Commodore W.P. Hayes, CD	1982
2652	Colonel A.B. Smith, MC, CD, QC	1983
RRA15	Brigadier-General B.A. Howard, CD	1984
7268	J.W. Brown	1985
5604	K.A. Smee	1986
3010	E.P. McLoughlin	1987
4100	J.J. Choquette	1988
3475	H.F. Champion-Demers	1989
3173	Major-General J.A. Stewart	1990

APPENDIX R

Secretary-Treasurers and Executive Directors of the RMC Club of Canada since 1940

College Number	Name	Year Elected
a) Secretary-Treasurers		
H5225	R.D. Williams	1940–56
H6888	Colonel T.F. Gelley	1957–68
2116	Brigadier-General J.S. Ross	1969–70
H2354	Colonel H.W.C. Stethem	1971–81
2499	Lieutenant-Colonel W.H.T. Wilson (acting secretary-treasurer)	1981
2480	Colonel D.W. Strong	1981–6
b) Executive Directors		
2480	Colonel D.W. Strong	1986–

Name Index*

*Numbers in italics indicate footnotes.

Subject Index*†

*Numbers in italics indicate footnotes.
†Due to the changing names within the Canadian Forces over the period covered in the book, references to the different services are found under "generic" names: Armed Forces, Air Force, Army, Navy, Permanent Force.

of Graduate Studies and Research, 198, 199
    working group on bilingualism, 164–65
Defence Council, 27, 31, 52, 54, 64, 107, 109, 124, 148
    and women in combat roles, 181, 188
Defence Research Board, 34, 50, 83, 160, 167, 192–98
    grants, 198
Defence Secretary, 27, 31, 52
Degree-granting powers, 32, 51, 52–54, 57, 59, 62, 65–66, 107, 116,
    148, 150, 153, 192, 193
    CMR, 162
    Royal Roads, 116
Department of National Defence, 31, 38, 52, 56, 59, 76, 105, 194
    administration of RMC by, 31
    development funds, 203
    grants new colours to CSCS, 67
    headquarters reorganization, 76, 80, 91
    use of French, 162
Depression (1929), 21
Deterrence, 112, 114
Director-general of individual training programs (DGITP), 95, 95, 106
Director-general of personnel research and development, 128
Director-general of postings and careers, 194
Director-general of recruiting, education, and training (DGRET), 123,
    129, 139, 140, 147, 173, 176, 182–83
Director-general of recruiting and training, 80
Division of Continuing Education. *See* Canadian Forces (Extension)
    College
Dominion cadetships, 25, 38, 39, 56
Dress regulations, 120
Drill, 46, 97, 126, 157, 175–76
Drugs, 119
Duntroon, 15

Educational (Academic) Council, CSC, 80, 83, 85, 108, 109, 111, 115,
    117, 137, 147, 150, 152, 163, 167, 169, 171, 174, 194, 199–200
Engineering Institute of Canada, 62
Ex-cadets. *See also* Royal Military College Club
    16, 18, 19
    in British Army, 16, 17
    in Korean War, 28
    role in campaign to re-open RMC, 23–25
    World War I, 20
Extension courses, 110, 149, 150–52, 153–54
    degree, 152
    RCAF, 149
    women in, 182

Faculty Board, 47, 51, 61–63, 65, 97, 100, 106, 108–109, 128, 141,
    157
    bilingualism and, 162, 172–73, 174
    Cadet-Wing education officer on, 121

Faculty Club. *See* Royal Military College, senior staff mess
Faculty Council, 45, 46, 47, 48, 52, 53, 55, 57, 61–62, 65, 66, 72, 83,
    109, 116, 121, 125, 140, 148, 150, 153–54, 157, 178
    bilingualism and, 161, 167, 169, 170, 171, 172, 175, 176, 177
    task force recommendations and, 134
Faculty Review Board, 65
Fieldtrips, 127
Fisher and Tedman report, 43–44
Fisheries' College, St. John's, Newfoundland, 141
Football team, 71, 92
Fort Champlain, 79, 80, 89
Fort Frederick, Martello Tower, 70, 170
Fort Frontenac, 102
Fort Haldimand, 43, 89
Fort Lasalle, 89
France, World War I, 19, 21
Francophones, 15, 26, 35–36, 43, 44, 47, 72, 116, 157–68, 169–75,
    177, 178, 179
    opposition to South African War, 18
    staff recruitment, 175, 178–79
    women cadets, 185
French as a working language, 159, 162

Garigue study, 136
General Staff, 23, 24, 27, 33
"Get Well Program", 141
Girouard Building, 173
Glassco Commission, 67–68, 72, 75, 78, 81, 82, 83
Globe and Mail, 101
Graduate studies, 65, 66, 96, 193–94
Great Britain. *See* Britain
Greenwich. *See* Royal Naval College, Greenwich
Gymnasium, 72, 107–108

Halifax Garrison, 19
Harvey Mudd College, 116
Honorary degrees, 65, 67, 130, 175
House of Commons, 17, 18, 22, 35, 36, 67, 91, 105, 181
    Special Committee on Defence (1963–65), 77–78, 80
Howard-English Commission, 4, 6–7
Human rights legislation, 183, 188

Initiation practices. *See* "Recruiting"
International peacekeeping, 29
Iron ring ceremony, 201

Keyham. *See* Royal Naval Technical College
Kingston, 130
Korean War, 29, 30, 32–33

Landymore Committee reports, 49–50, 61, 78–79